THE TWO-STROKE CYCLE ENGINE

Combustion: An International Series
Norman Chigier, *Editor*

DATE DUE

040704			
OCT 0 8 2002			
APR 1 8 2008			

THE TWO-STROKE CYCLE ENGINE

Its Development, Operation, and Design

John B. Heywood
Massachusetts Institute of Technology
Cambridge, MA

Eran Sher
Ben-Gurion University
Negav, Israel

USA	Publishing Office:	TAYLOR & FRANCIS 325 Chestnut Street Philadelphia, PA 19106 Tel: (215) 625-8900 Fax: (215) 625-2940
	Distribution Center:	TAYLOR & FRANCIS 47 Runway Road, Suite G Levittown, PA 19057-4700 Tel: (215) 269-0400 Fax: (215) 269-0363
UK		TAYLOR & FRANCIS 1 Gunpowder Square London EC4A 3DE Tel: +44 171 583 0490 Fax: +44 171 583 0581

THE TWO-STROKE CYCLE ENGINE: Its Development, Operation, and Design

1 2 3 4 5 6 7 8 9 0

Printed by Braun-Brumfield, Ann Arbor, MI 1999.

A CIP catalog record for this book is available from the British Library.
∞ The paper in this publication meets the requirements of the ANSI Standard Z39.48-1984 (Permanence of Paper).

Library of Congress Cataloging-in-Publication Data

Heywood, John B.
       The two-stroke cycle engine: its development, operation, and
    design / John B. Heywood, Eran Sher.
          p.   cm. -- (Combustion: an international series)
       Includes bibliographical references and index.
       ISBN 1-56032-831-2 (alk. paper)
                1.   Two-stroke cycle engine. I. Sher, Eran. II. Title.
    III. Series: Combustion (New York, N.Y. : 1989)
    TJ790.H49   1999
       621.43--dc21   98-52365
                                                                                  CIP

Taylor and Francis, Inc. ISBN: 1-56032-831-2 (case)

Published in North America by the Society of Automotive Engineers, Inc., 400 Commonwealth Drive, Warrendale, PA 15096-0001, USA.

Society of Automotive Engineers ISBN: 0-7680-0323-7
Library of Congress No. 99-60257

SAE Order number: R-267

CONTENTS

PREFACE

The internal combustion engine is the most widely used power-producing device in the world today. It has reached that position through its compatibility with cheap and readily available hydrocarbon fuels, and its combination of low cost, ruggedness and reliability, high power output for a given engine weight and size, and high efficiency. There are two basic internal combustion engine operating cycles: the four-stroke and the two-stroke cycle. They differ primarily in how the gas exchange process is effected. The four-stroke cycle uses more than one full revolution of the crankshaft—almost one-third the total cycle—to replace the burned gases that fill the cylinder at the end of the power-producing expansion process with fresh charge. The two-stroke gas exchange process uses about one-third of a revolution (and cycle) to scavenge the cylinder. The obvious trade-off is twice the number of power strokes in the two-stroke engine at a given speed, but much less effective scavenging.

This book is about the two-stroke cycle internal combustion engine. The two-stroke engine is widely used at the small-size and very large size ends of the engine market. In its small-size spark-ignition engine forms, the two-stroke cycle engine is relatively cheap, compact and light, simple, and robust. This is the basis of its market appeal in mopeds, scooters, motorcycles, and snowmobiles, in portable devices such as chainsaws and bush cutters, in agricultural and construction devices such as lawn mowers, disc saws, and snow blowers, in the outboard marine engine arena, and in light and in remotely piloted aircraft. The very large diesel engines used in marine and power-generation applications are also two-stroke cycle engines. These large internal combustion engines are the most efficient and cost effective prime movers currently available, and can achieve energy conversion efficiencies (useful work output relative to fuel energy input) of up to 55%. The two-stroke diesel has also been used in the locomotive and in parts of the truck market. The passenger-car and truck engine markets are, however, dominated by the four-stroke cycle engine.

This historical market distribution may change as a result of our need to reduce the environmental impact and energy consumption of vehicles and devices powered by internal combustion engines, in an increasingly competitive global marketplace. The two-stroke engine's small size and low weight for a given maximum power are likely to become more significant as vehicle weights must be reduced. One important opportunity here is the invention and development over the past decade of new gasoline fuel injection technology, which produces finer and more controllable fuel sprays. Used for injecting fuel directly into the engine cylinder, this technology may expand the market opportunities for the two-stroke spark-ignition engine, because direct fuel injection solves two of the engine's major problems: excessive fuel consumption and high hydrocarbon emissions. A second opportunity is the continuing development of better exhaust gas catalyst technology, which promises ever decreasing pollutant emission levels. It seemed appropriate to us, therefore, to write this book: a comprehensive professional reference on the history, the operating and performance characteristics, and the design of two-stroke cycle engines.

The technical literature on two-stroke cycle engines is neither readily available nor well organized. Thus a primary objective of this text is to collect together and organize this technical literature and present a coherent and up-to-date summary of our knowledge and understanding of this widely used and promising internal combustion engine concept. The book develops a fundamental understanding of the processes essential to the successful operation of the two-stroke cycle engine. It describes and evaluates the various types of models that have been developed to predict aspects of two-stroke engine operation. It reviews and explains the experimental methods that are used in two-stroke engine development and testing. It examines the major issues involved in the engine's design. It covers smaller, gasoline-fueled, spark-ignition engines as well as diesels (in both smaller sizes for potential automotive applications and larger sizes for marine and power-generating applications). It attempts to bring together, within the limits of our current knowledge base, the theory and practice of this important engine field.

We anticipate that our book will serve as a reference for professionals who work on or are interested in engines. These include researchers, developers, designers, inventors, students, faculty, and the many different users of two-stroke cycle engines. We hope that a reference text such as this will be widely used both as an introduction to this field and as a resource on "what is known" about this engine concept.

The book is organized as follows. First, in Chapter 1, we provide an overview of the two-stroke cycle concept, and develop the basic engine terminology needed to quantify and understand engine operation. We next review the history of the two-stroke cycle engine and how it has evolved over several hundred years to its present form. Chapters 2–5 then focus on the engine's scavenging process, which is both its Achilles heel and its yet to be fully realized opportunity. Effective scavenging of the burned gases in the cylinder after expansion, by the fresh charge, is essential to achieving high power densities and a robust combustion process. The critical aspects of the scavenging process are described and the parameters that quantify the process are developed in Chapter 2. The next chapter reviews the available experimental methods used to quantify the scavenging process. Chapter 4 then describes the scavenging models (both phenomenological models and more fundamentally based computational fluid dynamic models) that have been developed and used. The design of the scavenging system—the intake, transfer, and exhaust ports (or valves), and the engine's intake and exhaust systems, are examined next in Chapter 5. Then Chapter 6 reviews engine combustion, in both spark-ignition and diesel engines, providing an introduction to engine combustion as well as details of each of these two-stroke engines' combustion processes. In Chapter 7 an extensive summary of how the major air pollutants form inside the two-stroke engine cylinder and of the more promising emission control techniques in the engine and in the exhaust with catalysts is presented, covering both spark-ignition and diesel processes. The Chapter 8 discusses the important interdependent topics of two-stroke engine friction, lubrication, lubricant properties, and ways to minimize wear. The book ends with Chapter 9, where the scavenging behavior, combustion and emissions control requirements, and friction aspects of two-stroke engine operation are integrated together to describe and explain the performance characteristics of the different types of engine. Each chapter is extensively illustrated and referenced.

Due to its inherently attractive attributes, the internal combustion engine will be of great value to mankind for many years into the future. It behooves us to understand its potential and its limitations to the full.

John Heywood and Eran Sher
August 18, 1998

ACKNOWLEDGEMENTS

Many people have helped in significant ways as we put together this book. Heartfelt thanks are due to Dr. Mark Dulger of the Department of Mechanical Engineering, Ben-Gurion University, for his generous and professional help with the preparation of the drawings, and to Ilai Sher for his elegant and diligent work preparing, and greatly improving most of the figures in the book. We are grateful to Professors Haruo Yoshiki of the Institute of Industrial Science, University of Tokyo, and Isaac Garbar of Ben-Gurion University, for their invaluable advice. Our special thanks go to Karla Stryker, John Heywood's assistant, who cheerfully, patiently, and faultlessly transferred countless revisions, extensions, and changes into the master version of the text, and also chased down many critical details. The MIT sabbatical leave fund and the Sun Jae Professorship (endowed by Daewoo Heavy Industries Corporation) supported John Heywood's writing during academic year 1997–98, and the High Temperature Gas Dynamics Laboratory in the Mechanical Engineering Department at Stanford University graciously acted as his host. Eran Sher is indebted to the Sloan Automotive Laboratory at MIT for providing him the support and the opportunity to spend several invaluable, fruitful, and unforgettable, periods at MIT during the past 10 years.

The authors are indebted to the following companies for providing valuable data and figures: Cox Hobbies, Inc., Santa Ana, CA, USA; Ishikawajima-Harima Heavy Industries Co., Ltd., Tokyo, Japan; Kolbenschmidt AG, Neckarsulm, Germany; MAN, Augsburg, Germany; MAN B&W Diesel A/S., Copenhagen, Denmark; New Sulzer Diesel Ltd., Winterthur, Switzerland; Orbital Engine Co. Ltd., Perth, Australia; STIHL, Waiblingen, Germany; Suzuki Motor Co., Hamanatsu, Japan; Wärtsilä NSD Switzerland Ltd., Winterthur, Switzerland. Also several organizations and individuals provided us with high quality drawings and photographs, as acknowledged in the text; we greatly appreciate their assistance. We are especially grateful to the Society of Automotive Engineers for permission to reproduce a large number of figures from their publications in this text.

The authors also wish to thank these organizations and publishers for permission to reproduce figures and tables from their publications: Carnot Press; The Combustion Institute; Combustion Science and Technology; Editions Technip; Elsevier Science; Haynes Publishing, The McGraw-Hill Companies; MAN and B&W Diesel A/S; Orbital Engine Company; Professional Engineering Publishing Ltd.; Council of the Institution of Mechanical Engineers; Sandia National Laboratories; Society of Automotive Engineers; St. Martin's Press; Wärtsilä NSD Switzerland Ltd.

Finally, we are deeply thankful to our truly admirable families, especially our wives, Peggy and Edith, as well as our children (James, Stephen and Ben Heywood, and Elite and Ilai Sher) for gracefully and patiently living with the fact that writing, and especially finishing off, a major book takes much time and energy that could have gone into other activities. Without their full and continuing support, and their encouragement, our book would not have been completed.

John Heywood and Eran Sher

SYMBOLS, SUBSCRIPTS, AND ABBREVIATIONS

1 SYMBOLS

a	coefficient; crank radius
A	area
A_c	cylinder cross-sectional area
A_C	valve curtain area
A_d	piston deflector area
A_E	effective area of flow restriction
A_p	piston crown area
A_P	port area
A_R	reference area
b	coefficient
B	cylinder bore
c	coeficient; connecting rod length; sound speed; specific heat
c_p	specific heat at constant pressure
c_r	relative charging ratio
c_v	specific heat at constant volume
C_D	discharge coefficient
C_f	engine speed fluctuation rate
d	coefficient
D	diameter
D_{AB}	diffusion coefficient

D_s	valve stem diameter
D_v	valve diameter
e	coefficient; specific energy
E	energy
f	coefficient of friction; frequency
f_c	contact area ratio
f_f	lost flame area fraction
F	force
h	specific enthalpy
h_c	heat-transfer coefficient
h_P	port open height
H	enthalpy; hardness
I	intensity of sound; moment of inertia
k	critical shear stress; thermal conductivity
k_i^+, k_i^-	forward and backward rate constants for ith reaction
K	constant
K_h	humidity correction
l	characteristic length scale
l_T	characteristic length scale of turbulent flame
L	piston stroke
L_I	sound intensity level
L_v	valve lift
m	mass; mass of cylinder charge
$\dot{m}$	mass flow rate
m_r	mass of residual gas
M	molecular weight
n	number of moles; polytropic exponent
n_c	number of cylinders
n_P	number of ports
n_R	number of crank revolutions per power stroke
N	crankshaft rotational speed
p	cylinder pressure; pressure
P	power
$\dot{q}$	heat-transfer rate per unit area
Q	heat transfer
$\dot{Q}$	heat-transfer rate
Q_{ch}	fuel chemical energy or gross heat release
Q_{HV}	fuel heating value
Q_n	net heat release
Q_w	heat transferred to wall
r	radius
r_c	compression ratio
R	ratio, connecting rod length to crank radius; gas constant; radius
R_d	piston deflector ratio
R_e	Reynolds number

R_{sc}	short-circuiting ratio
s	distance from crank axis to piston pin; specific entropy
S	entropy
S_L	laminar flame speed
S_p	piston speed
t	time
T	temperature; torque
T_{loss}	sound transmission loss
u	specific internal energy; velocity
u'	turbulence intensity
u_s	sensible specific internal energy
U	fluid velocity; internal energy
v	specific volume; velocity
V	cylinder volume; velocity; volume
V_{cc}	crankcase clearance volume
$V_{cc,max}$	maximum crankcase volume
V_d	displaced cylinder volume
W	work transfer
W_c	work per cycle
W_p	pumping work
x, y, z	spatial coordinates
x	mass fraction
$\tilde{x}$	mole fraction
x_b	burned mass fraction
x_r	residual mass fraction
y	hydrogen-to-carbon (H/C) ratio of fuel; mass fraction; mole fraction; volume fraction
Y	maximum port height
α	angle; angular acceleration; boundary friction fraction; thermal diffusivity $k/(\rho c)$
β	exhaust gas purity
γ	specific heat ratio c_p/c_v
δp	root-mean-square fluctuating sound pressure
$\Delta\theta$	burn angle
$\Delta\theta_b$	rapid burning angle
$\Delta\theta_d$	flame development angle
η_{cat}	catalyst conversion efficiency
η_{ch}	charging efficiency
η_f	fuel conversion efficiency
η_m	mechanical efficiency
η_{rt}	retaining efficiency
η_{sc}	scavenging efficiency
η_{tr}	trapping efficiency
θ	crank angle
λ	relative air/fuel ratio; wavelength

Λ	delivery ratio
μ	dynamic viscosity
ν	kinematic viscosity μ/ρ
ξ	flow friction coefficient
ρ	density
$\rho_{a,0},\ \rho_{a,i}$	air density at standard and inlet conditions
σ	normal stress or loading; standard deviation; Stefan–Boltzman constant
τ	characteristic time; dimensionless time; shear stress
ϕ	fuel/air equivalence ratio
ω	angular velocity

2 SUBSCRIPTS

a	air
b	burned gas; brake value
c	cylinder
del	delivered
short	short-circuiting
e	engine; exhaust
f	end of combustion; friction; fuel
fcz	fresh charge zone
i	indicated; intake; species I
ign	at ignition
m	mixing zone; model
n	cycle number
p	piston
P	port; prechamber
r	reed valve parameter
rt	retained
R	reference value
s	isentropic; stoichiometric
sc	scavenge
t	total
tr	trapped
u	unburned gas
v	valve
w	wall; water vapor
0	reference value (usually atmospheric conditions); stagnation value

3 NOTATION

$\delta,\ \Delta$	difference
$\overline{x}$	average or mean value

$\tilde{x}$	value per mole
[ ]	concentration, moles per volume
{ }	mass fraction
$\dot{x}$	rate of change with time

4 ABBREVIATIONS

A/F	air/fuel ratio
bmep	brake mean effective pressure
BC, ABC, BBC	bottom-center crank position, after BC, before BC
COV	coefficient of variation
dB	decibels (dBA decibels, A-weighting)
del	delivered
EGR	exhaust gas recycle
EI	emission index
EPC, EPO	exhaust port closing, exhaust port opening
equil	equilibrium value
EVC, EVO	exhaust valve closing, exhaust valve opening
F/A	fuel/air ratio
FID	flame ionization detector (HC)
imep	indicated mean effective pressure
IPC, IPO	inlet port closing, inlet port opening
IVC, IVO	inlet valve closing, inlet valve opening
max	maximum value
mep	mean effective pressure
NDIR	nondispersive infrared analyzer
SC, SO	scavenge port closing, scavenge port opening
sCO, sHC, sNO_x	specific CO, HC, and NO_x emissions
sfc	specific fuel consumption
TC, ATC, BTC	top-center crank position, after TC, before TC
tp	tailpipe value

CHAPTER ONE

OVERVIEW, BACKGROUND, AND HISTORY

Internal combustion engines are used to produce mechanical power from the chemical energy contained in hydrocarbon fuels. The power-producing part of the engine's operating cycle starts inside the engine's cylinders with a compression process. Following this compression, the burning of the fuel–air mixture then releases the fuel's chemical energy and produces high-temperature, high-pressure combustion products. These gases then expand within each cylinder and transfer work to the piston. Thus, as the engine is operated continuously, mechanical power is produced.

Almost all internal combustion engines utilize the reciprocating piston-in-cylinder geometry shown in Fig. 1-1. The oscillating motion of the piston is converted to the rotary motion needed to transmit mechanical power, through the connecting rod and crankshaft arrangement shown. Each upward or downward movement of the piston is called a stroke. There are two commonly used internal combustion engine cycles: the two-stroke cycle and the four-stroke cycle. Both cycles are based on the preceding description of the essential power-producing sequence: compression of the unburned mixture, its combustion, and then expansion as burned gases.

The fundamental difference between these two cycles is in their gas exchange process: that is, the removal of burned gases at the end of each expansion process and the induction of fresh mixture for the next cycle. The two-stroke has an expansion, or power, stroke in each cylinder during each revolution of the crankshaft, and the exhaust and the charging processes occur simultaneously as the piston moves through its lowest, or bottom center (BC), position shown in Fig. 1-1. In the four-stroke cycle, the burned gases are first displaced by the piston during an upward exhaust stroke and the fresh charge then enters the cylinder during the following downward intake stroke. It is the two-stroke cycle engine that is the subject of this book.

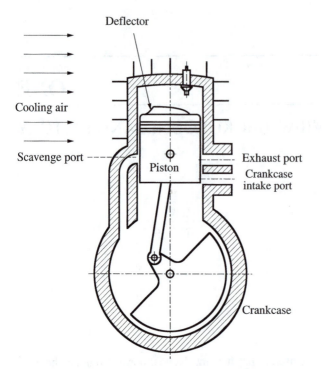

Figure 1-1 A crankcase-scavenged two-stroke cycle engine.

There are two basic types of internal combustion engine: the spark-ignition engine (sometimes called the Otto-cycle engine), and the compression-ignition engine, which is often called the diesel. In the spark-ignition engine, the fuel–air mixture is essentially premixed prior to combustion, which is initiated with a spark discharge. In the diesel, air alone is inducted and compressed; fuel is injected into the cylinder just before combustion commences. The injected fuel vaporizes and mixes rapidly with the high-temperature compressed air and spontaneously ignites, burning as it mixes. The two-stroke cycle is used in both spark-ignition and diesel engines.

The two-stroke cycle is hardly a new concept; in fact, all internal combustion engines developed before Nicolaus Otto invented the four-stroke cycle engine in 1876 operated on a two-stroke cycle. Today, the two-stroke engine is a refined power unit that offers high performance while being compact, simple, and lightweight. The largest ($\sim$ 30,000 MW) and the smallest ($\sim$ 15 W) reciprocating piston engines in the world use the two-stroke cycle.

If it were possible to run a two-stroke cycle engine at the same crankshaft rotational speed and work production per unit of displaced cylinder volume as a four-stroke cycle engine, the two-stroke would develop twice as much power as a four-stroke of the same size because the number of cycles per unit time would be twice as large. However, because in the two-stroke cycle engine the burned gases are pushed out from the cylinder by the entering fresh charge, losses due to mixing of fresh charge and burned gases are unavoidable, and the theoretical power advantage of the two-stroke engine over its four-stroke counterpart cannot be fully realized.

1-1 PRINCIPLES OF THE TWO-STROKE CYCLE ENGINE

As the two-stroke cycle lacks separate intake and exhaust strokes, a scavenging pump is required to drive the fresh charge into the cylinder. In one of the simplest and most frequently used types of two-stroke engine designs, the bottom surface of the piston in conjunction with that portion of the crankcase beneath each cylinder is used as the scavenging pump. Figure 1-2 shows the typical sequence of cycle events in this type of engine, and Fig. 1-3 relates these events to a cylinder-pressure versus cylinder-volume trace and the port positions along the liner. The cycle begins while the piston is travelling upward toward the top center (TC) crank position, and the crankcase intake port is uncovered by the piston (IO in Fig. 1-2). Fresh charge (either air or fuel–air mixture) en-

Two stroke gas exchange process with crankcase compression

E	Exhaust port	S	Scavenge port
EO	Exhaust opens	SO	Scavenge opens
EC	Exhaust closes	SC	Scavenge closes
I	Intake port	TC	Top center
IO	Intake opens	BC	Bottom center
IC	Intake closes		
IG	Ignition point		

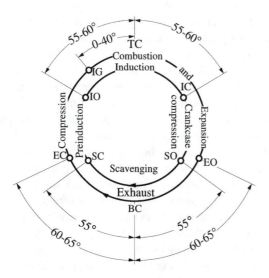

Figure 1-2 Typical sequence of two-stroke cycle events: The outer circle shows the processes occurring inside the cylinder as a function of crank angle; the inner circle shows those occurring in the crankcase. Developed from *Bosch Automotive Handbook*.[1]

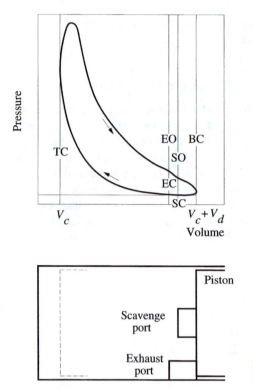

Figure 1-3 Cylinder-pressure versus cylinder-volume trace of a two-stroke cycle engine.

ters into the crankcase through the intake manifold while the charge within the cylinder continues to be compressed by the upper part of the piston. The charge is then ignited (either by an electrical discharge in a spark-ignition engine or by a spontaneous ignition process in a diesel), combustion occurs and the burned gases in the cylinder expand as the piston travels toward bottom center (BC). At the same time, as the crankcase volume decreases and the intake port is still open, some of the fresh charge may escape to the atmosphere through the intake manifold in a reverse flow. Approximately 60° after TC, the inlet port closes (IC), and the fresh mixture in the crankcase is then compressed. The in-cylinder gas exchange process begins as the exhaust port is opened (EO). As the piston continues its downward travel, it then opens the scavenge or transfer ports. When both the scavenge and exhaust ports are open, the cylinder is subjected to a pressure gradient that simultaneously governs the inflow and outflow streams through the open ports. During this period—the scavenging period—the compressed fresh charge in the crankcase flows through the transfer ducts into the cylinder and scavenges the burned combustion products out of the cylinder through the exhaust port. The ports and the projection on the piston (the deflector) are shaped so that most of the fresh charge will sweep up to the top of the cylinder before flowing to the exhaust port. This is done to scavenge the combustion products more completely from the upper part of the cylinder and prevent significant amounts of the fresh charge from flowing directly to the exhaust port, a process called short-circuiting. In the second half of this period, the piston travels upward, the crankcase volume increases, and a reverse flow from the cylinder to the crankcase through the scavenge ports may occur depending on the charging pressure and engine speed. The gas exchange process is completed when the piston covers up and closes the exhaust port (EC).

An alternative to using the crankcase to compress the fresh charge prior to scavenging is to employ an external pump. A positive displacement Roots blower can be used, or a centrifugal compressor, driven from the crankshaft. In larger two-stroke cycle engines a blower and a turbocharger can be combined together. The crankshaft-driven blower provides compression for starting, and at lower speed; at higher speeds, the turbocharger provides higher air flow rates, and hence higher power for a given size engine.

The most efficient gas exchange process will completely replace the products of combustion by fresh charge, at charge pressure and temperature, without wasting any fresh charge through the exhaust. In practice, the gas exchange process is far from this perfect displacement process; although part of the fresh charge does displace combustion products without mixing or loss, another part mixes with the combustion products, and other portions short-circuit directly to the exhaust port. The success of the scavenging process is a function not only of the geometry of the cylinder and port assembly, but also of factors such as how the fresh charge is introduced into the cylinder (external blower, crankcase, surge tank, etc.), engine speed, engine load, and atmospheric conditions.

1-2 CHARACTERISTICS OF THE TWO-STROKE CYCLE ENGINE

One power stroke per revolution.   Doubling the number of power strokes per unit time relative to the four-stroke cycle increases the power output per unit displaced volume. It does not, however, increase by a factor of 2. The outputs of two-stroke engines range from only 20% to 60% above those of equivalent-size four-stroke units. This lower increase in practice is a result of the poorer than ideal charging efficiency: that is, incomplete filling of the cylinder volume with fresh air. Doubling the number of power strokes per unit time also halves the intervals between combustion-generated pressure impulses. This results in a smoother crankshaft torque versus time diagram, and because it doubles the basic frequency of the torsional vibrations transmitted to the vehicle structure, it makes vehicle resonances a less significant problem. Consequently, a smaller lighter-weight flywheel can be used. There is, however, a major disadvantage: the higher frequency of combustion events in the two-stroke cycle engine results in higher average heat-transfer rates from the hot burned gases to the engine's combustion chamber walls. Higher temperatures and higher thermal stresses in the cylinder head, and especially the piston crown, result.

Scavenging with fresh charge.   Inherent in the two-stroke cycle is the process of scavenging the burned gases from the engine cylinder with fresh charge. This gas exchange process has several consequences. First, charging losses are inevitable. Under normal operating conditions in a typical two-stroke engine, about 20% of the fresh charge that enters the cylinder is lost due to short-circuiting to the exhaust. In carbureted spark-ignition engines, this process results in very high hydrocarbon emissions and poor fuel economy compared with the four-stroke cycle engine. However, as both exhaust and charging occur around BC, the exhaust and intake ports can be situated near the bottom end of the cylinder and can be covered and uncovered by a long-skirt piston (Fig. 1-1). This simplest geometric two-stroke cycle configuration obviates the need for valves and their actuating gear. This substantially simplifies the engine structure and the production process, and significantly reduces engine cost. However, because the opening and closing of the ports are symmetrical around BC, the intake closes before the exhaust and a simple supercharging process is therefore not possible. For this and other reasons, externally blown two-stroke scavenging systems are also in common use that, because they normally use valves to control the exhaust (and sometimes inlet) flow(s), lose the simplicity of the crankcase-scavenged design.

The importance and complexity of the gas exchange process in two-stroke engines should already be apparent. It will be the focus of the next four chapters. There is the obvious complexity of the in-cylinder flow as the fresh charge displaces and also mixes with the burned gases, and partially short-circuits the cylinder by flowing directly into the exhaust. In addition, a reverse flow of fresh charge through the intake manifold occurs at low engine speeds, and a reverse flow of cylinder charge through the scavenge duct occurs at high engine speeds. These flows are partly responsible for the deterioration of the cylinder charging process at off-design engine speeds, which, in turn, results in a falloff in torque. Furthermore, the effectiveness of the scavenging process is sensitive to the exhaust manifold pressure, which makes it difficult to employ a turbocharging system, a catalytic converter, noise-reduction devices, and an engine brake system. Throttling the intake of the two-stroke cycle engine to reduce airflow at part load, as is done in four-stroke cycle spark-ignition engines, further complicates the scavenging process. Throttling of the exhaust flow is also sometimes used to control the scavenging process.

Direct fuel injection. Because there is a substantial (of order 20%) loss of fresh air during the scavenging process due to short-circuiting and mixing with exhausting burned gases, direct injection of fuel into the cylinder is especially attractive with the two-stroke cycle. In this case, the gas exchange process is performed with fresh air alone, and the required amount of fuel is injected into the cylinder after the exhaust ports and valves have closed. Thus no fuel is lost with the exhausted air, and the severe fuel consumption and hydrocarbon emission penalties of premixed-charge two-stroke engines are avoided. The two-stroke diesel has always operated with direct fuel injection and has been competitive with four-stroke cycle diesels in large sizes in marine applications. Direct-injection two-stroke spark-ignition engines also show promise of competing with conventional port-fuel-injected spark-ignition engines in several applications. They obviously avoid the fuel short-circuiting of conventional premixed two-stroke cycle engines. Because the injected fuel spray at light and intermediate loads only partially mixes with the air in the cylinder, these engines operate in a stratified-charge mode which provides satisfactory combustion when operating with substantial excess air: that is, well lean of the stoichiometric or chemically correct fuel–air mixture composition.

Emissions.   The emissions characteristics of two-stroke cycle engines are obviously important and are different from those of four-stroke engine. The emissions requirements that must be met depend on the application and are especially strict in the transportation arena. Conventional two-stroke spark-ignition engines with carburetors (or port fuel injection), which are scavenged with a premixed fuel–air mixture, have extremely high unburned hydrocarbon emissions due to the significant amount of short-circuiting fresh charge. Direct-injection two-stroke spark-ignition engines remove this problem. However, at part load they operate lean, with excess air. They cannot, therefore, utilize the three-way catalyst technology that is widely used with stoichiometric operating four-stroke cycle spark-ignition engines to reduce engine-out emissions of hydrocarbons (HC), carbon monoxide (CO), and oxides of nitrogen (NO_x) before the exhaust gases enter the atmosphere. An oxidation catalyst can be used to reduce engine emissions of HC and CO, although the wide range of two-stroke spark-ignition engine exhaust gas temperatures (which are higher than the four-stroke engine's at high load and lower than the four stroke at low load) makes catalyst durability more challenging. However, the relatively poor scavenging of the two-stroke engine cylinder results in lower NO_x emissions. Dilution of the cylinder charge with burned gases by retaining (effectively recycling) a larger fraction of the exhaust gases lowers the burned-gas temperatures and therefore rates of formation of nitric oxide (NO). The two-stroke spark-ignition engine has inherently high dilution at light load. This high dilution at light loads and idle can, however, result in poor combustion and uneven operation.

Lubrication.   Another challenge of the two-stroke engine is effective lubrication. Crankcase-scavenged two-stroke engines cannot use conventional four-stroke wet-sump lubrication systems because large quantities of oil would be drawn into the combustion chamber. The mixing of oil and fuel with the incoming air is the method commonly used to lubricate the piston rings and skirt in small carbureted two-stroke spark-ignition engines. Direct-injection two-stroke gasoline engines spray oil into the air stream. Large two-stroke diesels bleed oil through holes in the cylinder liner. These are reasons why scavenging systems with an external blower are often the preferred choice: a conventional wet-sump lubrication system can be used.

To highlight the essential points in the preceding discussion, Table 1-1 summarizes the main advantages and disadvantages of a simple crankcase-scavenged two-stroke engine over an equivalent four-stroke cycle engine.

1-3 ENGINE DESIGN AND OPERATING PARAMETERS

In this section, the basic parameters commonly used to characterize the operation of a reciprocating internal combustion engine are defined. These include geometrical relationships, torque and power, normalized quantities such as specific fuel consumption, and the important dimensionless efficiencies. The definitions and notation used in this book are compatible with those developed by Heywood.[2] Definitions and terminology that are specifically related to the gas exchange process in two-stroke cycle engines are given in Chapter 2.

Table 1-1 Advantages and drawbacks of a simple crankcase-scavenged two-stroke engine over a four-stroke engine

Peculiarity	Advantage	Drawback
1. Products of combustion are scavenged by the fresh charge.	Exhaust and intake strokes are removed, thus doubling the number of power strokes per unit time.	Loss of fresh charge to exhaust results in fuel losses in pre-mixed (carbureted) engines; increases fuel consumption and HC emissions.
	Mixing between the two gases results in inherent exhaust gas recycling (EGR), reducing NO_x emissions.	Mixing between the two gases results in a low charging efficiency (low torque) and poor operation at idle. Scavenge pressure must be slightly above exhaust pressure.
2. Doubles number of power strokes per unit time.	Higher engine power/weight ratio and power/swept-volume ratio; smoother torque-versus-time profile; smaller flywheel (lower moment of inertia)	Higher thermal stresses and component temperatures (especially piston)
3. Intake pressure must be slightly above exhaust pressure.		Charge compression is needed prior to scavenging; there are difficulties in employing turbocharging system, catalytic converter, noise reduction and engine brake systems
4. Piston-controlled ports	Simple structure; low production cost; small bulk volume	Symmetrical timing around BC, which results in a loss of torque at low and high speeds; long-skirt piston, high piston inertia
5. Piston lubrication by oil-in-air mixture	Simple maintenance	Higher HC and smoke emissions

1-3-1 Geometrical Relationships

The basic geometrical parameters of a reciprocating engine are (see Fig. 1-4):

The *displaced volume* (V_d) is the volume swept by the piston while travelling from bottom center crank position (BC) to top center (TC). Per cylinder, V_d is

$$V_d = \frac{\pi}{4} B^2 L \tag{1-1}$$

where B is the cylinder bore and L is the piston stroke.

The *clearance volume* (V_c) is the cylinder volume at TC: that is, the volume enclosed by the cylinder head, the cylinder walls, and the piston head surfaces at TC.

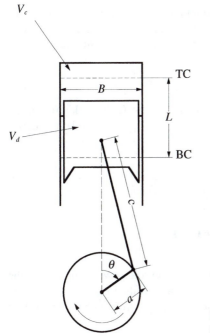

V_c

B

TC

L

V_d

BC

c

θ

a

Figure 1-4 Reciprocating engine geometry arrangement and notation.

Table 1-2 Typical values of major geometric ratios for two-stroke cycle engines

Engine type	Size	r_c	B/L	c/a	$V_{cc}/(V_d + V_c)$
Spark ignition		8–12	0.8–1.2	3–4	2.5–4.5
Compression ignition	Small	12–24	0.8–1.2	3–4	2.5–4.5
	Large	12–16	0.3–0.7	5–9	—

The *compression ratio* (r_c) of an engine is defined as

$$r_c = (V_d + V_c)/V_c \tag{1-2}$$

Two other important geometric relations are the ratio of cylinder bore to piston stroke (B/L) and the ratio of connecting rod length to crank radius (c/a). Note that the crank radius is related to the stroke via the relation $L = 2a$. In crankcase-compression two-stroke engines, there is another important ratio that relates the *maximum crankcase volume* (V_{cc}) to the maximum cylinder volume: $V_{cc}/(V_d + V_c)$. Typical values of these ratios are given in Table 1-2.

From geometrical relationships, the ratio of the *trapped cylinder volume* (V), at any *crank angle* (θ), and the clearance volume is

$$\frac{V}{V_c} = 1 + \frac{1}{2}(r_c - 1)\left[\frac{c}{a} + 1 - \cos\theta - \sqrt{\left(\frac{c}{a}\right)^2 - \sin^2\theta}\right] \tag{1-3}$$

Because several engine phenomena relate to piston motion, it is useful to define a characteristic speed based on piston speed rather than the rotational speed of the crankshaft. If the rotational speed is N, an appropriate characteristic speed is the *mean piston speed* $(\overline{S}_p)$:

$$\overline{S}_p = 2LN \qquad (1\text{-}4)$$

Typical maximum values are 8–16 m/s for small engines and 5–10 m/s for large engines. The *instantaneous piston velocity* (S_p), relative to the mean piston speed, is obtained by differentiating Eq. (1-3) to yield

$$\frac{S_p}{\overline{S}_p} = \frac{\pi}{2}\sin\theta\left(1 + \frac{\cos\theta}{\sqrt{(c/a)^2 - \sin^2\theta}}\right) \qquad (1\text{-}5)$$

1-3-2 Torque, Power, Indicated Work per Cycle, and Mechanical Efficiency

The power delivered by the engine crankshaft (P_b) is the usable power delivered by the engine to the load, and is termed the engine's *brake power*. If the torque exerted by the driveshaft is T, then the brake power is related to the engine torque and the crankshaft rotational speed N by

$$P_b = 2\pi NT \qquad (1\text{-}6)$$

Engine torque is usually measured with the engine clamped on a test bed with its driveshaft connected to a dynamometer. The brake power may then be obtained from Eq. (1-6). It is important to note that the brake power is a product of the engine torque and the crankshaft angular speed, and therefore the maximum engine power is developed at an engine speed that is higher than that which gives the maximum torque.

The work transfer from the gas in a cylinder to the piston during a single cycle $(W_{c,i})$ is termed the *indicated work per cycle*. Pressure data for the gas in the cylinder when integrated with respect to cylinder volume over the operating cycle of the engine (as plotted in Fig. 1-3) yield this indicated work per cycle. Thus,

$$W_{c,i} = \int p\,dV \qquad (1\text{-}7)$$

For two-stroke cycle engines, this integral is calculated over the full cycle. For four-stroke cycle engines that are naturally aspirated, the integral is usually calculated over the compression and expansion strokes. For turbocharged four-stroke cycle engines, the integral is calculated over the full four strokes of the cycle (the full cycle).

The indicted power of an engine having n_c cylinders is consequently defined as

$$P_i = n_c W_{c,i} N/n_R \qquad (1\text{-}8)$$

where n_R is the number of crank revolutions for each power stroke per cylinder. For two-stroke cycle engines, $n_R = 1$; for four-stroke cycles, $n_R = 2$.

The indicated power differs from the brake power by the power lost in overcoming engine mechanical or rubbing friction, driving essential engine accessories, and the pumping power. The total power loss is termed the friction power (P_f):

$$P_f = P_i - P_b \tag{1-9}$$

The ratio of the brake power delivered by the engine to the indicated power is termed the *mechanical efficiency* (η_m):

$$\eta_m = P_b/P_i = 1 - P_f/P_i \tag{1-10}$$

The mechanical efficiency of an engine depends strongly on its geometrical design, materials from which it is made, surface treatments, lubrication, engine speed, and throttle position (in throttled engines) or load. In practice, the mechanical efficiency varies from about 90% at low engine speed and wide-open throttle, to zero at idle.

1-3-3 Mean Effective Pressure

Torque and power depend on engine size. A useful relative measure of an engine's performance is obtained by dividing the work per cycle by the cylinder volume displaced per cycle. This normalized performance parameter is termed the *mean effective pressure* (mep):

$$\text{mep} = \frac{W_c}{V_d} \tag{1-11}$$

It has the dimensions of pressure. It can be expressed in terms of indicated or brake quantities to give imep or bmep, respectively. In terms of power [Eq. (1-8)] and torque [Eq. (1-6)]

$$\text{mep} = \frac{n_R P}{n_c V_d N} = \frac{2\pi n_R T}{n_c V_d} \tag{1-12}$$

where mep, P, and T can be brake or indicated values. With units,

$$\text{mep(kPa)} = \frac{P(\text{kW}) \times 10^3}{V_d(\text{dm}^3) N(\text{rev/s})} \tag{1-13a}$$

$$\text{mep(lb/in}^2) = \frac{P(\text{hp}) \times 396{,}000}{V_d(\text{in}^3) N(\text{rev/min})} \tag{1-13b}$$

and

$$\text{mep(kPa)} = \frac{6.28 T(\text{N-m})}{V_d(\text{dm}^3)} \tag{1-14a}$$

$$\text{mep(lb/in}^2) = \frac{75.4 T(\text{lb}_f\text{-ft})}{V_d(\text{in}^3)} \tag{1-14b}$$

Brake mean effective pressure is a measure of specific engine torque (i.e., the torque per unit of displacement volume), and its maximum value is obtained when the engine torque reaches a maximum. Typical maximum values of bmep for two-stroke spark-ignition engines are 650–1000 kPa, and for large diesels are 1500–1750 kPa. For four-stroke cycle engines, typical maximum values are 950–1200 kPa for naturally aspirated spark-ignition engines, 750–900 kPa for naturally aspirated diesels, and 1250–1700 kPa for turbocharged automotive diesel engines.

The engine's *specific power* is sometimes used as another measure for the relative output of the engine. This is defined as the engine power per unit piston projected area. Thus, if $V_d = A_c L$, where A_c is the cylinder cross-sectional area, Eqs. (1-4) and (1-12) may be solved for the engine power per unit piston projected area (assumed equal to the cylinder cross-sectioned area) to yield

$$\text{specific power} = \frac{P_b}{n_c A_c} = \frac{(\text{bmep})\overline{S}_p}{2n_R} \tag{1-15}$$

1-3-4 Specific Fuel Consumption and Fuel Conversion Efficiency

In engine performance tests, fuel consumption is measured as the fuel mass flow rate. A more useful normalized parameter quantifying fuel usage is the *specific fuel consumption* (sfc). The specific fuel consumption of an engine is a measure of how effectively an engine is using the fuel supplied to produce power, and is defined as the ratio of fuel mass flow rate ($\dot{m}_f$) to power, at given operating conditions:

$$\text{sfc} = \dot{m}_f / P \tag{1-16}$$

A lower sfc implies better energy conversion. Typical best values are 270 g/kW-h for spark-ignition engines, 200 kg/kW-h for small and medium size compression-ignition engines, and as low as 160 g/kW-h for the largest size compression-ignition engines.

A more fundamental measure of the engine's energy conversion performance is the *fuel conversion efficiency* (η_f). This dimensionless parameter is defined as the ratio of the work per cycle to the fuel energy supplied per cycle which can be released by combustion. Thus, the fuel conversion efficiency is

$$\eta_f = \frac{W_c}{m_f Q_{\text{HV}}} = \frac{P}{\dot{m}_f Q_{\text{HV}}} \tag{1-17}$$

where Q_{HV} is the heating value of the fuel. The fuel's heating value is the chemical energy released when unit mass of fuel is fully burned with sufficient air, at atmospheric pressure and temperature. With fuels containing hydrogen, whether the water in the products of combustion is liquid or vapor affects Q_{HV} (giving higher or lower heating values, respectively). By convention, the lower heating value is usually used in Eq. (1-17).

The indicated and brake values of the fuel conversion efficiency are related by the mechanical efficiency:

$$\eta_{f,b} = \eta_m \eta_{f,i} \tag{1-18}$$

Combining Eqs. (1-16) and (1-17) shows that

$$\eta_f = \frac{1}{(\text{sfc}) Q_{\text{HV}}} \tag{1-19a}$$

or, with units,

$$\eta_f = \frac{3600}{\text{sfc}(\text{g/kW-h}) Q_{\text{HV}}(\text{MJ/kg})} \tag{1-19b}$$

$$\eta_f = \frac{2545}{\text{sfc}(\text{lb}_m/\text{hp-h})\,Q_{HV}(\text{Btu/lb}_m)} \tag{1-19c}$$

Thus, specific fuel consumption is inversely proportional to fuel conversion efficiency for normal hydrocarbon fuels (which have closely comparable heating values). Typical heating values for the commerical hydrocarbon fuels used in engines are in the range 42–44 MJ/kg (18,000–19,000 Btu/lb).

1-3-5 Relative Proportions of Air and Fuel

The *air/fuel ratio* (A/F) and the *fuel/air ratio* (F/A) at which the engine operates are defined as follows:

$$A/F = \dot{m}_a/\dot{m}_f \tag{1-20a}$$

and

$$F/A = \dot{m}_f/\dot{m}_a \tag{1-20b}$$

where $\dot{m}_a$ is the mass flow rate of the air flowing into the engine.

The appropriate proportions of air and fuel are best defined in relation to the minimum amount of air required to fully oxidize the fuel. This is called the stoichiometric or chemically correct amount of air. For a hydrocarbon fuel of composition $(CH_y)_n$— gasoline, for example, has a typical composition of C_7H_{13}—the stoichiometric combustion equation is obtained from an oxygen balance:

$$(CH_y)_n + \left(1 + \frac{y}{4}\right)(O_2 + 3.773N_2) \rightarrow$$

$$nCO_2 + n\frac{y}{2}H_2O + n\left(1 + \frac{y}{4}\right)3.773N_2 \tag{1-21}$$

Air is 21% oxygen, 78% nitrogen, and 1% argon, and is usually approximated by an oxygen–nitrogen mixture in molar proportion of 1:3.773.[2] By converting the molar proportions of Eq. (1-21) to mass proportions, we obtain the stoichiometric A/F and F/A ratios: for gasoline these are typically 14.6 and 0.0685, respectively. For hydrocarbon fuels, these stoichiometric ratios are a weak function of the C:H ratio of the fuel. For other fuels they can be substantially different: for example, for methanol, CH_4O, these ratios are 6.47 and 0.155.[2]

Air–fuel mixtures can have more or less air than this stoichiometric requirement. We call such mixtures *fuel-lean* or *fuel-rich*, respectively. The composition of the combustion products formed by burning fuel-lean or fuel-rich mixtures is significantly different, so it is a common practice to use two normalized mixture composition parameters: the fuel/air equivalence ratio ϕ, and the relative air/fuel ratio $\lambda (= 1/\phi)$, where

$$\phi = (F/A)_{\text{actual}}/(F/A)_s \tag{1-22a}$$

and

$$\lambda = (A/F)_{\text{actual}}/(A/F)_s \tag{1-22b}$$

The subscript s stands for stoichiometric conditions ($\phi = \lambda = 1$), where there is just enough air in the fuel–air mixture to burn the fuel fully. For fuel-lean (excess air) mixtures, $\phi < 1$ and $\lambda > 1$, and for fuel-rich (air deficient) mixtures, $\phi > 1$ and $\lambda < 1$. For additional discussion of the properties of such mixtures, see Heywood.[2]

1-4 TWO-STROKE ENGINE APPLICATIONS

1-4-1 Common Designs and Applications

The two-stroke cycle can be used with both spark-ignition and compression-ignition engines. Common applications are where low weight, small bulk, simple construction, and high power output are the prime considerations. It is interesting to note that the largest and smallest internal combustion engines in use today are two-stroke cycle engines. The smallest engines are those used in models that employ carbureted compression-ignition engines having an output power well below 100 W. These are crankcase loop- or cross-scavenged engines† in which a glow plug is usually installed to assist cold starting (Fig. 1-5). The largest engines are those used in stationary power plants. These are turbocharged compression-ignition engines having a uniflow (or along the cylinder axis) scavenging system.

Two-stroke cycle engines may be divided into three major categories according to their power output: small engines (0.5–50 kW), medium-size engines (50 kW–1 MW) and large engines (above 1 MW). Small two-stroke engines are customarily naturally aspirated spark-ignition engines with a crankcase loop-scavenged system. They are mainly used when compactness, light weight, and ability to run at unusual altitudes

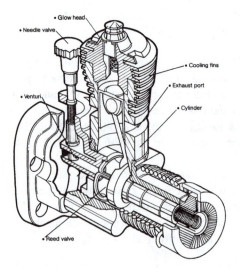

Figure 1-5 Small carbureted compression-ignition Cox reed-valve two-stroke cycle engine. (Courtesy Cox Hobbies, Inc.)

†Loop- and cross-scavenged refer to the overall flow pattern of the scavenging fresh charge within the cylinder.

Figure 1-6 Low-pressure die-cast aluminum cylinder and piston for chain saw two-stroke spark-ignition engine. (Courtesy Kolbenschmidt AG.)

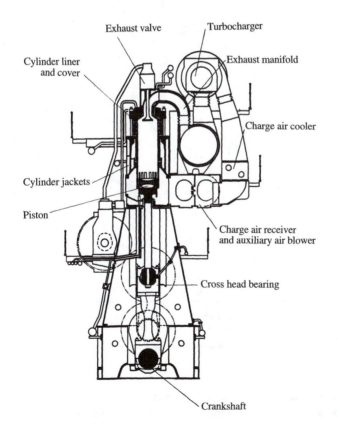

Figure 1-7 Cross section of Sulzer RTA 84 cross-head marine diesel engine. (Courtesy Wärtsilä NSD Switzerland Ltd.)

Table 1-3 Applications of loop-scavenged spark-ignition two-stroke engines (developed from Laimbock[3])

	Engine displacement (cm³)	Cylinder size (cm³)	Specific power (MW/m³)	Exhaust system†	Intake system‡
PORTABLE DEVICES					
Chainsaw	30–110	30–110	35–60	I	P(R)
Bushcutter	25–50	25–50	35–60	I	P
Tree sprayer	25–50	25–50	35–60	I	P
Pump	100–300	100–300	30	I, II	P
Generator	30–100	30–100	20–25	I	P
VEHICLES (homologated)					
Two-wheeler:					
Bicycle auxil. engine	15–50	15–50	20–25	II	P(R)
Moped	50–75	50–75	18–75	II, III	P, R
Scooter	50–200	50–200	40–75	II, III	R
Motorcycle	50–550	50–550	75–170	III, IV	R
Passenger car	500–900	250–300	28–32	II	P, R
VEHICLES (non homologated)					
Model engine two-wheeler	1–25	1–25	75–280	I, III, IV	D
Road racing	80–500	80–167	200–270	IV	R, D, P
Motocross	50–550	50–550	100–250	IV	R
ATV	125–350	125–175	150–170	IV	R
Karting	100–250	100–125	200–250	III, IV	D
Snowmobile	250–900	125–375	100–150	III, IV, V	D, R, P
AIRCRAFT					
Model engine	1–25	1–25	75–280	I, III, IV	D
Ultralight	350–900	300–350	50–60	II	R, D, P
Remotely piloted aircraft	440–600	110–300	50	I	P
MARINE					
Model engine	1–25	1–25	75–280	I, III, IV	D
Boat racing	350–700	87–175	100–125	IV	D
Wetbike	400–650	200–325	100–150	II, V	R
Outboard engine	125–3600	125–450	30–80	I, II, V	R
Sail drive	200–600	200–250	30	I, II	R
AGRICULTURAL & CONSTRUCTION DEVICES					
Lawn mower	85–150	85–150	20	I, II	P
Mower	150–300	150–300	25–35	II	P
Swath rake	150–300	150–300	25–35	II	P
Jolter (vibratory)	50–150	50–150	25–35	II	P
Jumping jack	75–95	75–95	25	I	P
Ground boring mach.	50–150	50–150	25–35	I, II	P
Ground milling mach.	150–300	150–300	25–35	II	P
Disc-saw	50–100	50–100	30–50	II	P
Snow-thrower	50–300	50–300	30	I, II	P
STATIONARY & MULTIPURPOSE DEVICES					
	20–600	20–300	25–40	I, II	P
Fire pump	185–600	185–300	30–45	I, II	P, R

† Exhaust system classification: I, short port and expansion into the open, or volume; II, manifold tube and volume (2-in-1 manifold and volume); III, manifold tube, diffuser, baffle plate (2-in-1 manifold and volume); IV, diffusor, cylindrical centerpart, baffle cone, tailpipe; V, multicylinder with merged, manifolded exhaust system.

‡ Intake system classificaton: R, reed valve intake; D, rotary intake; P, piston controlled intake; P + R, combined piston and reed valve intake.

are required. Portable chainsaws, lawn mowers, snowmobiles, outboards, small electrical generators, mopeds, motor scooters, motorcycles, and unmanned aircraft vehicles are examples of applications of these small engines. Examples of this class are a chainsaw engine (Fig. 1-6) of 111-cm³ displacement which develops 5.2 kW at 7000 rev/min,

Table 1-4 Examples of large-bore CI two-stroke cycle engines

Type	Units	MAN–B&W K70MC-S	Mitsubishi 4UECGOL	MAN–B&W K90MC	Sulzer-IHI RTA84
No. of cylinders	—	4	4	4	4
Bore	m	0.70	0.60	0.90	0.84
Stroke	m	1.96	1.90	2.55	2.40
Displaced volume	liter	3,016	2.150	6,486	5,320
Max. output	MW	10.2	6.25	17.3	13.2
at speed	rev/min	120	110	94	90
Brake spec. fuel cons.	kg/kW-h	0.160	0.170	0.162	0.171
Max. mean piston speed	m/s	7.8	7.0	8.0	7.2
Max. bmep	MPa	1.70	1.60	1.70	1.55
Max. combustion pressure	MPa	—	13.0	—	—
Power/gross weight	kW/kg	—	0.023	0.023	0.022
Power/displacement volume	MW/m^3	3.40	2.91	2.66	2.49
Type of scavenging	—	Uniflow	Uniflow	Uniflow	Uniflow
Major use		Stationary	Marine	Marine	Marine

and a Yamaha motorcycle engine of 248-cm^3 displacement which develops 20 kW at 7500 rev/min. A summary of typical applications of loop-scavenged spark-ignition engines is given in Table 1-3.

Medium-size two-stroke cycle engines are usually turbocharged compression-ignition engines having a uniflow scavenged system. They are frequently used in armored vehicles and military tanks.

Large two-stroke engines are well suited to marine propulsion. Figure 1-7 shows a cross section of such an engine. These low-speed compression-ignition uniflow-scavenged engines are able to match the power–speed requirements of any ship with relatively few cylinders in a simple direct-drive arrangement. The engine shown has a bore of 840 mm and stroke of 2900 mm. It produces 3.46 MW per cylinder at 78 rev/min. Such engines are turbocharged, with auxiliary blowers for low-load operation. These very large engines are available with from 4 to 12 cylinders. They can also readily follow the propeller law without adjustment, and are easily reversed.[4] These large engines can achieve brake fuel conversion efficiencies up to 54%.[5]

Some examples of large-bore compression-ignition engines and small-bore spark-ignition engines are listed in Tables 1-4 and 1-5, respectively.

1-4-2 Special Designs and Applications of the Two-Stroke Engine

Owing to the simple structure, light weight, and low production cost of the two-stroke engine, it is often attractive for various special applications. Free-piston compressor engines, gas producers, and pulse-jet engines are good examples.

Table 1-5 Examples of small-bore carbureted SI two-stroke engines

Type	Units	Yamaha	Yamaha	Kawasaki	Suzuki	Suzuki	Daihatsu	Stihl	Sachs	Rotax
No. of cylinders	—	1	3	1	3	2	2	1	2	2
Bore	cm	5.6	6.7	5.4	8.4	5.4	6.2	5.9	5.0	7.6
Stroke	cm	5.0	6.6	4.8	7.2	5.4	5.9	4.2	6.7	6.4
Displaced volume	cm^3	123	698	110	1197	247	356	111	352	581
Effective compression ratio	—	7.0	6.0	7.0	6.5	7.5	6.9	5.2	—	5.9
Max. output	kW	11.9	29.8	3.2	55.9	33.6	17.9	—	20	47
at speed	rev/min	7,000	5,500	5,000	5,300	8,500	5,000	7,000	7,250	6,500
Max. torque	N-m	16.7	64.7	6.47	110.8	37.3	36.3	—	25.0	74.0
at speed	rev/min	6,500	3,100	3,700	4,500	8,000	4,000	—	6,500	5,800
Bsfc	kg/kW-h	0.375	0.361	0.496	0.408	0.321	0.375	—	0.350	0.236
Mean piston speed	m/s	11.6	12.1	8.0	12.7	15.3	10.3	13.8	16.2	13.9
Max bmep	kPa	870	590	370	590	970	650	—	446	800
Power/ gross weight	kW/kg	0.42	0.43	0.22	0.49	0.98	0.29	0.45	2.17	1.65
Power/displaced volume	MW/m^3	96.7	42.7	29.0	46.7	136	52.2	46.8	56.8	80.9
Type of scavenging†	—	C	S	S	S	S	S	S	S	S
Major use‡	—	G	G	G	G	A	A	Chainsaw	Aero	Aero

† Abbreviations: C, cross; S, Schnürle;
‡ Abbreviations: G, general, A, automotive.

The free-piston engine is a combined engine and compressor unit whose power piston is directly connected to the compressor piston without any intermediate crankshaft and connecting rod linkage. The free-piston engine may be used either as a compressed air supplier for external use (the free-piston air compressor) or as a gas generator that supplies hot gas at high pressure (the free-piston gasifier). In the latter, the air supply to the engine must be supercharged to maintain the receiver pressure upstream of the gas user. The essential features of a basic unit are illustrated in Fig. 1-8. The cylinder consists of three parts: the power, the compressor, and the bounce compartments. The piston oscillates between its bottom center (BC) and top center (TC) positions, controlling the scavenge and exhaust ports of the power cylinder as it does so. The operating cycle the working fluid follows is either an SI or a CI two-stroke cycle. When the piston travels downward toward BC, fresh air from the atmosphere enters into the compressor through the inlet valve while the contents of the bounce cylinder are compressed. When the piston travels upward toward TC, the fresh air is compressed and then delivered to the receiver through the exit valve. During this period, the piston is driven by the high pressure in the bounce cylinder, which effectively acts like a flywheel, absorbing energy from the piston on the downward stroke and returning it on the upward stroke.

In its classical form, the free-piston engine is an opposed-piston two-stroke cycle concept containing two free pistons that operate between the combustion chamber and the bounce chamber. There are two basic versions, which differ from each other by their compression method. In the inward compression version, compression takes place on the inward stroke of the pistons (i.e., when the pistons move toward the spark plug in Fig. 1-8). Compressed air is delivered to an air receiver where it is stored until needed for the next inward stroke. The energy for compression is supplied by the bounce cylinders, which are, therefore, large. In the outward compression version, compression takes place on the outward stroke (i.e., during the power stroke). If compressed air is not needed for external use, the compressed air may be delivered directly from the compressor cylinder via a transfer duct to the power cylinder. Because the bounce cylinders only have to store energy for compression in the power cylinder, they can be much smaller in diameter. The inward compressing version, however, has the advantages of compactness and mechanical simplicity.

The history of the free-piston engine began in the mid-1920s with Marquis R. de Pescara, who was interested in using a free-piston engine for helicopter propulsion.[6] In the early 1940s, Sulzer announced a 7000-hp compound power plant in which the free-piston hot-gas generator was proposed as a substitute for the conventional engine (Fig. 1-9).[7,8] The main elements of the free-piston engine were the two opposed pistons, four-compartment cylinder, main turbine, auxiliary turbine, charging compressor, and electric generator. On the power stroke, energy is stored in the bounce cylinders while air is compressed in the outer compressor chamber (outward compression version) and delivered into one end of the power chamber, which is a uniflow-scavenged diesel engine. With this arrangement, the stroke and compression ratio are not fixed but are determined by the mass of the reciprocating parts and the amount of fuel injected. Excessive compression and combustion pressures are avoided by controlling the compression in the inner compressor chambers by the location of the spill ports. In some other free-piston engines, the pistons are constrained by a synchronizing shaft and rock-

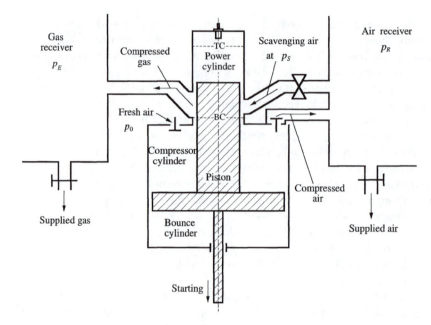

Figure 1-8   Essential features of free-piston hot-gas generator or air compressor.

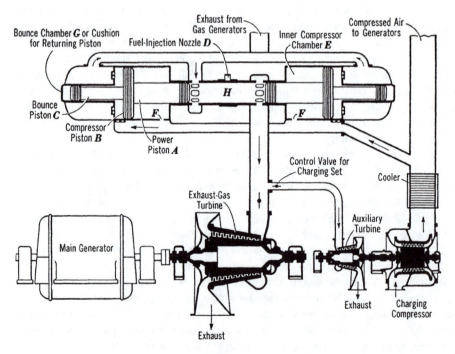

Figure 1-9   Free-piston hot-gas generator with gas turbine.[7]

ing arm to ensure that the two piston assemblies remain in phase. For a working pressure of 400 kPa, the Sulzer free-piston hot-gas generator oscillates at 560 cycles per minute, delivers hot gas at a temperature of 450°C, and each hot-gas generator supplies a power of 1.05 MW at a thermal or fuel conversion efficiency of 43%.

Another more recently proposed application for the free-piston engine is as a power source in the propulsion system of an automotive vehicle.[9,10] The propulsion system uses a two-stroke spark-ignited free-piston engine coupled to a hydraulic pump compartment and an accumulator where high-pressure hydraulic fluid is stored for transmission of power. The energy in the accumulator is transmitted to hydraulic motors that provide the tractive effort.

The pulse-jet engine is another special design of the two-stroke engine. In its simplest form it consists of nothing more than a suitably shaped duct and a fuel supply system. It has no moving parts and may be used either as a propulsion engine or as a hot-gas generator. Figure 1-10 shows schematically the operating cycle of an idealized valveless pulse-jet engine. The engine is constructed from an air inlet duct, a combustion chamber, and a resonance tube for exhausting the products of combustion. Following Putman et al.,[11] its operation can be divided into four phases (see Fig. 1-10). Phase 1 starts with ignition of the combustible mixture in the combustion chamber at point A of the cycle. Ignition occurs by the hot gases flowing backward, or by the residuals of the previous burning phase, or by a hot spot in the burner. The burning region expands and moves outward in both directions but, owing to the engine configuration, moves mainly toward the exit of the combustion chamber. In some designs, a one-way valve is installed at the inlet duct to avoid the backflow. In phase 2 (B–C), the gases expand and move outward while the pressure in the combustion chamber falls below atmospheric pressure due to momentum effects. Consequently, the outward flow slows down (point C). In phase 3 (C–D), some of the products of combustion move back through the inlet duct. In phase 4 (D–A), gases move strongly back through both ends and momentum effects cause the pressure to rise above ambient. The main inward flow is through the inlet, and most of the gas enters the combustion chamber as fresh combustible mixture. With the rapid burning of this new charge, the cycle repeats. The operating cycle may be either an SI or CI two-stroke cycle.

The first combustion-driven oscillator was reported by Byron Higgins in 1802.[11] A hundred years later, attempts were made to use the pulse combustor as a hot-gas source for gas turbines. In 1906, Marconnet proposed using the pulse-jet combustor for propulsion. The compactness and remarkable simplicity of the pulse-jet combustor have intrigued designers for years. However, apart from the large-scale use of pulse combustors during the Second World War as thrust producers for propelling German V-1 cruise missiles, and subsequently for various French prototype engines, the use of thrust-producing pulse combustors has been relatively rare. This is because of their poor specific fuel consumption and low specific thrust when compared with turbojet and turbofan engines. Despite their shortcomings, some pulse jets may still be viable as heaters and hot-gas producers,[11] and also as power units for expendable low-cost short-duration-mission subsonic vehicles such as target drones and remotely piloted vehicles.[12]

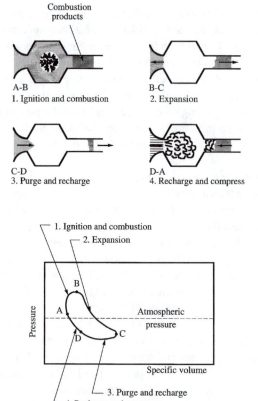

Figure 1-10 The phases in the operation of an idealized pulse-jet engine. Developed from Putman et al.[11]

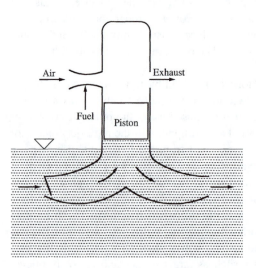

Figure 1-11 Schematic diagram of a piston pulse-jet engine.

Among the wide variety of types of pulse-jet engine,[11] the piston pulse-jet engine for water propulsion[12] is of interest (Fig. 1-11). The engine is constructed from a cylinder, a bifurcated duct, a piston, a simple carburetor, and a glow plug. The cycle begins as the vessel advances to the left and water enters through the reed valve, pushing the piston upward. The fresh charge inside the cylinder is compressed and ignition occurs. Combustion takes place, the piston travels downward, and a water jet is directed to the right (the reed valve is closed). Meanwhile the cylinder exhaust port is exposed and the burned gases blow down into the atmosphere. Owing to the high inertia of the water and the piston, the pressure inside the cylinder falls below atmospheric levels and a new charge is drawn into the cylinder. When the pressure in the duct falls below the stagnation pressure at the duct entrance, the reed valve opens and a new cycle begins. In some versions, an externally heated flash boiler replaces the power cylinder and the piston is not needed: instead, the water acts as the piston. This pulse-jet engine is popular in "Putt-Putt" toy boats.

1-5 HISTORICAL DEVELOPMENT OF TWO-STROKE CYCLE ENGINES

The two-stroke cycle engine is as old as the concept of the heat engine, which has served mankind for over 450 years. All internal combustion engines developed prior to the appearance of the Otto four-stroke cycle engine in 1876 operated on a two-stroke cycle. During the earlier part of this period, engines were designed to operate on gunpowder, and many efforts even during the first half of the 19th century were aimed at applications with this type of fuel. Charging the cylinder with fresh mixture and scavenging the products of combustion have always been a major concern of the engine designer. In this section, we present a brief historical review of some pre-1900 landmarks in the development of two-stroke engines, with special emphasis on the various solutions proposed for improving the gas exchange process.

1-5-1 Exploration Period (1508–1860)

As far back as 1508, Leonardo da Vinci sketched the "fire engine" for "raising heavy weight by fire".[13] A strong 1.5-m^3 cylindrical vessel, 60 cm in diameter and 6 m long, was installed with a piston having a leather surround for sealing. It was proposed that about 0.2 kg of gunpowder be introduced into the cylinder while the piston was situated at TC. The charge was to be ignited through a narrow hole, the gunpowder would then explode, driving the piston downward. The ensuing expansion would decrease the temperature and pressure of the gases in the cylinder. Then atmospheric pressure would press the piston up and raise a load. This machine would have been able to raise 1600 kg to a height of about 3 m. Although he did not build the engine, Leonardo da Vinci could be considered the inventor of the atmospheric reciprocating internal combustion engine and his concept was, without doubt, a two-stroke cycle engine.

In 1673 Christian Huygens, a Dutch physicist, was the first to build a gunpowder engine. In fact he built several engines of the same design with different cylinder dimensions.[13] The diameters ranged from 68 to 325 mm and the cylinder lengths

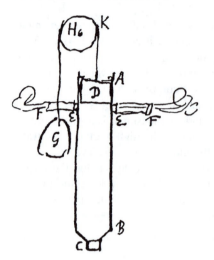

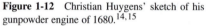

Figure 1-12 Christian Huygens' sketch of his gunpowder engine of 1680.[14,15]

from 54 to 120 mm. Figure 1-12 shows one of Huygens' later designs, from 1680. Hardenberg[13] describes these engines as follows. "The cylinders were made of tin-plated sheet metal, lined with a thick layer of plaster. The piston was sealed with leather pressed onto the liner surface by means of sponges which were kept wet with water contained in a cavity in the piston. Wetted flexible leather hoses at the sides of the top part of the cylinder served as valves; these hang limp after gunpowder combustion and thus shut the exit ports in the cylinder wall, which were tapered to prevent the hoses from being drawn in. A small vessel charged with gunpowder was screwed into the bottom of the cylinder immediately after a piece of tinder with one end reaching into the powder had been ignited." Evidently, the working principle was the same as Leonardo's engine, although the design differed in several respects. Huygens found that the engine cylinder could be emptied by only about 80%, but he did not report any measured data with regard to the loads lifted and the heights to which they had been raised.

Denis Papin, a French physician, who had been Huygens' assistant and who continued his experiments, improved the original design of his teacher. One of his major design features was locating the exhaust valve in the piston. Papin was the first who explicitly noted, in 1688, the problem of scavenging the cylinder of burned gas at the end of the cycle. He found that with his design, about one-fifth of the charge remained in the cylinder as residuals. In fact the final pressure was slightly higher than atmospheric because of the weight of the exhaust valve. Papin's two-stroke cycle engine achieved a thermal efficiency of 4% with a bmep of 20 kPa, very low by today's standards.[13]

The first recorded engine that used combustion of a vaporized-fuel–air mixture within a cylinder to produce continuous reciprocating piston movement, and did not rely on atmospheric pressure to do the work, was developed by Robert Street in England in 1794 (Fig. 1-13). The engine had a cast-iron open-top cylinder that was heated underneath by a solid-fuel stove. A heavy solid cast-iron piston worked a water pump. Turpentine was poured through a funnel onto the hot bottom of the cylinder, where it vaporized and mixed with the air drawn into the cylinder by raising the piston manually.

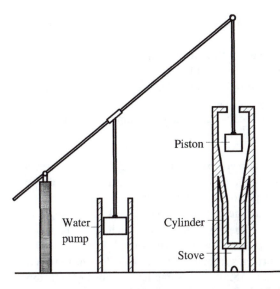

Figure 1-13 Robert Street's turpentine-fueled engine of 1794. Developed from Sher.[16]

A flame was introduced to a touch hole in the cylinder wall, which ignited the turpentine vapor, which then blew the piston to the end of its stroke. At the end of the stroke, the piston protruded out of the cylinder, allowed a part of the burned gas to exhaust, and then fell back into the cylinder under gravity with guidance from slides. Then the whole cycle could start again. In this machine, no attempt was made to scavenge the burned gas that was trapped in the cylinder as the piston fell back into the cylinder.

George Medhurts, an English engineer, in 1800 was the first to propose a two-stroke engine with a controlled cyclic repetition of the working processes. In his direct-acting expansion machine, with a cylinder capacity of 437 cm^3, the connecting rod acted through a parallel-motion mechanism on a crankshaft.[13] To produce a power output of one horsepower, about 1 cm^3 of gunpowder should be introduced from a funnel by means of a stopcock, as the piston approached its top center position. This stopcock was to be rotated by the crankshaft at half its speed by means of gearing. Ignition would be effected by a spark-producing flint wheel mechanism, actuated by the piston when reaching TC. The air enclosed between the piston and the bottom of the cylinder would act as a spring to return the piston. A small spring-loaded inwardly opening valve in the bottom was expected to be opened by atmospheric pressure at the TC position of the piston. With the piston in its bottom center position, the pressure in the working cylinder should be lower than atmospheric, which would cause the valve in the cylinder head to be opened and permit the combustion gases to be exhausted until the piston would shut the valve when approaching TC.

An interesting mechanism for scavenging burned gases from the cylinder was proposed and tested by Isaac de Rivaz of Switzerland in 1805. The Rivaz engine, which was installed in one of his later vehicle models, used a methane–air mixture that was produced on board from coal. The operation of the engine was initiated by movement of a hand-controlled lever (Fig. 1-14) that performed three simultaneous functions[15]:

a. A lower piston was pulled down and away from the power piston, creating a vacuum in the combustion chamber to suck in a combustible mixture.
b. A valve opened to allow the premeasured and mixed charge of gas and air to flow into the combustion chamber.
c. Contacts in an electric circuit closed, causing a spark to jump a gap between two wire terminals in the combustion chamber.

A pressure-operated exhaust valve allowed part of the burning gases to escape during the expansion process. The valve was closed when the cylinder pressure dropped below atmospheric as the piston continued to travel upward. When the internal pressure once again exceeded atmospheric, toward the end of the inward working stroke, the exhaust valve was opened again. As the two pistons came close to each other, the residual exhaust gases were almost completely purged from the cylinder.

The problem of scavenging the burnt gas from the cylinder was also one of the main concerns of William Cecil in England in 1820. Cecil's engine was constructed from a vertical cylinder and two horizontal expansion chambers (Fig. 1-15). A mixture of air and hydrogen was introduced into the cylinder as its piston travelled downward (owing

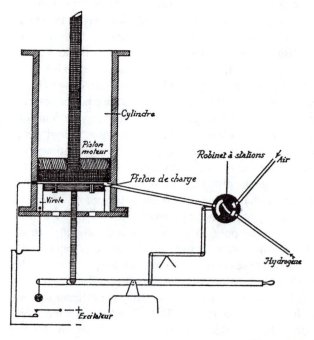

Figure 1-14   Isaac de Rivaz's engine of 1805.[15]

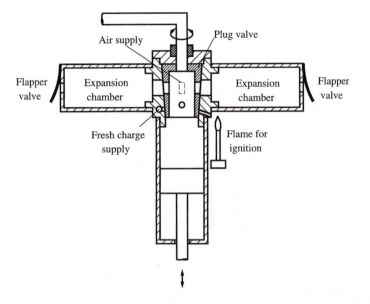

Figure 1-15   William Cecil's engine of 1820. Developed from Cummins.[15]

to the inertia of the flywheel). During the intake stroke, the plug valve was positioned so that the expansion chambers were exposed to the atmosphere while the cylinder was sealed. When the piston approached BC, the plug valve was rotated quickly to a position where a small valved port allowed the continuously burning flame to ignite the cylinder contents. After ignition, the cylinder was sealed again by rotating the valve back. At this position, wide ports in the plug valve interconnected the two expansion chambers and the cylinder. As the mixture burned and expanded, air and residual gases from the previous cycle in the expansion chambers were pushed out through the flapper valves. The rapidly cooling gases in the cylinder created a vacuum, which caused the flapper valves to close. Atmospheric pressure acting on the underside of the piston pushed it upwards for the power stroke. When the piston reached TC, the plug valve was rotated back to its intake position to begin the next cycle.

A different solution for the gas exchange problem was proposed by Samuel Morey of New Hampshire in 1826.[15] Morey's engine (Fig. 1-16) consisted of two pistons in a single cylinder. The cycle starts when the connecting rod lifts the two pistons while the two are attached to each other. Meanwhile air is drawn into the cylinder through the air valve. As the lower piston reaches its TC position, the main piston continues to travel upward and fresh mixture is drawn through the mixture port into the space between the pistons. At a certain point, the ignition port opens and the cylinder contents are ignited. The main piston continues the expansion stroke while the combustion gases escape through the perforated lower piston and are forced out through the pressure-operated leather valve. The working stroke begins as atmospheric pressure pushes down the main piston and ends when the main piston makes contact with the lower piston at BC. By these means, the space between the two pistons is fully scavenged.

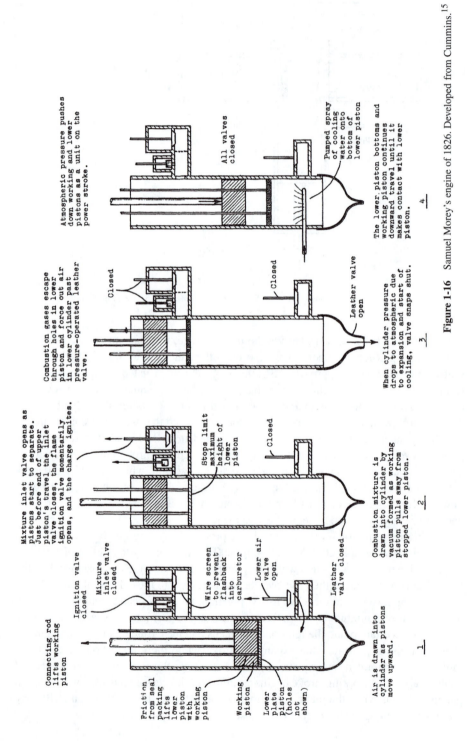

Connecting rod lifts working piston

Ignition valve closed

Mixture inlet valve closed

Wire screen to prevent flashback into carburetor

Lower air valve open

Friction from seal packing lifts lower piston with working piston

Working piston

Lower plate piston (holes not shown)

Leather valve closed

Air is drawn into cylinder as pistons move upward.

_1_

Mixture inlet valve opens as pistons start to separate. Just before end of upper piston's travel the inlet valve closes, the flame ignition valve momentarily opens, and the charge ignites.

Stops limit maximum height of lower piston

Closed

Combustion mixture is drawn into cylinder by vacuum formed as working piston pulls away from stopped lower piston.

_2_

Combustion gases escape through holes in lower piston and force out air in lower cylinder past pressure-operated leather valve.

Closed

Leather valve open

When cylinder pressure drops to atmospheric due to expansion and start of cooling, valve snaps shut.

_3_

Atmospheric pressure pushes down working and lower pistons as a unit on the power stroke.

All valves closed

Pumped spray of cooling water onto bottom of lower piston

The lower piston bottoms and working piston continues downward travel until it makes contact with lower piston.

_4_

Figure 1-16 Samuel Morey's engine of 1826. Developed from Cummins.[15]

28

The first internal combustion engine that employed a compression stroke prior to ignition was the two-stroke engine developed by William Barnett in England in 1838. The Barnett engine consisted of a double-acting piston and an auxiliary pump. The fresh charge was pumped into the upper chamber about 30° before TC and then compressed by the piston up to TC. At TC, the mixture was ignited and burned while the piston travelled downward. At the same time, the auxiliary pump began pulling exhaust gas from the lower chamber through the exhaust port situated midway down the cylinder. When the piston covered the exhaust port, the auxiliary pump began to pump new fresh mixture into the lower chamber (through a port in the chamber head), where it was compressed by the lower face of the piston and ignited. Cummins[15] points out that by using only pressure-controlled valves to admit the fresh charge into the cylinder, the charge would flow into the chamber having the lowest pressure. If the entire charge was to enter the desired chamber, the pressure in the opposite chamber had to remain higher—a condition not possible in this configuration. Although Barnett's proposed gas exchange process seems inefficient, his basic idea of scavenging the cylinder from one side to the other by using an external pump and a piston-controlled port has later been applied successfully to many types of engines.

1-5-2 Industrialization Period (1860–1900)

The Frenchman Lenoir's engine (1860) was the world's first production gas engine: over 400 units were sold. The Lenoir engine was a two-stroke cycle engine operating without compression. Figure 1-17 shows several cycles from an operating engine in a diagram of cylinder pressure versus cylinder volume.[15] Combustible mixture was drawn into the cylinder during the first half of the downward stroke, and then after the inlet valve had closed, it was ignited. The second half of the downward stroke generated power. The upward stroke was used for exhaust, and the scavenging efficiency† was equal to the volumetric expansion ratio. The engine looked very much like a horizontal steam engine and had double-acting cylinders (giving two power strokes per revolution) with a separate intake and exhaust valve. These were plate-type slide valves, one on each side of the cylinder, reciprocated by eccentrics. The high consumption of expensive town gas (a mixture of hydrogen and carbon monoxide), and the frequent need to clean the spark plugs and charge the batteries, led to the eventual stoppage of production within 10 years.

In 1872, Brayton marketed an engine in which combustible mixture was compressed in a separate cylinder to 500–600 kPa and delivered to an intermediate chamber. From the latter it passed into the working cylinder through a valve and wire-gauze screen to prevent backfiring into the intermediate chamber. A small flame was kept burning in the combustion chamber, and the charge burned at the rate at which it entered, thus maintaining an almost constant pressure throughout a considerable part of the power stroke. Although a large part of the burnt gas was displaced by the moving piston (while the remaining part was mixed with the fresh charge inside the cylinder), the

---

†The scavenging efficiency is the ratio of mass of fresh charge (or air) trapped in the cylinder to the total mass of trapped charge; see Section 2-1.

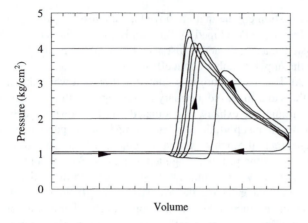

Figure 1-17 Cylinder pressure versus cylinder volume for several cycles, from Lenoir engine of 1860. Developed from Cummins.[15]

engine had one power stroke for each revolution of the crankshaft and therefore can be thought of as a two-stroke cycle engine. Figure 1-18 shows pressure–volume diagrams for both the motor (or working) cylinder and the pump (or compression) cylinder, illustrating how this engine operated.[15] Despite the use of a wire-gauze screen, backfiring was a constant source of trouble with the city gas fuel, and therefore in his later engines Brayton used kerosene. Both the Lenoir and Brayton engines had very low efficiencies: the Lenoir because it operated without compression, and the Brayton because the greater part of the power generated in the working cylinder was consumed operating the compression cylinder.

While Nicolaus Otto was developing the four-stroke cycle engine, the Scotsman Clerk, Robson and Atkinson in England, and Koerting in Germany, were hard at work on the two-stroke cycle engine. Clerk began work on a gas engine at the end of 1876, and his first patent was taken out at the beginning of 1877. The patent dealt with an engine of the air-pressure vacuum type, in which the explosion compressed air into a reservoir and caused a partial vacuum in both the explosion chamber and a vessel connected with it. Clerk's next patent was taken out in 1878. In this engine an external pump compressed a mixture of fuel gas and air into a reservoir at the full pressure required for compression; the mixture under compression was admitted to the cylinder during the first part of its upward stroke, cut off, and then ignited by a platinum igniter. The piston is driven downward by the expanding high-pressure gases, and exhaust takes place at the end of the downward stroke close to bottom center. The engine produced a brake power of 2.25 kW.

The engine usually known as the Clerk-cycle engine was patented in 1881 (Fig. 1-19). It employed the same sequence of operations as the Otto engine, yet it performed these in two complete strokes of the piston instead of four. It was a single-cylinder single-acting engine with a one-way inlet valve situated in the combustion chamber head, but with no mechanical exhaust valve as in the Otto engine. Exhaust was effected through ports cut in the cylinder wall, controlled by the motion of the piston

Motor or Working Cylinder

Mean Pressure 30.2 lbs. per sq. in.   8 ins. dia. Cylinder   Stroke 12 ins.   200 revs. per min.

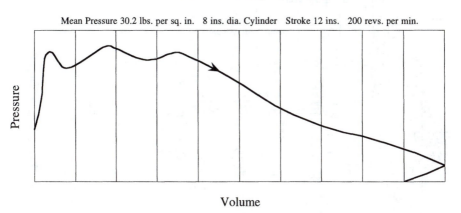

(a)

Pump or Compression Cylinder

Mean Pressure 27.6 lbs. per sq. in.   8 ins. dia. Cylinder   Stroke 6 ins.   200 revs. per min.

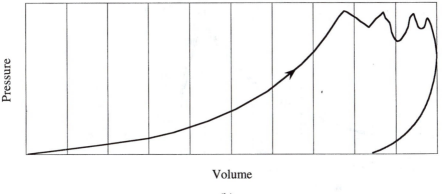

(b)

Figure 1-18   Pressure–volume diagrams for (a) the motor cylinder and (b) the pump cylinder of Brayton's 1872 engine. Data from Clerk's tests of 1878 (Clerk, *The Gas and Oil Engine*, 1896).[15]

toward the bottom of its stroke. This was the first engine where the exhaust discharge was timed and controlled by the main piston only (at about the same time, Robson used a separate exhaust valve that was opened and closed at the proper time by an external means). In Clerk's engine, an auxiliary pumping cylinder was incorporated to provide a positive means of charging the cylinder. This pumping cylinder forced the fresh charge under pressure (some 14 kPa above atmospheric pressure) through a one-way inlet valve into the combustion chamber of the working cylinder. A slide valve controlling the admission of the ignition flame to the combustion chamber was located on top of the cylinder head, and the whole cylinder was jacketed for water cooling. It is interesting

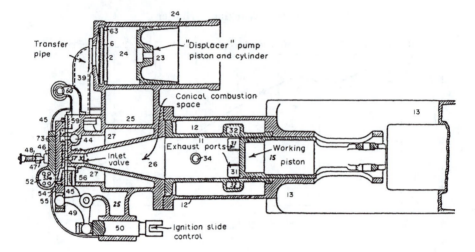

Figure 1-19   Layout of the Clerk engine of 1881.[17]

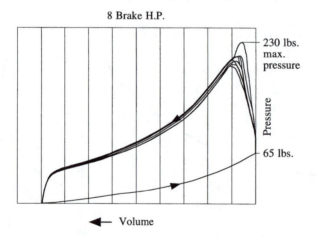

Figure 1-20   Indicator diagram of one of the first two-stroke engines made by Sterne and Co. in 1881 (Science Museum, London).

to note that the shape of the cylinder head was designed to assist the scavenging effectiveness of the fresh charge entering the cylinder, and also to preserve some degree of fresh-charge, burned-gas stratification. Figure 1-20 shows an indicator diagram (cylinder pressure versus volume) of one of the first of these engines. Clerk's engine, built by Sterne & Co., the Campbell Gas Engine Co., and many other makers, was widely used.

Further important developments of the two-stroke cycle engine came in 1891 when Joseph Day, an English engineer, produced his simpler engine design. In Day's design, the pumping cylinder was dispensed with and use was made of the pumping capabilities of the underside of the working piston. To arrange this, the crankcase was made gas tight, and the carburetor fed fresh fuel–air mixture into the crankcase through a one-way inlet valve that opened on the upward or suction stroke of the underside of the piston, and closed under the pressure of the descending piston. Admission to the cylinder occurred through a port in the cylinder wall that was opened by the piston near the bottom of the stroke. The descending piston then compressed the gas in the crankcase. Next, the gas entered the working cylinder through the scavenge duct and scavenge port, and was directed up toward the top of the cylinder by a special deflector on the piston. This deflector served two functions: first, to direct the fresh gas up into the combustion space so that it would displace the residual gas and drive it down toward the exhaust port; second, to prevent as far as possible the fresh gas escaping through the exhaust port on the opposite side of the cylinder. In its simplest form, the Day engine had only three moving parts for each cylinder. The Day engine provided two further steps in the development of the three-port two-stroke cycle engine: substitution of a closed crankcase for the closed inner end of the cylinder, and piston-controlled scavenge ports and the one-way valve for flow into the crankcase. The flow into the crankcase from the carburetor is controlled by the third port in the cylinder wall, which is uncovered by the bottom of the piston skirt at the top of its upward stroke. The descending piston then closes the crankcase inlet port, and the entrapped gas in the crankcase is then compressed for delivery to the cylinder through the scavenge port. This was an important simplification, and this scavenging approach has been used, successfully, more than any other.

There were many more who subsequently contributed important detailed refinements that have made the two-stroke cycle engine into the effective and reliable machine it is today. However, many of the basic principles were developed by the pioneers of the last two decades of the 19th century.

1-6 DEVELOPMENTS IN THE 20TH CENTURY

Large-bore two-stroke diesel engines were initially used exclusively for stationary applications but soon the advantages of high thermal efficiency, low maintenance, high power to engine-weight ratio, and its power–speed characteristics began to be recognized. The low-speed diesel engine has become the natural choice of propulsion plant for ocean going ships, and engineering progress in the marine industry and engine technology has been strongly coupled. Figure 1-21 lays out the chronology of marine two-stroke cycle diesel engine developments during this century.[4] The first two-stroke engine installation for marine use was in 1905. This Sulzer engine was of the trunk-piston type having four cylinders of 175-mm bore and 250-mm stroke; it produced 19 kW per cylinder at a speed of 375 rev/min. This pioneering engine employed uniflow scavenging with scavenge air valves in the cylinder head and exhaust ports in the lower part of the liner.

In 1909, Sulzer switched to cross-flow port scavenging, with exhaust and scavenging ports on opposite sides of the cylinder. As this dispensed with cylinder-head valves, a much simpler cover design with better reliability was achieved. A double row of scavenge ports provided a degree of aftercharging through the upper row of ports, after the exhaust ports had been covered by the rising piston. In early designs, the secondary air flow was controlled by mechanically operated valves, then by rotary valves, and later by automatic nonreturn reed valves. Cross scavenging was employed until the adop-

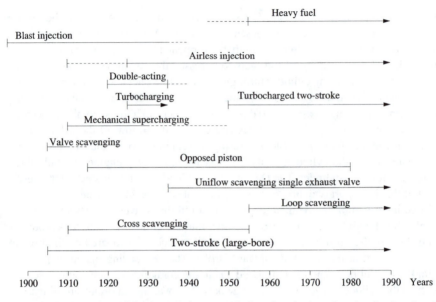

Figure 1-21 Chronology of development for two-stroke large-bore low-speed marine diesel engines (from Brown[4]).

tion in 1956 of loop scavenging (where the scavenging ports are designed to direct the incoming fresh air toward the cylinder liner wall opposite the exhaust ports).

In the 1920s, several engine builders introduced double-acting engines as a means of increasing unit outputs (Fig. 1-21). The largest merchant ship installation of a double-acting engine was completed in 1924 by MAN. The engine had six cylinders, each of 70-cm bore and 120-cm stroke, which produced a maximum output of 21 MW at 125 rev/min. These two-stroke double-acting engines had problems such as the cracking of piston rods, poor combustion in the lower cylinder, and poor reliability of the piston rod stuffing box. In addition, they were very large and maintenance was inconvenient. Eventually, all double-acting diesel designs were abandoned; in fact turbocharging did far more for the development of high-powered marine diesels than use of double-action engines could ever have achieved.

A major drawback of early marine diesel engines was the air-blast injection system. Practical difficulties in constructing a liquid fuel pump that metered precisely and injected the fuel rapidly, however, made air-blast injection expedient. This technology remained in use well into the 1930s, but the air compressors required caused many operating problems, required additional maintenance, and added considerably to the parasitic power losses. In 1910, McKechnie introduced an accumulating system for airless fuel injection, a concept that was eventually adopted by all manufacturers of large marine diesel engines during the 1930s. The high peak pressures generated in the airless injection system required a stronger camshaft drive, cams, and tappet rollers, so suitable materials and manufacturing techniques had to be developed.

Turbocharging has been the greatest single technical contribution to diesel engine progress. It removed the constraints of natural aspiration (i.e., drawing in air directly from the atmosphere), enabling large increases in power output and significant reductions in engine size and weight. Figure 1-22 shows how two critical engine performance parameters, the brake mean effective pressure [work per cycle per unit displaced cylinder volume, see Eq. (1-11)] and brake fuel conversion efficiency [Eq. (1-17)], have increased over the years, especially since about 1950. Turbocharging increases the amount of air trapped in the cylinder and, hence, the amount of fuel that can be burned. It also increases the engine's efficiency. Both these factors increase bmep and power density. Turbocharging played a major role in helping the diesel engine displace the steam turbine, thus beoming the most compact, economical, reliable prime mover of today. But it was almost 50 years after Büchi's first turbocharging patent in 1905 before this technology could be applied to large marine two-stroke engines. Considerable attention has been given to solving the greater problems of turbocharging two-stroke engines, compared with four-stroke engines. The larger air flow requirement of the two-stroke reduces the exhaust temperature, and consequently the exhaust energy available to drive the turbocharger turbine. At the same time, scavenging efficiency is very sensitive to exhaust back pressure. The overall result is that turbochargers must have a much higher efficiency for use with two-stroke diesel engines. The necessary improvements in turbocharger design were significantly aided by the gas turbine developments of the 1940s.

The concept of pressure charging to increase the mass of air charge in the cylinder had been known before diesel engine development even began. Most of these engine-builders tried various pressure-charging techniques in the 1920s and 1930s. All naturally aspirated two-stroke engines required some supplementary pumping to ensure an adequate flow of scavenging air: double-acting reciprocating air pumps driven off the crankshaft, changing later to a more compact arrangement of side-mounted pumps driven by levers off crossheads, attached Roots-type blowers, or, in some installations, independently driven pumps or blowers. In the early cross-scavenged engines, as mentioned above, additional air was supplied from an auxiliary reciprocating pump through an upper row of scavenge ports at a slightly higher pressure than the scavenge air.

The use of cheaper heavier fuel oils has become a crucial element in the remarkable operating economy of these low-speed diesel engines. Fuels with poor combustion characteristics, such as coal tar oil and, especially, residual fuel oils, have been used extensively in stationary engines and also in marine engines from the earliest days. However, much greater wear and fouling of pistons and cylinder liners detracted from the benefit of a cheaper fuel.[4] Consequently, devices such as a diaphragm separating the cylinder scavenge space from the crankcase, to protect the running gear from harmful combustion residues, and provision for hot fuel to be circulated, to keep the fuel pumps warm, were introduced. High-pressure pipes were steam traced, clearances of the pump plunger and injector needle were increased to avoid sticking over the wider temperature range experienced, and fuel nozzles were more effectively cooled to prevent carbon deposit buildup.

Analyses of anticipated trends in ship types, ship sizes, service speeds, fuel costs, engine prices, engine dimensions, and propeller design indicated the need for lower shaft speeds.[4] Lengthening the stroke/bore ratio is the optimum solution for low-speed engines, permitting a simple cost-effective direct propeller drive, while maintaining appropriate values for the mean piston speed. The stroke/bore ratio was thus increased up

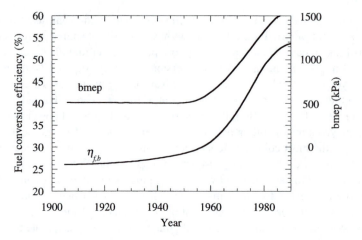

Figure 1-22 Evolution of brake fuel conversion efficiency and brake mean effective pressure of large-bore two-stroke diesel engines (modified from Brown[4]).

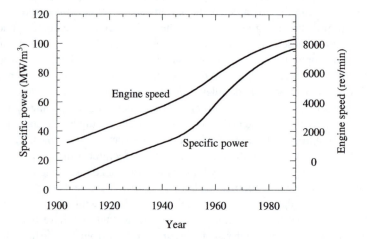

Figure 1-23 Evolution of specific power and maximum engine speed of small SI two-stroke engines (modified from Sher[16]).

to 2.9; see Table 1-4. At this substantially increased ratio, the traditional loop-scavenged system is at a disadvantage in terms of scavenging effectiveness, and the transition to uniflow scavenging with poppet-type exhaust valves in the cylinder head was, therefore, a logical development.

Engine measurement technology has been greatly improved in recent years. Computers permit on-line data acquisition and processing so that comprehensive performance and emissions measurements can now be taken and analyzed. Development of new two-stroke cycle diesel engines entails a combination of scale-model tests, laboratory tests with actual components, and extensive trials on a full-scale test engine. Much of the theoretical background to diesel engine design has been well known for many

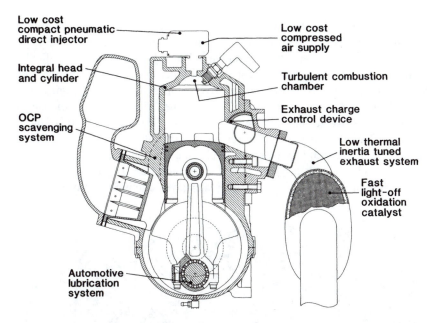

Figure 1-24 Cross section of Orbital combustion process direct-injection crankcase-scavenged two-stroke cycle spark-ignition engine. (Courtesy Orbital Engine Co.)

years. Computer data processing, however, has brought about a breakthrough in theoretical analysis, permitting the solution of complex problems with many unknowns. Detailed patterns of stress, temperature, and deformation can now be accurately calculated even for very complicated systems, particularly in regions of stress concentration and high heat-transfer rates.[2]

While the evolution of the large-bore two-stroke diesel engine has been dominated by engineering developments in scavenging methods, supercharging machines, and fuel-injection systems, the evolution of the small-bore spark-ignition engine has followed the development of its four-stroke counterpart. For many years, however, the two-stroke spark-ignition engine lagged behind the four-stroke engine in power output and fuel consumption because of design limitations associated with the complicated gas exchange process (a subject we will develop in detail in Chapter 2). Ninety years of research and development have resulted in a gradual increase in the specific power of these engines from a few megawatts per cubic meter to close to 100. Figure 1-23 summarizes these trends. Advances in design, materials, and manufacturing have opened the way to increase the maximum engine speed from several hundred rev/min to above 8000 rev/min (sometimes to 12,000 rev/min in specially designed engines).

In recent years, new fuel-injection technology has been developed that makes feasible direct injection of the fuel—gasoline—into the two-stroke cycle engine cylinder. With in-cylinder injection while the exhaust ports (or valves) are closed, the fresh air lost during the scavenging process is no longer accompanied by fuel, and substantial improvements in fuel consumption and emissions result. Furthermore, the engine's combustion and fuel-injection systems can now be designed to produce a nonuniform fuel distribution within the combustion chamber, which permits very lean engine operation (or highly dilute operation with residual burned gases) at part load. Accordingly, the scavenging constraints on such direct-injection engines are different from those of carbureted premixed gasoline–air two-stroke cycle engines, and the scavenging performance can be further improved. Figure 1-24 shows the Oribital Engine Company's direct-injection two-stroke gasoline engine system, one of the more successful of these recent designs.[18] The novel features of the engine are identified: the air-assist direct-fuel-injection system, the compact combustion chamber for the stratified combustion process (at part load), the exhaust port with a flow control device, the low-thermal-inertia exhaust system and close-coupled oxidation catalyst to remove hydrocarbon and

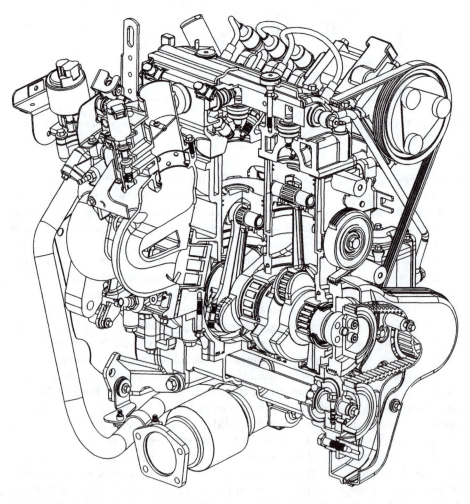

Figure 1-25 Cutaway drawing of Orbital 1.2-liter three-cylinder direct-injection crankcase-scavenged two-stroke spark-ignition engine used in 100-car fleet test. (Courtesy Orbital Engine Co.)

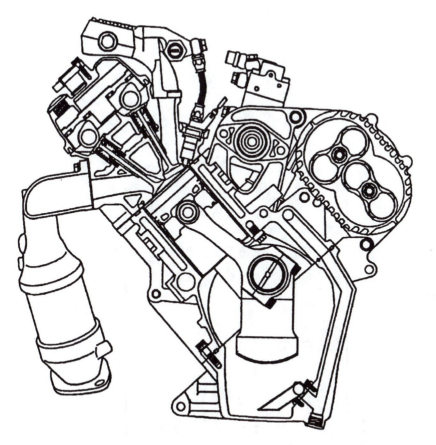

Figure 1-26 Cross section of prototype Toyota S-2 direct-injection two-stroke valved gasoline engine. The scavenging air is externally compressed in a belt-driven roots-type blower.[20]

carbon monoxide emissions from the exhaust gases effectively and quickly after engine start up, and the crankcase scavenging system with its dry sump, grease-lubricated roller crankshaft bearings, with oil sprayed into the fresh air as it is transferred to the cylinder.

Table 1-6 compares some major features of two-stroke cycle automotive engines that have evolved during this century with those of these direct-injection spark-ignited two-stroke engines now being developed. Note that the scavenging process has improved remarkably (as evidenced by the increases in specific power), while the overall geometry—the bore-to-stroke ratio of the engine—has not changed at all. The scavenging process has been primarily improved by refining the geometry of the engine ports. One of the more recent engines listed in Table 1-6 is valved rather than ported (Toyota) and three (Toyota, Subaru, and Orbital IES) use an external blower. Accordingly they can use a conventional four-stroke cycle engine crankshaft design and wet-sump lubrication system.

Figures 1-25 and 1-26 show actual layout drawings of two of the more recent direct-injection two-stroke cycle spark-ignition engines, the in-line three-cylinder crankcase-

Table 1-6 Development of the two-stroke cycle SI engines[19]

Engine	Year	Displacement (cm^3)	Config-uration	Bore/stroke	Power (kW–rpm)	Specific power† (MW/m^3)	Torque–speed* (N-m–rpm)	Oil sump
GM Olds	1906	3218	O2	1.00	17.9–1000	6	—	Dry
Zoller	1930	723	L4	—	22.4–2600	31.3	—	Wet
GM	1932	4160	L4	—	131 –3000	31.3	528–1500	Wet
Trabant	1934	594	L2	0.99	19.4–4200	32.8	54–3000	Dry
Wartburg	1935	993	L3	0.94	37.3–4250	37.3	98–3000	Dry
SAAB	1956	842	L3	0.96	403 –4250	35.8	81–3000	Dry
GMX4	1968	2470	X4	—	75.3–4000	30.6	271–2000	Wet
Subaru	1989	1500	V4	1.08	123 –6000	82.1	220–3000	Wet
Toyota	1989	3000	L6	0.91	179 –3600	59.7	490–2800	Wet
Orbital X	1986	1200	L3	1.04	65.6–5500	55.2	125–3500	Dry
Orbital IES	1989	1000	L3	1.17	67.9–6000	67.9	132–4000	Wet
GMCDS2	1990	1500	L3	1.00	83.6–5500	55.2	176–2500	Dry

†Maximum values.

scavenged Orbital engine and the Toyota S-2 roots-type blower, valved, two-stroke direct-injection engine, respectively. The 1.2-liter Orbital engine uses Orbital's air-assist fuel-injection technology, which produces a fuel spray with very small drop sizes ($\sim$ 10-μm diameter), and roller bearings on the connecting rod and crankshaft as a consequence of the dry crankcase required by the choice of crankcase scavenging. The piston rings and skirt are lubricated primarily by a fine oil spray into the air flow between the crankcase and the cylinder. The Toyota engine uses a belt-driven external air blower and poppet valves rather than ports, and, therefore, can use conventional connecting rods, crankshaft, and crankcase sump-based lubrication system. The engine has five valves in each cylinder's head: two for intake and three for exhaust. A high-pressure liquid fuel-injection system provides the controlled spray development inside the cylinder for stratified combustion at light loads (droplet sizes of about 25 μm). In this version of the engine, a second injector in the cylinder head is used to achieve a more homogeneous fuel distribution inside the cylinder at higher loads.

Crankcase-scavenged two-stroke engines like that in Fig. 1-25 are superior in terms of smaller engine bulk volume, weight and maximum engine speed, and hence have higher power density, lower friction, and lower engine cost. Poppet-valve externally blown two-stroke engines such as that in Fig. 1-26 are superior in oil consumption, durability, and scavenging efficiency, but worse in maximum engine speed, engine size and weight, and engine cost.[20]

REFERENCES

1. Bosch, *Automotive Handbook*, 3rd English ed., Robert Bosch GmbH, distributed by SAE, Warrendale, PA, 1993.
2. Heywood, J.B., *Internal Combustion Engine Fundamentals*, McGraw-Hill, New York, 1988.
3. Laimbock, F.J., "The Potential of Small Loop-Scavenged Spark-Ignition Single Cylinder Two-Stroke Engines," SAE Paper 910675, SAE SP-847, 1991.
4. Brown, D.T., "A History of the Sulzer, Low-Speed Marine Diesel Engine," Special report, Sulzer, Winterthur, 1984.
5. "New Developments in Two-Stroke Low Speed Diesel Engines for Stationary Applications," Special report, MAN and B&W, Copenhagen, 1986.
6. London, A.L., "Free Piston and Turbine Compound Engine—Status of the Development," *SAE Trans.*, vol. 62, pp. 426–435, 1954.
7. Meyer, A., "Recent Developments in Gas Turbines: The Free-Piston Engine," *Mech. Eng.*, vol. 69, no. 4, pp. 273–277, April, 1947.
8. Obert, E.F., *Internal Combustion Engines and Air Pollution*, Harper and Row, New York, 1973.
9. Hibi, A., "Hydraulic Free Piston Internal Combustion Engine," *Hydraulic Pneumatic Mechanical Power*, pp. 57–59, March, 1984.
10. Hibi, A., and Kumagai, S., "Hydraulic Free Piston Internal Combustion Engine—Test Results," *Hydraulic Pneumatic Mechanical Power*, pp. 244–249, Sept., 1984.
11. Putman, A.A., Belles, F.E., and Kentfield, J.A.C., "Pulse Combustion," *Prog. Energy Comb. Sci.*, vol. 12, no. 1, pp. 43–79, 1986.
12. Payne, P.R., "A Progress Report on Pulsejets," In *10th IECEC*, pp. 535–547, August, 1975.
13. Hardenberg, H.O., "An Historical Overview of Gunpowder Engine Development—1508–1868," presented at the second Annual Fall Technical Conference of the ASME Internal Combustion Engine Division, Michigan, October, 1989.
14. Day, J., *Engines—The Search for Power*, St. Martin's Press, New York, 1980.
15. Cummins, L., *Internal Fire*, SAE R-100, 1989.

16. Sher, E., "Scavenging the Two-Stroke Engine," *Prog. Energy Comb. Sci.*, vol. 16, pp. 95–124, 1990.

17. Counter, C.F., *The Two-Cycle Engine*, Pitman, Lonodn, 1932.

18. Schlunke, K., "The Orbital Combustion Process Engine," In *10th Vienna Motorsymposium*, VDI No. 122, pp. 63–68, 1989.

19. Wyczalek, F.A., "Two-Stroke Engine Technology in the 1990's," SAE Paper 910663, SAE SP-849, pp. 1–8, 1991.

20. Nomura, K., and Nakamura, N., "Development of a New Two-Stroke Engine with Poppet Valves: Toyota S-2 Engine," In *A New Generation of Two-Stroke Engine for the Future?* (ed. P. Duret), Proceedings of International Seminar held at IFP, Rueil-Malmaison, France, Nov., 1993, pp. 53–62, Editions Technip, Paris, 1993.

TWO

GAS EXCHANGE FUNDAMENTALS

The first section of this chapter develops the fundamentals and terminology of the gas exchange process in two-stroke cycle engines (an introductory description of this process is given in Sections 1-1 and 1-2). We then review the various scavenging and charging methods used in two-stroke engines, including supercharging, turbocharging, and charge stratification. The basic geometrical relationships and the parameters commonly used to characterize other aspects of engine operation are developed in Section 1-3.

2-1 FUNDAMENTALS AND TERMINOLOGY

In two-stroke cycle engines, every outward (or downward) stroke of the piston is a power stroke. To achieve this, the burned gases have to be replaced by fresh charge while the piston moves through its bottom center (BC) position. The period during which the products of combustion are exhausted from the cylinder and the fresh charge enters is called the *gas exchange period*. This period begins at the time the exhaust port (or valve) is first exposed or opens (EO in Fig. 1-2) and is completed at the time when both ports (exhaust and scavenge) close (EC in Fig. 1-2). It is appropriate, and convenient, to subdivide the gas exchange process into two periods: the *exhaust blowdown period*, from exhaust opening (EO) to scavenge port opening (SO); and the *scavenging period*, which is the remainder.

A quantitative discussion of the two-stroke cycle scavenging process requires precise terminology and an appropriate set of parameters. The following terminology is used throughout this book:

ambient conditions—atmospheric conditions;

standard conditions—conditions at standard atmospheric pressure and temperature;

scavenge conditions—conditions prevailing at the inlet of the scavenge duct;

fresh charge—charge entering the scavenge ports (pure air when the fuel is injected directly into the cylinder, or fuel–air mixture for carburetted or port-injected engines);

cylinder charge—cylinder content at any given instant (during the gas exchange process, the cylinder charge amount and composition vary with time; at any instant, the cylinder charge is always equal to the fresh charge retained plus the residual);

residual—cylinder charge retained from the previous cycle (may contain burned gas, air, and fuel);

retained—within the cylinder volume at any given instant (at the end of the gas exchange process, the retained conditions become the trapped conditions);

trapped—retained in the cylinder at the end of the gas exchange process.

The following notation is used:

m—mass of cylinder charge (varies with time during the gas exchange process);

m_{rt}—mass of retained cylinder charge;

m_{tr}—mass of trapped cylinder charge (at the end of the gas exchange process);

m_0—reference mass = displaced volume × fresh-charge density at ambient conditions. The reference volume should be defined as the cylinder trapped volume. However, as the trapped volume is not usually given, the displaced volume, which can easily be evaluated, will be used.

The following parameters are used to describe the progress of the gas exchange process (note that, during the gas exchange process, these parameters are time-dependent quantities and that, at the end of the process, the retained conditions become the trapped conditions).

The *delivery ratio*

$$\Lambda = \frac{\text{mass of fresh charge delivered}}{m_0} \qquad (2\text{-}1)$$

compares the actual mass of delivered fresh charge at any given instant to the total amount required in an ideal charging process, the reference mass.

The *charging efficiency*

$$\eta_{\text{ch}} = \frac{\text{mass of fresh charge retained}}{m_0} \qquad (2\text{-}2)$$

indicates how effectively the cylinder volume has been filled with fresh charge at any given instant.

The *scavenging efficiency*

$$\eta_{\text{sc}} = \frac{\text{mass of fresh charge retained}}{m} \qquad (2\text{-}3)$$

indicates to what extent the burnt residuals have been replaced with fresh charge at any given instant.

The *retaining efficiency*

$$\eta_{rt} = \frac{\text{mass of fresh charge retained}}{\text{mass of fresh charge delivered}} \qquad (2\text{-}4)$$

indicates what fraction of the fresh charge delivered to the cylinder is retained at any given instant. At the end of the gas exchange process, the retaining efficiency becomes the *trapping efficiency*.

$$\eta_{tr} = \frac{\text{mass of fresh charge trapped}}{\text{mass of fresh charge delivered}} \qquad (2\text{-}5)$$

The *exhaust gas purity*

$$\beta = \text{mass fraction of fresh charge in the exhaust gas}$$
$$\text{at any given instant} \qquad (2\text{-}6)$$

The charging efficiency, retaining efficiency, and delivery ratio are related by

$$\eta_{ch} = \Lambda \eta_{rt} \qquad (2\text{-}7)$$

At the end of the gas exchange process,

$$\eta_{ch} = \Lambda \eta_{tr} \qquad (2\text{-}8)$$

In some references, an additional property, the *purity*, is introduced. This is defined as the ratio between the mass of pure air retained at the end of the scavenging process and the reference mass.

In the real gas exchange process in two-stroke engines, mixing occurs as the fresh charge displaces the burned gases, and some of the fresh charge is expelled. The relation between these gas exchange parameters is, therefore, complex. Two idealized models of the scavenging process are useful for illustrating the relationships between these gas exchange parameters: the isothermal pure-displacement model, and the isothermal perfect-mixing model. The formal analysis of the general cases where these two ideal processes occur under nonisothermal conditions will be given in Section 4-1. Here we describe the isothermal versions of these models.

For *pure isothermal displacement*, it is assumed that the entering fresh charge pushes out the burnt gases by a pure displacement mechanism, that the process occurs at a constant cylinder volume, pressure, and temperature, that no mass or heat is allowed to cross the interface between the fresh charge and the in-cylinder burned gases, and that the cylinder walls are adiabatic. Thus, $m = m_{tr} = m_0$ and, by definition,

$$\begin{array}{llllll} \text{for } \Lambda \leq 1 & \eta_{ch} = \Lambda & \eta_{sc} = \Lambda & \eta_{rt} = 1 & \text{and} & \beta = 0 \\ \text{for } \Lambda > 1 & \eta_{ch} = 1 & \eta_{sc} = 1 & \eta_{rt} = 1/\Lambda & \text{and} & \beta = 1 \end{array} \qquad (2\text{-}9)$$

The pure-displacement model is the upper bound for the scavenging process and is never achieved in practice. Figure 2-1 shows the behavior of this model. Note that if, at the end of the gas exchange process, the total delivery ratio $\Lambda \leq 1$, then $\eta_{tr} = 1$, and if the total delivery ratio > 1, then $\eta_{tr} = 1/\Lambda$.

For *perfect isothermal mixing*, it is assumed that as the fresh charge enters the cylinder it mixes instantaneously with the cylinder charge to form a homogeneous mixture.

The instantaneous composition of the gas leaving the cylinder through the exhaust port is, therefore, a mixture of fresh charge and burned gases of the same mixture composition as the composition in the cylinder ($\beta = \eta_{sc}$). The process is assumed to occur at a constant cylinder volume, pressure, and temperature, the cylinder walls are adiabatic, the entering charge has the thermodynamic properties of the ambient, and the two gases involved follow the ideal gas law and have the same molecular weights with identical and constant specific heats. Thus, $m = m_{tr} = m_0$.

Following these assumptions, between time t and $t + dt$, a mass element dm_i of fresh charge is delivered to the cylinder and is uniformly mixed throughout the cylinder volume. An equal amount of fluid, with the same composition, leaves the cylinder during this time interval. Thus, the mass of fresh charge delivered between t and $t + dt$ that

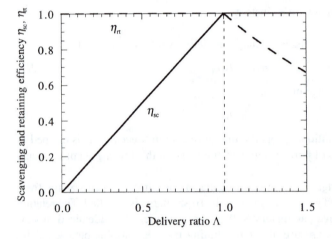

Figure 2-1 Results of the pure isothermal displacement model.

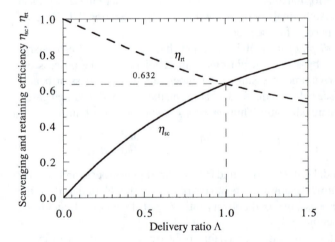

Figure 2-2 Results of the perfect isothermal mixing model.

is retained, dm_{rt}, is given by

$$dm_{rt} = dm_i(1 - m_{rt}/m_0) \tag{2-10}$$

This integrates over the duration of the scavenging process to give

$$m_{rt}/m_0 = 1 - \exp(-m_i/m_0) \tag{2-11}$$

The results are therefore

$$
\begin{aligned}
\eta_{ch} &= 1 - e^{-\Lambda} & \eta_{sc} &= 1 - e^{-\Lambda} \\
\eta_{rt} &= (1 - e^{-\Lambda})/\Lambda, & \beta &= 1 - e^{-\Lambda}
\end{aligned}
\tag{2-12}
$$

If at the end of the gas exchange process the total delivery ratio is Λ, then the trapping efficiency $\eta_{tr} = (1 - e^{-\Lambda})/\Lambda$. In modern engines, the gas exchange performance is significantly more efficient than the perfect isothermal mixing approximation and this ideal model, therefore, forms the lower bound for the practical scavenging process. Figure 2-2 shows the behavior of the perfect mixing model.

2-2 SCAVENGING METHODS

There are a number of port layouts commonly used in two-stroke spark-ignition and diesel engines. The basic types of arrangements, which are usually termed methods of scavenging, are illustrated in Fig. 2-3. The general advantages, drawbacks, and applications of each arrangement are summarized in Table 2-1. Cross- and loop-scavenging

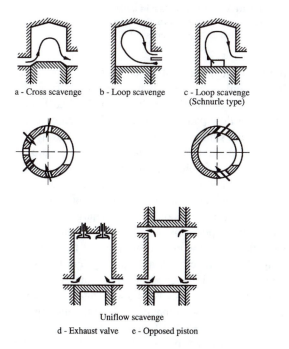

a - Cross scavenge   b - Loop scavenge   c - Loop scavenge
(Schnurle type)

Uniflow scavenge
d - Exhaust valve   e - Opposed piston

Figure 2-3 Basic types of scavenging arrangements.

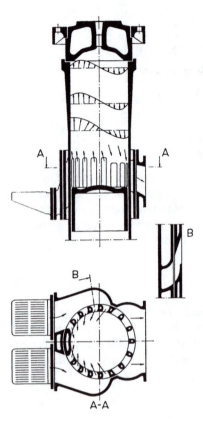

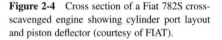

Figure 2-4   Cross section of a Fiat 782S cross-scavenged engine showing cylinder port layout and piston deflector (courtesy of FIAT).

systems use exhaust and scavenge ports in the cylinder wall, uncovered by the piston as it moves toward bottom center (BC). The uniflow system may use scavenge ports in the cylinder liner with exhaust valves in the cylinder head, or scavenge and exhaust ports with opposed pistons. Despite the different flow patterns obtained with each cylinder geometry, the general operating principles are similar. The fresh charge must always be supplied to the scavenge ports at a pressure higher than the exhaust-system pressure.

In early designs of two-stroke engines, the scavenge or transfer ports were usually opposite the exhaust port. The fresh charge entered one side of the cylinder, directed by the inclined scavenge ducts toward the cylinder head, and crossed to the other side (Fig. 2-3a). With this layout, a significant part of the incoming fresh charge passed directly across the cylinder and out of the exhaust port: this is called short-circuiting. The piston was, therefore, shaped with a deflector on the crown to direct the fresh charge upward toward the cylinder head (Fig. 2-4). This cross-scavenging system has good scavenging characteristics when operated at partially opened throttle (lighter loads) and low engine speeds. At higher throttle openings (higher loads), the scavenging efficiency deteriorates due mainly to the poorly scavenged region that exists at high entering velocities along the cylinder wall above the scavenge ports (opposite the exhaust port). Another problem of this arrangement is the hot edges of the deflector that increase the engine's tendency to knock and preignition. The compression ratio of an engine having

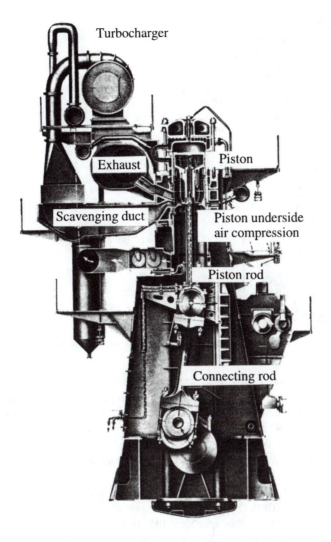

Turbocharger

Exhaust

Piston

Scavenging duct

Piston underside
air compression

Piston rod

Connecting rod

Figure 2-5   Cross section of a MAN KZ105/180 loop-scavenged engine, developing 3000 kW (4000 hp) per cylinder at 106 rev/min (courtesy of MAN).

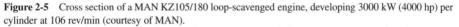

this kind of port layout is therefore limited. On the other hand, the cross-scavenged engine has substantial packaging and manufacturing advantages: the cylinder-to-cylinder spacing in a multicylinder configuration can be as close as is practical for intercylinder cooling considerations with this simple port arrangement. Further, from the manufacturing perspective, it is possible to drill the scavenge and exhaust ports directly, in-situ and in one operation, from the exhaust port side and thereby reduce manufacturing costs.[1]

A different arrangement, where the exhaust ports are above the scavenge ports (MAN-type loop-scavenging system), is shown schematically in Fig. 2-3b. A typical example is shown in Fig. 2-5. In this design, the fresh charge stream is directed toward

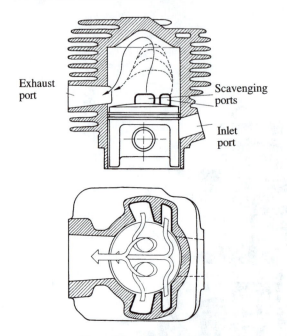

Exhaust port

Scavenging ports

Inlet port

Figure 2-6   A small Schnürle-type loop-scavenged engine (courtesy of Kolbenschmidt AG).

the unported wall, flows toward the cylinder head, changes its direction, and continues toward the exhaust port. The long path of the entering charge requires high-momentum jets, and one would expect, therefore, that this type of engine performs better at wide-open throttle (WOT). For this reason, this arrangement is well suited to diesel engines where engine load is controlled by the amount of fuel injected rather than by a throttle valve. Because of the relatively high velocity of the incoming charge, the mixing between the fresh charge and the residuals is quite extensive, and although good scavenging efficiency achieved, it is still inferior to that of uniflow-scavenged engines.

Another method that avoids the use of the troublesome deflector piston was developed by Schnürle in the 1920s. In this approach, the fresh charge is directed toward the opposite side of the cylinder to the exhaust port, across a piston with an essentially flat top. Instead of the single scavenge port placed diametrically opposite the exhaust port, a pair of scavenge ports were located symmetrically around the exhaust port on the same level as the exhaust port, as shown in Fig. 2-3c. A development of this approach is the Curtis-type loop-scavenged engine, in which more scavenge ports are added opposite the exhaust port. In these two arrangements, the fresh charge path is shorter than in the MAN-type loop-scavenging system (Fig. 2-3b), thus eliminating the requirement for high-momentum jets. Consequently, the scavenging performance of the Schnürle loop-scavenging system (and its Curtis derivative) is better at throttled conditions, and mixing between the fresh charge and the burned gases is reduced. This type of scavenging system is widely used in small-bore SI engines; an example is shown in Fig. 2-6.

Figures 2-3d and e show two types of uniflow scavenging (sometimes called through scavenging) systems. Here the in-cylinder fresh charge flows in one direction

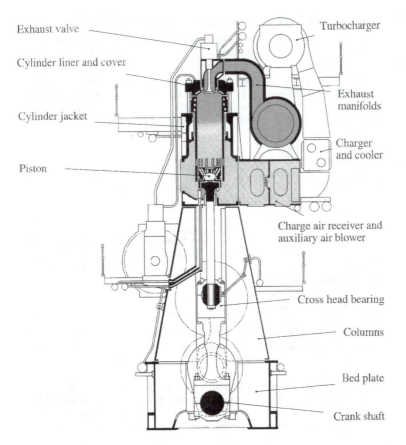

Exhaust valve

Cylinder liner and cover

Cylinder jacket

Piston

Turbocharger

Exhaust manifolds

Charger and cooler

Charge air receiver and auxiliary air blower

Cross head bearing

Columns

Bed plate

Crank shaft

Figure 2-7 Cross section of an IHI-Sulzer RTASS uniflow-scavenged engine, developing 1590 kW per cylinder at 127 rev/min.

only; the cylinder fills with fresh charge from one end, and the burned gases are expelled from the other. With this arrangement, a poorly scavenged annular-shaped region close to the cylinder walls may result. To overcome this problem, the scavenge ports in some layouts are designed to produce a controlled swirling motion, thus using centrifugal forces to push the burned gases out of this annular region (Junker method). In this case, the turbulence intensity increases and the combustible mixture is better prepared (this is an important feature, especially for diesel engines). However, as the fresh charge then adheres to the cylinder wall, a small volume in the central core is sometimes poorly scavenged.[2] Figure 2-7 shows a typical example of this type of engine. In fact, the most efficient prime movers ever made are the low-speed uniflow-scavenged two-stroke diesel engines with a high stroke-to-bore ratio of about 2.5. A brake specific fuel consumption of as low as 170 g/kW-h has been achieved (see Table 1-3), which is equivalent to a brake fuel conversion efficiency above 50%. For the higher stroke-to-bore ratios, the pressure inside the cylinder at the end of the expansion stroke may fall to too low a value for an efficient blowdown process unless the engine speed is further

Table 2-1 Classification of different scavenging methods and their applications

Method	Advantages	Drawbacks	Applications
Cross	Good scavenging at partial throttling and low speeds Low engine volume for multicylinder arrangements Low manufacturing cost	High bsfc at high throttle opening and high speeds High tendency to knock limits compression ratio	Small outboard engines, and some other specific applications (see Table 1-4)
Loop, MAN-type	Good scavenging at WOT Low surface-to-volume ratio combustion chamber Low manufacturing cost	Poor scavenging at part-throttle operation	Large-bore marine CI engines
Loop, Schnürle-type	Good scavenging at WOT and medium engine speed Fair scavenging at part throttle and other than medium engine speeds Low manufacturing cost	High bsfc at part throttle operation	SI engines for a large variety of applications (see Tables 1-2 and 1-4)
Uniflow, exhaust valve	Very good scavenging at WOT for high stroke-to-bore ratio Excellent bsfc	Need for exhaust valves; thus more complex and higher manufacturing cost	Large-bore low-speed CI marine and stationary engines (Table 1-4)
Uniflow, opposed piston	Very good scavenging at WOT for high stroke-to-bore ratio	Need for mechanical coupling between two crankshafts	Sometimes used in large-bore low-speed CI marine engines

reduced. It should be noted that engine speed is often limited by various mechanical constraints. For most engines used in today's motorcycles and outboards, where the stroke-to-bore ratio is between 0.8 and 1.1, the data indicate that uniflow scavenging is not significantly better than the best of the loop-scavenged designs.[1] Because uniflow scavenging entails considerable additional mechanical complexity relative to a port-controlled engine, a marginal improvement in engine efficiency would not necessarily offset this extra complexity.

2-3 CHARGING METHODS

Because the two-stroke cycle gas exchange process occurs when both the exhaust and the scavenge ports are open, the pressure inside the cylinder is normally above atmospheric pressure. This gas exchange or scavenging process requires that the fresh charge be supplied to the engine cylinder at a high enough pressure to displace the burned gases from the cylinder. At the same time, the pressure should be low enough to minimize the scavenging air pumping work. For carburetted and port-fuel-injected engines, an additional restriction is imposed on the scavenge pressure; the pressure should result in a stable flow stream and a low enough delivery ratio to minimize direct fuel losses to the exhaust port (short-circuiting). In such cases, and also when lean operation is required, stratification of the charge—nonuniform distribution of the fuel within the cylinder—can be an advantage, and the scavenging charge supplier must be designed accordingly. Raising the pressure of the fresh charge is done in a separate pump, blower, or compressor. Several methods for charging the cylinder have been proposed. Many of these are in use today and others are useful as a source of ideas for possible future developments. We now review the fundamentals of these charging methods.

2-3-1 Crankcase-Compression Engines

In a crankcase-compression engine, the fresh charge is compressed in the crankcase by the underside of the working piston, prior to its admission to the cylinder through the scavenge ducts. The closing and opening of the inlet, scavenge, and exhaust ports are controlled by the piston itself, and thus in its simplest form (which is similar to the original Day engine of 1891, see Section 1-5-2), the present engine requires only three moving parts for each cylinder. This engine concept benefits greatly from this simplicity and has been used successfully as a spark-ignition prime mover for more applications than any other two-stroke engine type (see Table 1-3).

The sequence of events in this engine's operating cycle and the characteristics of its gas exchange process have already been introduced in Sections 1-1 and 1-2. The efficiency of the gas exchange process in cross- and loop-scavenged engines is far from that of the pure-displacement model: while a part of the fresh charge displaces combustion products without mixing or loss, another part mixes with the products of combustion, and other portions short-circuit directly to the exhaust port. The relative magnitude of these various portions is a function not only of the geometry of the cylinder, crankcase, and port assembly, but also of operational factors such as the rate of fresh charge introduction into the cylinder (a time-dependent quantity), engine speed, engine load or throttle position, and atmospheric conditions. The throttle is used to control the amount of air that flows into the crankcase each cycle. Crankcase-compression engines, in their multicylinder form, require that the crankcase regions beneath each cylinder be separated and sealed. This necessitates a dry sump, lubrication of piston skirt and ring by oil sprayed into the scavenging air, and roller crankshaft and connecting-rod bearings.

Figure 2-8 shows the effect of the engine speed on the fresh charge velocity at the scavenge port, together with the pressure variation in the crankcase (p_c), the scavenge port (p_s), and the exhaust port (p_e), at wide-open-throttle conditions (WOT). The engine layout is shown in Fig. 2-9, and its specifications are given in Table 2-2. Note that the inlet port in this engine is installed with a reed valve—a device that prevents backflow from the crankcase to the atmosphere through the intake system. This subject is discussed in more detail in Chapter 5. At an engine speed of 3000 rev/min, the pressure inside the crankcase increases as the piston travels downward toward BC. At scavenge port opening (SO), the pressure reaches a maximum value of about 130 kPa, and at the same time the pressure at the exhaust port falls to about atmospheric pressure. As the scavenge port opens, the fresh charge is driven into the cylinder by the pressure difference and its velocity increases to about 65 m/s. Meantime, the pressure in the crankcase

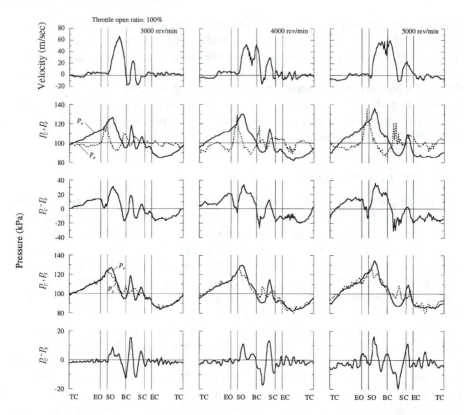

Figure 2-8 The effect of engine speed on the fresh-charge velocity at the scavenge port, and the pressure variation in the crankcase (p_c), the exhaust port (p_e), and scavenge port (p_s) at wide-open-throttle conditions.[3] Engine details are given in Table 2-2.

decreases, a compression wave reaches the exhaust port, and the pressure difference becomes negative just before BC. A backflow from the cylinder to the crankcase appears to follow this change in pressure difference and reaches a value of 15 m/s. At this stage, the exposed area of the scavenge port is at its maximum and therefore the mass of fluid that flows back is important. Oscillations of the exhaust and crankcase pressure follow, which result in oscillations in the velocity of the incoming fresh charge until the scavenge port closes (SC). For all engine speeds, the pressure at the scavenge port follows the pressure inside the crankcase. At a higher engine speed of 5000 rev/min, the time available for the blowdown period is too short, and at SO the pressure at the exhaust port has not fallen below the crankcase pressure. As a result, a backflow of residuals from the cylinder to the crankcase can be observed. The pressure inside the crankcase increases further. Simultaneously, the exhaust pressure decreases and the crankcase contents are driven into the cylinder. The maximum velocity of the incoming charge again reaches a value of about 65 m/s. As time advances, the pressure difference becomes negative, and another backflow occurs. However, now the open area of the scavenge port is small and the amount of fluid that flows back is also small.

Table 2-2 Specifications of the engine shown in Fig 2-9

Type of engine	Two-stroke crankcase-compression loop-scavenged SI engine of Schnürle type; single cylinder; reed valve installed at the inlet port
Displaced volume	98.2 cm^3
Bore	5.0 cm
Stroke	5.0 cm
Compression ratio	6.59
Exhaust port	EO, 83° BBC; EC, 82° ABC
Scavenge port	SO, 57° BBC; SC, 60° ABC

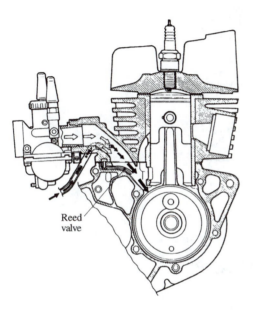

Reed
valve

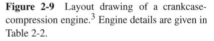

Figure 2-9 Layout drawing of a crankcase-compression engine.[3] Engine details are given in Table 2-2.

The difference between the progress of the gas exchange process under firing and motoring conditions is demonstrated in Fig. 2-10. It shows that, although under motoring conditions the fresh charge enters the cylinder relatively evenly during the scavenging period with a characteristic velocity of about 20 m/s, an important backflow occurs under firing conditions at BC. Also, the peak scavenge flow velocity of the fired engine is about 60% higher than that of the motored engine. Beause the path of the entering charge inside the cylinder and the mixing process between the fresh charge and the cylinder contents are both strongly dependent on the velocity of the incoming jet, these differences between motoring and firing conditions are very significant. The differences are mainly attributed to the difference in in-cylinder pressure history, in which the cylinder pressure under firing conditions is much higher due to combustion. Consequently, the cylinder pressure is much higher during the blowdown phase and results in high-amplitude pressure waves in the exhaust system. These pressure waves

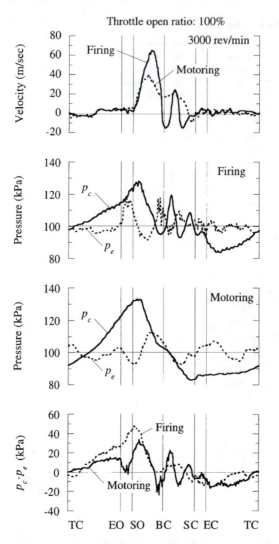

Figure 2-10 Comparison of the fresh-charge velocity at the scavenge port, the crankcase pressure (p_c), the exhaust port pressure (p_e), and the pressure difference ($p_c - p_e$), between a motored and a fired engine.[3] Engine details are given in Table 2-2.

directly affect the flow through the cylinder during the scavenging period. Figure 2-11 shows the total volume of fresh charge delivered during a single cycle versus engine speed for motored and fired operation. The displaced volume of the engine is 98.2 cm^3. Although the total volume delivered in the firing engine decreases moderately from 28 cm^3/cycle at 3000 rev/min to 22 cm^3/cycle at 5000 rev/min, it falls more steeply from 41 to 28 cm^3/cycle in the motored engine. The motored engine shows about a 40% better delivery ratio. Nevertheless, the trends are similar and, as a first approximation, measurements in the motored engine can be useful.

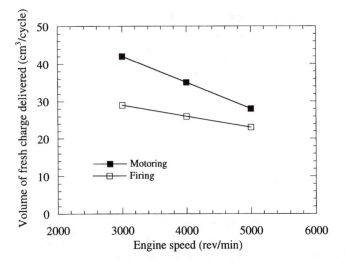

Figure 2-11 Volume of fresh charge delivered each cycle versus engine speed for motored and fired engines.[4] Engine details are given in Table 2-2. The displaced volume of the engine is 98.2 cm^3.

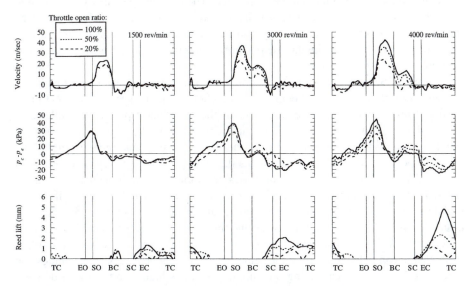

Figure 2-12 The effect of engine speed and throttling on the variation in fresh-charge velocity at the scavenge port, the pressure difference between the crankcase and the exhaust port ($p_c - p_e$), and the reed valve lift, through a motored engine cycle (developed from Ikeda et al.[4]): TOA, throttle open area; 100% is wide-open throttle. Engine details are given in Table 2-2.

Bearing these reservations in mind, we now examine the effect of the throttle position on the velocity of the incoming fresh charge in a motored engine (Fig. 2-12). The throttle is used to control the amount of air that flows into the crankcase each cycle. At low engine speed, closing the throttle to 20% of its wide-open area does not

Table 2-3 Specifications of the engine used in Fig. 2-13

Type of engine	Two-stroke CI, crankcase-compression, single cylinder, loop-scavenged engine of Schnürle type
Displaced volume	567 cm^3
Bore	8.5 cm
Stroke	10.0 cm
Compression ratio	23
Exhaust port	EO, 63° BBC; EC, 63° ABC
Scavenge port	SO, 54° BBC; SC, 54° ABC
Inlet port	IO, 53° BTC; IC: 53° ATC

significantly affect the mean air velocity. At high engine speed, the maximum velocity decreases by about 50% when the throttle is closed to 20% of its WOT area. Backflows at low engine speed occur during a large part of the gas exchange period, but only to a limited extent at high engine speed (with the throttle partially closed). At low speed, the time available for the crankcase to be filled is long enough at these throttle positions for the total cylinder trapped mass to be unaffected. The throttle open area would have to be further reduced to affect the trapped charge mass. At high speed, the velocity of the incoming charge through the intake system is high and so is the pressure drop across the throttle. Consequently, the throttle position affects the crankcase pressure, and thus the total cylinder trapped mass.

Figure 2-13 shows the effect of the crankcase volume and scavenge port width on the delivery ratio of a motored engine as a function of engine speed. Table 2-3 gives details of the engine. In these experiments neither an exhaust pipe nor an inlet duct was used, thus eliminating any possible pipe-tuning effects. The figure shows that for small port widths (when the ratio between the effective-area crank-angle integral of the scavenge port and a reference value equals 0.25 and 0.3), the delivery ratio decreases monotonically as the engine speed increases or as the port width decreases. In these cases, the pressure difference between the crankcase and the cylinder is not high enough to deliver the required amounts of fresh charge through a narrow port during the time available. As the engine speed increases, the time available decreases, and less charge is delivered. As the port becomes narrower, the effective port area is smaller and less fresh charge can delivered during the same time. For a higher crankcase compression ratio (Fig. 2-13b), the scavenging pressure difference is higher and a higher delivery ratio would be expected at a given engine speed.

For the cases when the scavenge port is wider still (when the ratio between the effective-area crank-angle integral of the scavenge port and a reference value equals 2.0), and for the same pressure difference between the crankcase and cylinder, the mass flow rate through the scavenge port is higher and the crankcase pressure falls to the cylinder pressure before scavenge port closing. As the piston continues its upward motion, the crankcase volume increases, the crankcase pressure decreases, and a backflow occurs from the cylinder to the crankcase through the scavenge port. As a result, the total delivery ratio decreases. At higher engine speeds, the time available is shorter and

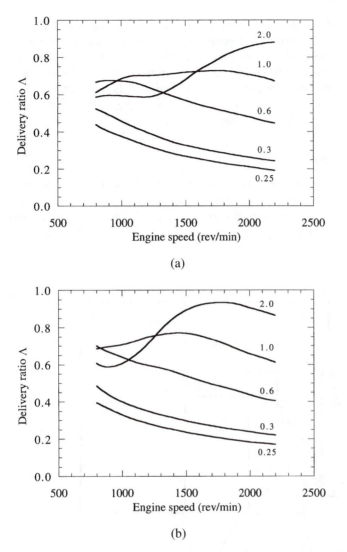

(a)

(b)

Figure 2-13   The effect of the crankcase volume and scavenge port width on the delivery ratio as a function of engine speed under motoring conditions (developed from Fukutani and Watanabe[5]; the engine details are given in Table 2-3): for crankcase compression ratio (a) $r_c = 1.35$ and (b) $r_c = 1.20$. The number by each curve is the ratio between the effective-area crank-angle integral of the scavenge port and a reference value of 0.0324 m^2-deg.

a smaller portion of the mass flows back. Thus an increase in the delivery ratio with increasing engine speed is expected. A trade-off between these effects results in the increasing–decreasing behavior of the delivery ratio with increasing engine speed. It seems that, for any particular port width, the peak delivery ratio is shifted to lower engine speeds with increasing crankcase volume. The value of the peak delivery ratio, however, decreases as the port width decreases. (See also Schweitzer.[6])

Following this discussion, it is important to note that the optimal dimensions of the scavenge port and the crankcase depend very much on the user's requirement for a particular engine torque-versus-speed curve at a given set of ambient conditions. It is also important to note that a higher delivery ratio does not necessarily result in a higher charging efficiency; the latter depends also on the degree of short-circuiting and mixing that occurs. Both of these are related in a complex way to the port geometry and to the time variation of the fresh charge flow rate.

2-3-2 Stepped-Piston Engines

The stepped-piston concept overcomes some of the drawbacks of crankcase-compression engines. Increasing the delivery ratio, isolation of the fresh charge from the crankcase, and improving engine performance at high altitude are some advantages. The essential features of the stepped-piston two-stroke engine are shown in Fig. 2-14.

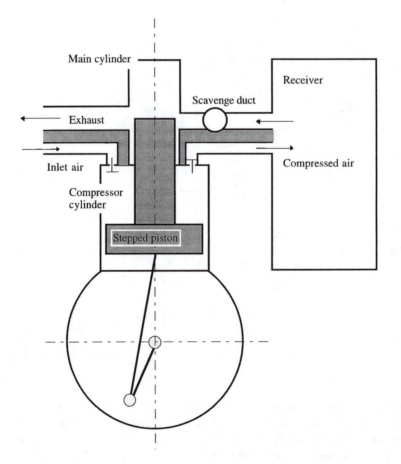

Figure 2-14   Schematic layout of a stepped-piston engine.

The engine is constructed of a stepped piston and a stepped cylinder, thus forming three compartments: a power, a compression, and a crankcase compartment. With this arrangement, the fresh charge is compressed in the compression compartment, delivered to a receiver, and introduced to the cylinder through the scavenge ports. In some designs, the fresh charge enters the crankcase compartment prior to its admission to the cylinder. As the piston travels downward, the volume of the compression compartment increases, the pressure thus decreases, and fresh charge is admitted. Meanwhile, the exhaust port of the power cylinder is exposed first, then the scavenging port, and the burnt gases inside the power cylinder are scavenged by the fresh charge, which previously was compressed in the crankcase compartment. As the piston ascends, both the exhaust and scavenge ports of the power cylinder are covered by the piston and, simultaneously, the intake reed valve of the compression compartment closes, the delivery reed valve opens, and fresh charge flows from the compression compartment to the crankcase. Just before top center (TC), the fresh charge in the power cylinder is ignited, combustion occurs, and the piston is pushed down for the power stroke.

Many designs incorporating stepped pistons were proposed in the early days of the internal combustion engine. Few of these apparently ever reached production, largely because of limitations of contemporary engine technology.[7] The British Dunelt motorcycle, produced from 1919 to 1930, employed a simple version of the stepped-piston engine to improve performance by increasing the displacement of the crankcase pump. This resulted in an excellent torque curve for which these machines became well known, at the expense, however, of high fuel consumption.[7] The Elmore car, produced in the United States with three or four cylinders from 1909 to 1913, employed another version of the stepped-piston engine. In this engine, an ordinary trunk piston was provided, at its lower end, with a circular flange carrying piston rings. This flange made a running fit in an enlarged bore concentric with the working cylinder. An annular space was thus formed between the small-diameter portion of the piston and its large bore, where the charge was compressed. A more effective pump than the crankcase resulted, as its clearance volume could be made as small as desired. This principle may be applied to both multicylinder and single-cylinder engines alike. For multicylinder engines, the charge compressed in the annular space in one cylinder during the downstroke of the piston is transferred to another cylinder in which the piston has simultaneously performed an upstroke. For single-cylinder operation, the fresh charge is compressed into a receiver, where it is stored during the next downstroke until the piston opens the scavenging ports.[7]

For applications at high altitudes, it has been found that a problem with the use of a crankcase-scavenged two-stroke engine is the sharp decrease in engine power with increase in altitude.[8] This was attributed not only to the low density of the ambient air, but also to deterioration of the efficiency of the gas exchange process due to the decrease in the delivery ratio. The main reason for the decrease in the delivery ratio at high altitudes is the inability of the crankcase volume to admit enough air when the pressure difference between the ambient and the crankcase volume is small.[9] Increasing the compression ratio of the crankcase is one possible solution; however, the fresh charge supply to the scavenge ducts will most likely take place during only a small part of the scavenging period and the scavenging process would probably be less efficient. Increasing the delivery ratio can easily be achieved with a stepped-piston engine.

Compared with the conventional crankcase-scavenged engine, the stepped-piston engine offers better scavenging, but higher pumping work, because both the compressor and the crankcase are required. If the increase in engine volume and weight are ignored, an optimal aspect ratio that gives the highest thermodynamic efficiency exists. The engine's bulk, however, is an important parameter that directly affects specific power, usually an important feature of the two-stroke engine. Adding a compressor with a high compression ratio at the inlet of the crankcase appears to be an attractive solution. This modification facilitates breathing at high altitude (higher pressure difference between ambient and the compressor compartment), does not require a high aspect ratio (low engine bulk), and allows slower fresh charge delivery to the power cylinder during the scavenging period.

2-3-3 Double-Piston Engines

In the past, one of the more popular types of two-stroke engines with piston-controlled ports, with nonsymmetrical timing, was one having two pistons in each cylinder with the combustion chamber between them (Fig. 2-15). With this arrangement, one piston controls the scavenging ports and the other the exhaust ports, the two sets of ports being located at opposite ends of a long cylinder. This results in what is known as the uniflow scavenging system; its scavenging process is expected to be much more efficient than

Figure 2-15 An opposed-piston uniflow-scavenged engine with an external rotary compressor.[10] The example shown is the 350-cm^3 Imperia engine giving approximately 22 kW.

in cross- or loop-scavenging engines, especially if the strokes are made relatively long and both scavenging and exhaust ports extend fully around the cylinder. Because of the nonsymmetric timing with this arrangement, the blowdown period can be made two to three times as long as in a conventional two-stroke engine. Also, because each set of ports can extend completely around the cylinder instead of less than halfway around, the port area can be made large enough to facilitate high-speed engine operation.

2-3-4 U-Cylinder Uniflow-Scavenged Engines

The U-cylinder uniflow-scavenged engine is a type of two-stroke cycle engine that has a one-way scavenging path in inverted U-form (Fig. 2-16). This approach, first employed in the Italian Garelli motorcycle, uses two parallel cylinders joined by a common combustion chamber; the exhaust port lies at the bottom of one cylinder and the scavenging ports in the other, while either or both pistons can be used to control the crankcase inlet ports.

In some of these designs, both pistons move exactly in phase, being coupled to a single connecting rod. The port timing is therefore symmetrical, but could be made slightly asymmetrical either by offsetting the cylinders or by coupling the pistons, using two connecting rods, to a single crankpin rotating in the plane of the cylinders. In this case, the pistons will move out of phase such that the exhaust port is exposed well before the scavenging port, whereas it can close much earlier than the scavenge port if desired. This system was used in the British Trojan car, using forked rods and spindle. The engine used in the Valveless car, a still earlier example, employed two crankshafts

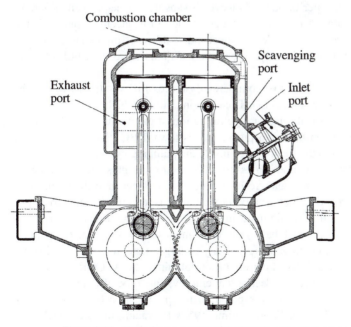

Figure 2-16 Vertical cross section of the Valveless U-cylinder-type uniflow-scavenged engine.

geared together and spaced at wider centers than those of the cylinders (Fig. 2-16). This arrangement provided nonsymmetrical timing with excellent balance. However, as one piston has started downward while the other is still rising, the enclosed volume remains substantially constant for several degrees and, inevitably, higher heat losses result. This can be overcome by coupling one piston to a specially designed four-rod system of suitable proportions (Puch and DKW motorcycles). In this way, the pistons can be made to move substantially in phase at TC, but will be out of phase at BC. This was especially necessary in racing DKW engines that, being supercharged, demanded nonsymmetrical port timing.

Still, whatever advantages the U-cylinder engine may have, the additional mechanical complications and the difficulty of cooling the space between the two cylinders weaken its advantages. Engines of this type were produced as commercial engines until about 1955 (in Japan), when they disappeared from production.[11]

2-3-5 Scavenging Pumps

The fresh charge must be supplied to the engine cylinder at a pressure higher than atmospheric. However, the pressure should be as low as possible, to minimize pumping work. The performance of a two-stroke cycle engine depends heavily on the characteristics of the fresh-charge pressure-raising system. Important types of pressurizing systems in current use include crankcase compression, centrifugal blowers, roots compressors, and piston pumps. Compared with the displacement pumps (roots and piston), the crankcase compression and centrifugal blower are simple, inexpensive, less bulky, and low-weight systems[12]; their respective performance, however, is sensitive to the pressure difference across the exhaust system. An important characteristic of displacement pumps is that the amount of air delivered per cycle is almost independent of the exhaust pressure; with these systems, mufflers and catalytic converters will, therefore, have little effect on the delivery ratio.

Clerk, in 1881, developed the first commercially successful two-stroke cycle engine with a separate piston pump (Fig. 2-17). The charging pump, which draws air through the carburetor and delivers it through a pipe to the top of the combustion chamber, is driven by the engine. When the main piston travels toward BC, the pumping piston moves upward to compress the fresh charge inside the pumping cylinder. When the exhaust ports are exposed, the cylinder pressure falls, the excess pressure above the top valve causes this valve to open, and fresh mixture is introduced into the main cylinder. One important purpose of such an auxiliary cylinder is to create asymmetrical port timing, relative to BC, and thus minimize any backflow through the scavenging ports. Engines using this Clerk approach have proved reliable over a wide range of operating conditions and have been employed successfully in large stationary engines (Körting and Oechelhauser) and small petrol car engines (Dolphin). The ALCOR engine[13] for motorcycle applications is a recent design that employs this type of pumping system. An interesting design that combines a separate piston pump and a U-cylinder engine (Fig. 2-18) was produced by DKW in the mid-1950s for racing motorcycles.[14] This engine achieved a spectacular specific power output of 105 W/cm^3.

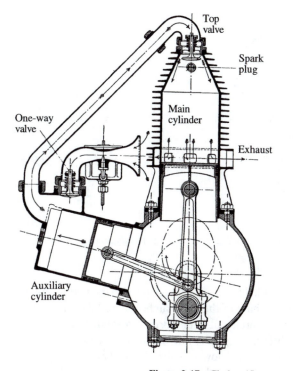

Figure 2-17 Clerk uniflow-scavenged engine, showing its auxiliary piston.

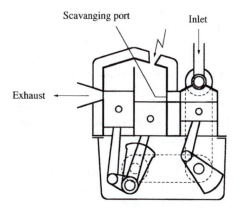

Figure 2-18 The DKW two-stroke racing motorcycle engine with a U-cylinder-type engine and a separate scavenging pump.[14]

2-3-6 Charge Stratification

For carburetted and port-fuel-injected engines, the fresh-charge pressure should result in a stable flow stream and a low enough delivery ratio to prevent significant short-circuiting losses of fuel to the exhaust port. In these cases, and also when lean operation is required, stratification of the charge can be an advantage. The potential benefits of charge stratification were recognized as early as 1905. In the Dolphin engine

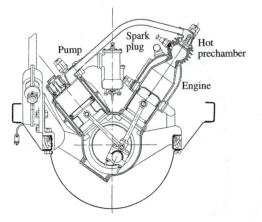

Figure 2-19 Vertical cross section through the Dolphin engine showing an early arrangement of charge stratification.

(Fig. 2-19), a hot prechamber was installed in the air passage prior to the main combustion chamber. The restricted neck at the base of this prechamber caused the gas to enter the cylinder at a high velocity and to separate out in the form of a solid cone sweeping the exhaust gases before it with a little mixing. It is interesting to note that the profile of the passage was designed with the aid of a flow-visualization technique using glass cylinders and smoke. This arrangement enables the engine to run when the throttle is almost closed because the main cylinder remains nearly full of inert exhaust gases while a small amount of fresh gas enters the prechamber and remains there until ignited. The neck meanwhile prevents mixing of the fresh and burned gases. The volume of gas in the pump cylinder is insufficient to fill the main cylinder. To increase the volume of this gas without increasing its quantity, it is expanded at the entrance to the main cylinder by being made to pass through the hot prechamber, which increases its specific volume. Dedeoglu and Kajina[15] have suggested modifing this idea by dividing the air into two portions. During the first stage, the cylinder is scavenged with an air portion warmed by exchanging heat with the exhaust gases. The unwarmed air portion, having higher density, is then introduced to fill the already well scavenged cylinder. For best results, the two portions are almost equal.[15]

Batoni[16] has developed a type of engine in which the gas exchange process progresses in two phases: scavenging and charging. The engine (Fig. 2-20) is of the opposing-piston type, having two cylinders of different displacements, 200 and 50 cm^3. Each cylinder is installed with a system of scavenging ports, while the small-displacement cylinder has the normal exhaust ports sealed off. During the first phase, the larger-displacement cylinder is charged with air alone, a part of which is lost through the exhaust port, while the intake charge of the smaller-displacement cylinder is rich mixture of approximately 4:1 air/fuel ratio. During the second phase, the rich mixture is delivered to the main cylinder and occupies its upper part. The compression stroke follows, and the rich mixture at the upper side is diluted by fresh air to stoichiometric or lean A/F values.

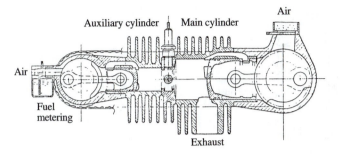

Figure 2-20 An opposed-piston stratified-charge engine.[16] The gas exchange process comprises a scavenging phase followed by a charging phase.

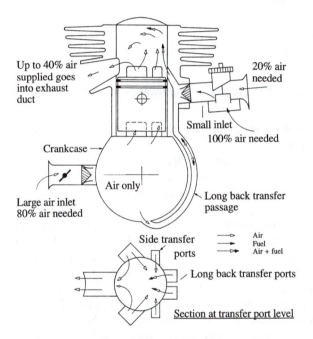

Figure 2-21 Air and fuel flow paths in the Blair crankcase-scavenged stratified-charge engine.[17]

Blair et al.[17] have presented a unique stratified crankcase-scavenged design in which the essential simplicity of the conventional ported two-stroke cycle engine is retained. The engine (Fig. 2-21) has a swept volume of 400 cm^3 and has two intakes to the crankcase, one of them being for a rich mixture and the other for air only. Both intakes are throttled and the throttles are linked mechanically. The air-only intake is controlled with a reed valve and inducts some 70–80% of the required air. A long back-scavenging port connects the lower part of the crankcase with the cylinder wall opposite the exhaust port. The rich mixture is introduced also via a reed valve and inducts some 20–30% of the required air through the long back-scavenging port. The charging process begins with air only in the proximity of the exhaust port, and with the rich mixture

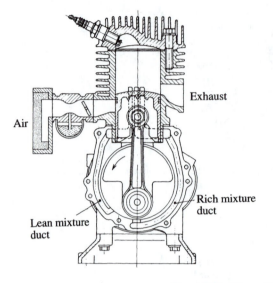

Air

Exhaust

Rich mixture
duct

Lean mixture
duct

Figure 2-22   The multilayer stratified scavenging (MULS) engine.[18]

retained as remotely as possible from the exhaust port. Hence most of the fresh charge
lost from the cylinder will be air alone.

Onishi et al.[18] realized that in an operating engine, due to combined centrifugal and
gravity forces, a part of the liquid fuel in the crankcase adheres to the inner wall of the
crankcase and flows downward along the wall. Based on this observation, they designed
an engine that has two sets of scavenging ports (Fig. 2-22). One set is connected to the
bottom part of the crankcase, thus collecting the liquid fuel to form a rich mixture, and
the other to the upper part of the crankcase volume, providing a lean mixture. The rich
mixture is introduced to the cylinder through a pair of scavenging ports opposite the
exhaust port, while the lean mixture is introduced to the cylinder through two pairs
of scavenging ports that are located on the exhaust-port side of the cylinder liner. The
scavenging flows are thus controlled by three pairs of scavenging ports. A rich mixture
flows into the cylinder toward the cylinder liner wall located opposite the exhaust port,
and a lean mixture flows through two pairs of scavenging ports so that the lean mixture
overlays the rich mixture. The rich mixture is thereby prevented from moving toward
the exhaust port by the lean mixture.

Another method of charge stratification has been suggested by Saxena et al.[19] They
used a dual-intake system consisting of primary and secondary branches (Fig. 2-23).
The primary branch is used to supply air–fuel mixture to the crankcase volume through
a carburetor. The secondary branch is used to supply pure air to the upper part of the
scavenging duct through a reed valve. In this arrangement, the scavenging duct is filled
with pure air. When the scavenging ports open, pure air enters first and then air–fuel
mixture follows. A modified version has been suggested by the same authors,[20] in
which the secondary branch is connected to the exhaust pipe rather than to the atmosphere (Fig. 2-24). During induction, a portion of the exhaust gases are recirculated to
the scavenging duct. At the same time, the fresh charge mixture is introduced to the

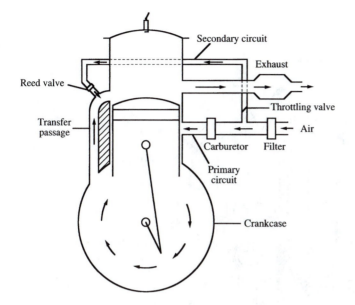

Figure 2-23   A dual-intake system for two-phase charge stratification.[19]

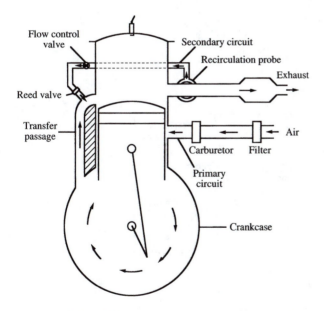

Figure 2-24   A selective exhaust gas recirculation (SEGR) system with two-phase charging.[20]

crankcase volume through the main branch. In the first phase of scavenging the recirculated exhaust gas enters the cylinder, and subsequently the air–fuel mixture enters from the crankcase. The engine power output can be controlled by the throttle valve provided

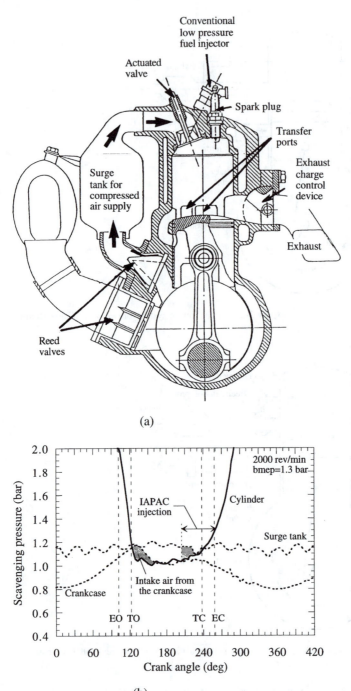

(a)

(b)

Figure 2-25 The "Injection Assistée Par-Air Comprime" (IAPAC) engine: (a) a cross section; (b) traces of cylinder, crankcase, and surge-tank pressures. (Developed from Duret and Moreau.[23])

in each branch. The principle of utilizing the exhaust gases to scavenge the cylinder prior to the admission of the fresh charge, thus delaying the charging process, has also been used though with a different configuration by Rochelle.[21]

Direct fuel-injection and intake fuel-injection systems can readily be designed to form a stratified charge inside the cylinder. Duret and co-workers[22,23] have proposed such an engine design, which employs an intake pneumatic fuel-injection system (Fig. 2-25a). The engine includes, in addition to its essential parts, a surge tank that serves as a compressed-air reservoir installed with a reed valve at its inlet and a valve at its outlet to the cylinder. During the expansion stroke, the surge-tank pressure increases while the fuel is introduced behind the closed valve. When the scavenging ports located at the bottom part of the cylinder open, the cylinder is scavenged with pure air for a period of about 60 crank-angle degrees (Fig. 2-25b). The surge tank outlet valve opens 30 crank-angle degrees before the scavenge port closing, and the fuel is pneumatically injected until about 10 crank-angle degrees before exhaust port closes (EC).

In another recent stratified direct-injection two-stroke engine design, developed by the Oribital Engine Company[24,25] (see Fig. 1-24), the cylinder is scavenged by pure air that is compressed in the crankcase and introduced through the scavenging ports in a conventional manner. The fuel is injected directly into the cylinder during the compression process. The gasoline is injected and finely atomized by an air-assisted atomizer with pulsed air at 350–700 kPa, so that the air–fuel mixture within the combustion chamber varies in composition, thus creating a stratified in-cylinder charge. These direct fuel-injection systems are a promising way to prevent any significant loss of fuel by short-circuiting during the gas exchange process. This fuel loss is a major problem with two-stroke cycle gasoline engines used in transportation.

REFERENCES

1. Blair, G.P., *The Basic Design of Two-Stroke Engines*, SAE, Warendale, PA, 1990.
2. Wallace, E.J., and Cave, P.R., "Experimental and Analytical Scavenging Studies on a Two-Stroke Opposed Piston Diesel Engine," SAE Paper 710175, 1971.
3. Ikeda, Y., Hikosaka, M., Nakajima, T., and Ohhira, T., "Scavenging Flow Measurements in a Fired Two-Stroke Engine by Fiber LDV," SAE Paper 910670, *SAE Trans.*, vol. 100, 1991.
4. Ikeda, Y., Hikosaka, M., and Nakajima, T., "Scavenging Flow Measurements in a Motored Two-Stroke Engine by Fiber LDV," SAE Paper 910699, 1991.
5. Fukutani, I., and Watanabe, E., "Limiting Delivery Ratio Curves of Crankcase-Scavenged 2-Stroke Cycle Engines by Throttling the Inlet Port," SAE Paper 860035, *SAE Trans.*, vol. 95, 1986.
6. Schweitzer, P.H., *Scavenging of Two-Stroke Cycle Diesel Engines*, Macmillan, New York, 1949.
7. Hooper, B., and Favill, J.E., "Modern Stepped Piston Diesel Engines," Inst. Mech. Engrs. Conference Pub. No. 1987-5, pp. 51–58, 1987.
8. Sher, E., "The Effect of Atmospheric Conditions on the Performance of an Air-Borne Two-Stroke Spark Ignition Engine," *Proc. Inst. Mech. Engrs., Part D*, vol. 198, pp. 239–251, 1984.
9. Sher, E., and Zeigerson, M., "A Stepped-Piston Two-Stroke Engine for High Altitude Applications," SAE Paper 940400, 1994.
10. Irving, P.E., *Two-Stroke Power Units, Their Design and Application*, Hart Publishing Co., New York, 1968.
11. Ishihara, S., "An Experimental Development of a New U-Cylinder Uniflow Scavenged Engine," SAE Paper 850181, 1981.

12. Cundari, D., and Nuti, M., "Appraisal of Regenerative Blowers for Scavenging of Small 2T S.I. Power-plants," SAE Paper 920781, 1992.

13. Sanz, M.S., "Two-Stroke Engine with Flow Induction Corrected at the Intake and Transfer Phases," SAE Paper 850184, 1985.

14. Humphries, J.D., *Automotive Supercharging and Turbocharging Manual*, Haynes, Sparkford, nr. Yoevil, Somerset, UK, 1992.

15. Dedeoglu, N., and Kajina, T., "Two-Stage Scavenging Improves Overall Engine Efficiency," SAE Paper 881263, 1985.

16. Batoni, G., "An Investigation Into the Future of Two Stroke Motorcycle Engine," SAE Paper 780710, *SAE Trans.*, vol. 87, 1978.

17. Blair, G.P., Hill, B.W., Miller, A.J., and Nickell, S.P., "Reduction of Fuel Consumption of a Spark-Ignition Two-Stroke Cycle Engine," SAE Paper 830093, *SAE Trans.*, vol. 92, 1983.

18. Onishi, S., Hong Jo, S., Do Jo, P., and Kato, S., "Multi-Layer Stratified Scavenging (MULS)—A New Scavenging Method for Two-Stroke Engine," SAE Paper 840420, *SAE Trans.*, vol. 93, 1984.

19. Saxena, M., Mathur, H.B., and Radzimirski, S., "A Stratified Charging Two-Stroke Engine for Reduction of Scavenged-Through Losses," SAE Paper 891805, 1989.

20. Saxena, M., Mathur, H.B., and Radzimirski, S., "Reduction of Fresh Charge Losses by Selective Exhaust Gas Recirculation (SEGR) in Two-Stroke Engines," SAE Paper 891806, *SAE Trans.*, vol. 98, 1989.

21. Rochelle, P.H.Ch., "Delayed Charging: A Means to Improve Two-Stroke Engine Characteristics," SAE Paper 941678, 1994.

22. Duret, P., Ecomard, A., and Audinet, M., "A New Two-Stroke Engine with Compressed-Air Assisted Fuel Injection for High Efficiency Low Emissions Applications," SAE Paper 880176, *SAE Trans.*, vol. 97, 1988.

23. Duret, P., and Moreau, J.F., "Reduction of Pollutant Emissions of the IAPAC Two-Stroke Engine with Compressed Air Assisted Fuel Injection," SAE Paper 900801, 1990.

24. Leighton, S.R.A, Southern, M.P., and Cebis, M.J., "The Orbital Combustion Process for Future Small Two-Stroke Engines," *A New Generation of Two-Stroke Engines for the Future?* (ed. P. Duret), Proceedings of the International Seminar held at IFP, Rueil-Malmaison, France, Nov. 1993, Editions Technip, Paris, 1993.

25. Houston, R., Archer, M., Moor, M., and Newman, R., "Development of Durable Emissions Control System for an Automotive Two-Stroke Engine," SAE Paper 960361, *SAE Trans.*, vol. 105, 1996.

THREE

EXPERIMENTAL METHODS FOR QUANTIFYING
THE SCAVENGING PROCESS

The performance of a two-stroke engine is strongly dependent on how well the burnt gases are scavenged from the cylinder volume and replaced with fresh charge: that is, on the efficiency of the scavenging process and on the amount of fresh charge trapped inside the cylinder at the end of the process. For the same pumping power input, a cylinder that is better scavenged will produce a higher engine brake power. Improved scavenging will minimize losses of fresh charge to the exhaust port through short-circuiting and, therefore, in engines with premixed charge will reduce fuel consumption and hydrocabon (HC) emissions. Lower fuel consumption at low engine load can also be achieved with a well-controlled scavenging process, often with appropriate stratification of the fresh charge within the cylinder to facilitate good combustion under lean and/or high residual gas operating conditions.

The characteristics of the scavenging process, that is, the scavenging flow details, its overall efficiency, and the distribution of the charge retained in the cylinder at the end of the process, depend very much on the operating conditions of the engine. Important parameters are the engine speed, engine load (especially in throttled engines), and ambient conditions. For example, the typical convex shape of the torque versus engine-speed curve is often attributed to the sensitivity of the scavenging efficiency to engine speed (see Section 3-1). Also, the scavenging flow behavior in two-stroke engines employing a Schnürle-type scavenging system, is strongly dependent on ambient conditions, which limits their suitability for high-altitude applications.

It is, therefore, important for the engine designer to obtain reliable information about the engine's scavenging progress over a wide range of operating conditions. Due to its complexity, the scavenging flow behavior cannot, at present, be accurately predicted. So experimental measurements, although difficult to obtain, are essential to engine development and design. The experimental methods used to quantify the scavenging process are the subject of this chapter. As there is no direct method for determining the mass of the fresh charge trapped inside the cylinder at the commencement of the compression process in a firing engine, several indirect methods have been developed. These may be classified into two main categories: measurements in motored engines; and measurements in fired engines.

3-1 MEASUREMENTS IN MOTORED ENGINES

With motored engine tests it is presumed that the scavenging characteristics are only weakly dependent on the combustion process. The engine itself, or a model of the engine, is driven by an external motor and the scavenging process is then analyzed. For a preliminary design of an engine geometry, it is sometimes easiest to evaluate the scavenging characteristics on the basis of a single cycle rather than several successive cycles. Static model tests, in which the cylinder is subjected to a steady air flow through its ports while the piston is locked at bottom center, are also used.

3-1-1 Static Model Tests

Static or steady flow model tests are mainly used to study the flow direction and distribution of the entering fresh charge, without introducing the effects of the moving piston and the unsteady nature of the process on the scavenging characteristics. Jante[1] found that scavenging ports can be designed to give an efficient scavenging process using a test on the engine with the cylinder head removed. Jante proposed that the fresh charge be introduced to the cylinder as a steady flow and that the velocity profile of the scavenging flow be measured just above the open cylinder. He argued that, to minimize short-circuiting, the upward flow from the scavenge ports toward the cylinder head should be concentrated on the wall opposite the exhaust port: "a stable closed rising current is obtained on the wall opposite to the exhaust ports." A zero-velocity interface between upward and downward streams should be located about halfway across the cylinder or slightly farther from the exhaust side. The area of the interface available for turbulent mixing between the two streams should be minimized if the zero-velocity contour appears on the diagram as a straight line perpendicular to the plane of geometrical symmetry. Figure 3-1 shows results obtained by Jante for four different port geometry designs.

The Jante method is popular among engine designers because it is relatively simple to carry out and can be performed on either the actual engine or a model cylinder. The method has been successfully employed in the development of two-stroke cycle engines over the past 25 years. Among the users are the Maschinenfabrik Augsburg-Nürnberg Aktiengesellschaft (MAN) of Germany, and Fiat of Italy. The method has been improved by Blair and co-workers,[2-4] who have developed an algorithm to analyze the recorded data and optimize the ports geometries accordingly. In addition, Blair and co-workers have extended its application to include piston-speed effects by using

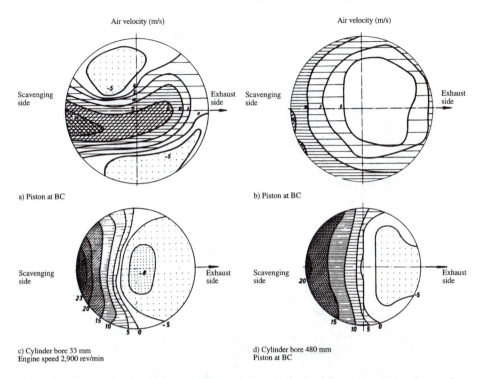

Figure 3-1 Scavenging flow patterns obtained with the cylinder head removed under steady-state flow conditions.[1] Figures show constant air velocity contours: (c) and (d) examples of satisfactory scavenging; (a) and (b) examples of an unsatisfactory scavenging flow.

a pulsating air flow. Figures 3-2 and 3-3 show the effect of engine speed and throttle position on the so-called Jante scavenging pictures of a loop-scavenged spark-ignition engine.[3] This particular engine apparently performs better at higher engine speeds and lower engine loads. The zero-velocity line, however, is too curved and wrinkled for the cases shown, which means that substantial fresh charge would be expected to mix with the burnt gas—a process that should be avoided as much as possible by optimizing the geometry of the cylinder and ports assembly. The Jante method has been further improved,[5-7] to consider the profiles of all three velocity components at the open end of the cylinder. Even in its simplest form, however, the Jante method is still a useful and easy-to-apply method for estimating the optimum geometrical features and dimensions of the scavenge ports.

It is interesting to note here that in earlier work, Percival[8] used a similar approach (in the sense that he used a steady-flow apparatus and measured the axial velocity profile) to optimize the geometry of the scavenge ports of a uniflow-scavenged engine. Percival suggested measuring the axial velocity profile of the flow pattern at an arbitrary cross section of the cylinder (without removing the cylinder head) and optimizing the port geometry to obtain as uniform a profile as possible.

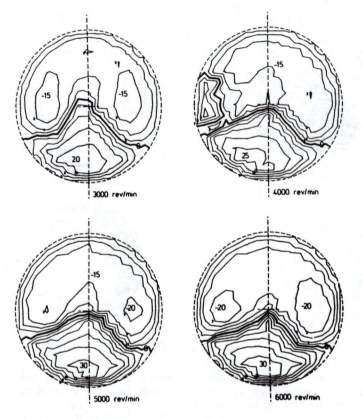

Figure 3-2 The effect of the engine speed on the Jante scavenging-flow patterns of a loop-scavenged spark-ignition engine, at full throttle.[3] The engine was motored with the cylinder head removed, while the delivery ratio was kept constant for all these cases. The zero-velocity line is drawn in heavily; the velocity contours are spaced in equal increments of 5 m/s.

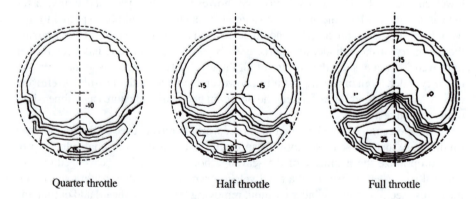

Figure 3-3 The effect of the throttle position on the Jante scavenging-flow patterns of a loop-scavenged spark-ignition engine, at a constant engine speed of 4500 rev/min, for the same conditions as shown in Fig. 3-2.[3]

Table 3-1 Summary of examples of steady flow scavenging tests

Performed by	Year	Ref.	Fresh charge	Burnt gas	Visualized by	Scavenge system	Flow field measurements
Percival	1955	8	Air	Air	—	Uniflow	2-D
Ferro	1958	11	Air	Air	Threads	Uniflow	3-D
Oggero	1968	9	Air	Air	—	Cross	3-D
Jante	1968	1	Air	Air	—	Schnürle	1-D
Oka and Ishihara	1971	12	Water + dye	Water	Color	Cross	None
Phatak	1979	15	Ammonia	Air	NH_3 sensor	Cross + Loop	None
Blair and Kenny	1980	3	Air	Air	—	Loop	1-D
Sung and Patterson	1982	6	Air	Air	—	Uniflow	2-D
Ishihara et al.	1988	5	Air	Air	—	Schnürle	3-D
Smyth et al.	1988	4	Air	Air	—	Schnürle	1-D
Smyth et al.	1990	16	Air	Air	—	Loop	3-D
Sher et al.	1991	7	Aair	Air	—	Uniflow	3-D
Sato et al.	1992	13	Water + ink	Water	Ink	U-type	None

Although the coefficient of discharge of the scavenge port as well as the inclination angle of the incoming charge depend on piston speed, piston position, and pressure difference across the cylinder (see Chapter 5), steady flow models have been found to be useful indicators of the effect of pressure ratio across the cylinder on the airflow rate through the cylinder.[9] Also, by comparing the profile of the axial velocity component at two distinct cross sections of the cylinder, steady flow models have also been found valuable for estimating the fresh charge losses to the exhaust ports (short-circuiting) in loop- and cross-scavenging engines.[9,10]

Steady flow methods are also a useful tool for visualizing key features of the scavenging flow. An impression of the flow inside the cylinder with a particular design of ports (and piston) may be obtained by using simple flow-visualization techniques: indicator strips can be mounted on threads to indicate the flow direction, its approximate magnitude, and the occurrence of vortices and instabilities. Ferro,[11] one of the pioneers of this technique, was able to identify vortex regions and flow separation from the cylinder walls. Ferro observed that in these flow situations the strips tended to flutter about their mean positions. The flow pattern inside the cylinder can also be visualized by using techniques such as coloring the incoming fresh charge[12,13] or by employing an indicator that is sensitive to a specific property of the incoming charge or an additive that is premixed with it. Acid indicators[14] and ammonia indicators[15] are two of many possible additives. Table 3-1 summarizes several examples of static model scavenging tests.

Although these steady flow engine and model tests have been found useful by many engine manufacturers as a first step in engine port geometry design, the conclusions drawn from such tests should be carefully examined for their applicability to real engines for the following reasons:

1. The phenomena associated with the scavenging process are unsteady and can only be partially simulated by a steady flow process.

2. In the engine, the movement of the piston and the blowdown process do affect various stages of the flow field inside the cylinder (see Section 3-3-1).
3. The velocity and the inclination angle of the incoming charge depend on the piston speed, the piston position, and the pressure difference across the cylinder (which may vary with time).

3-1-2 Dynamic Model Tests

Dynamic model tests are mainly used to investigate possible ways of modifying a particular port geometry to improve its scavenging characteristics. With this approach, the engine itself or a model of the engine is driven by an external motor, while the scavenging characteristics are evaluated by sampling, visualizing, or other measurement techniques. The experimental apparatus is usually constructed so it provides an inexpensive means to obtain valuable design information relative to the expense of a full prototype.

To obtain useful conclusions from an experimental simulation, the experimental apparatus should be designed so that the proper relationships between the relevant variables as well as the relative forces driving the flow inside the cylinder in both the model and the engine are the same. Among these are the geometric proportions, piston speed, pressure difference across the cylinder, fluid densities, and thermodynamic properties. Dimensional analysis of the governing conservation equations provides the desired similarity laws between the simulated and actual engines.[14] The following simplifying assumptions are appropriate:

1. Although the fresh charge and the burnt gas are different substances, they both follow the ideal gas law and are Newtonian fluids.
2. No chemical reactions are involved.
3. The transport coefficients molecular diffusivity, thermal diffusivity, and kinematic viscosity are independent of the temperature and pressure.
4. Surface tension, Duffor, and Soret effects may be neglected.

Based on these general (yet plausible) assumptions, the following similarity laws between the relevant dimensionless numbers in the model (m) and engine (e) can be derived[14]:

Geometric similarity—the geometric proportions of the engine are duplicated in the model, for example,

$$(B/L)_m = (B/L)_e \qquad (3\text{-}1)$$

Reynolds similarity—(the Reynolds number, $\rho u l / \mu$ or $u l / \nu$, is the ratio of fluid interial forces to viscous shear forces)

$$(u l / \nu)_m = (u l / \nu)_e \qquad (3\text{-}2)$$

Prandtl similarity—(the Prandtl number is the ratio of momentum diffusivity to thermal diffusivity)

$$(\nu / \alpha)_m = (\nu / \alpha)_e \qquad (3\text{-}3)$$

Schmidt similarity—(the Schmidt number is the ratio of momentum diffusivity to mass or species diffusivity)

$$(v/D_{AB})_m = (v/D_{AB})_e \qquad (3\text{-}4)$$

Eckert similarity—(the Eckert number is the ratio of fluid kinetic energy to enthalpy)

$$(u^2/h)_m = (u^2/h)_e \qquad (3\text{-}5)$$

Euler similarity—(the Euler number is the ratio of pressure forces to inertia forces in the flow)

$$(\Delta p/\rho u^2)_m = (\Delta p/\rho u^2)_e \qquad (3\text{-}6)$$

Molecular weight similarity law—

$$(M_u/M_b)_m = (M_u/M_b)_e \qquad (3\text{-}7)$$

The boundary and initial conditions must also match. Here,

l is length, for example, the piston stroke L or bore B;
M_b is the molecular weight of the burnt gas;
M_u is the molecular weight of the fresh charge;
u is the velocity, for example, the mean piston speed;
Δp is the pressure difference across the cylinder ports;
h is the enthalpy;
ρ is the density;
v is the kinematic viscosity;
α is the thermal diffusivity;
D_{AB} is the mass diffusivity between fresh charge and residual burnt gas.

Practically, it is very difficult, and may not be possible at all, to design an experiment for which the complete set of similarity laws is satisfied. However, if a particular phenomenon is of special interest, such as the mixing process or the blowdown effect, it is often possible to simplify the set further without significantly affecting the particular phenomenon to be investigated.

Dynamic model simulations have been found invaluable for visualizing the scavenging process using relatively simple and inexpensive techniques. A careful analysis of scavenging-flow pictures often reveals important information about poorly scavenged regions, the rate of mixing between the two fluids, short-circuiting, and so on. Using a pair of liquid fluids rather than a pair of gases[11,14,17–20] makes it possible to operate the model engine at a much slower speed while maintaining the similarity laws between the model and the real engine. The scavenging process may then easily be photographed. However, it is obvious that some important effects such as the blowdown, compression, and expansion processes cannot be simulated. Table 3-2 shows an example of typical relationships between the characteristic dimensions and variables of a real engine scavenging flow and its model equivalent. For this experiment, three similarity laws are conserved: geometric ratios, Reynolds similarity, and Euler similarity. The effects of

Table 3-2 Relationship between characteristic parameters of a real engine and its water model engine[14]

Characteristic dimension	Units	Real engine	Water model	Model-to-engine ratio
Physical constants of the fluids				
Density	kg/m^3	0.505	10^3	2000
Viscosity	kg/m-s	33.3×10^{-6}	10^{-3}	30
Similarity laws to be maintained				
Geometrical (bore/stroke)	—	1.25	1.25	1
Reynolds number	—	5380	5380	1
Euler number	—	445	445	1
Design parameters				
Bore	mm	53	80	3/2
Stroke	mm	67	100	3/2
Piston speed	m/s	6.7	0.067	1/100
Engine speed	rev/min	3000	20	1/150
Pressure difference	kPa	10	2	1/5

the blowdown, compression, and expansion processes, heat transfer and molecular diffusion between the two media, and their compressibilities have not been considered. An additional important factor is also usually omitted. If a crankcase-scavenged engine is to be studied, the effect of the variation of the crankcase pressure with time on the engine scavenging process is important. However, for simplicity, in all the published examples of dynamic model tests listed in Table 3-3, the pressure of the scavenging charge was kept constant, thus ignoring this effect.

Many experimenters prefer to make the incoming fluid visible by coloring it with a dye. However, because of the intense mixing between the two fluids, the visible interface does not necessarily indicate the real interface between the fluids, and only limited information can be obtained about the mixing process itself. It should be noted that the darkness of the color observed is not only a function of the mixing ratio, but also of the depth of the photographed field. In order to get more insight into the mixing process itself, the incoming fluid can be made visible by using a color-change pH indicator.[14] In this case the acidity of the liquid that simulates the burnt gas is prepared so that the visible interface between the two solutions will represent the front of a prescribed mixture ratio between the two fluids. A set of isoconcentration profiles can be obtained by this technique.

Figure 3-4 shows interface visualization results from a Schnürle loop-scavenged model engine. In this experiment the image distortion due to the curvature of the cylinder wall was minimized by constructing a square transparent box around the cylinder and filling its volume with the in-cylinder liquid. The three top pictures show the progress of the scavenging process observed by coloring the incoming fresh charge with a dye. The other two sets of pictures (the middle and bottom) show the same process observed by coloring the incoming charge with a pH indicator. Alkali ions were added to the water that represents the fresh charge, to form a dilute solution of sodium hydroxide

Table 3-3 Summary of examples of dynamic model scavenging tests

Performed by	Year	Ref. no.	Fresh charge simulation	Burnt gas simul.	Visual. facilities	Density ratio simul.	Blow-down simul.	Scavenge system	Measurement of
Sammons	1949	21	Various + powder	Various gases	+	+	+	Loop	Density analysis
Rizk	1958	22	Water	Water + dye	-	-	-	Loop	Dye conc.
Ohigashi	1960	23	NH_3	Air	-	-	-	Uniflow	NH_3 conc.
Ohigashi	1961	24	CO_2 + Freon 22	CO_2	-	-	-	Cross + loop	Freon 22 conc.
Ohigashi	1966	25	$NH_3 + SiClH_4$	NH_3 + air	+	+	-	Schnürle	NH_3 conc.
Martini	1971	26	Air	Air + H_2	+	+	+	Cross	Chem. anal.
Oka	1971	12	Water + dye	Water	+	-	-	Schnürle	Dens. + Temp.
Dedeoglu	1971	18	Tetrachloroethylene	Cyclo-hexane	+	+	-	Schnürle	Density analysis
Kannapan	1974	27	Air + CO_2	Air	-	+	+	Schnürle	CO_2 in exhaust gas
Phatak	1979	15	Air + NH_3	Air	+	-	-	Cross + loop	NH_3 sensor
Sanborn	1980	17	Water + sugar	Water	-	+	-	Schnürle	Sugar analys.
Sher	1982	14	Water + NaOH	Water + H_2SO_4	+	-	-	Schnürle	PH indicator
Sweeney	1985	28	CO_2	Air	-	+	-	Schnürle	CO_2 analysis

Table 3-3 (Continued).

Performed by	Year	Ref. no.	Fresh charge simulation	Burnt gas simul.	Visual. facilities	Density ratio simul.	Blow-down simul.	Scavenge system	Measurement of
Sanborn	1985	19	Water + dye	Deionized water	+	-	-	Schnürle	Electrical conductivity
Mirko	1985	20	Tetrachloroethylene + dye	Cyclohexane	+	+	-	Schnürle	Photography
Reddy	1986	29	Air	Air	-	-	+	Schnürle	Hot-wire anemometry
Sanborn	1988	30	Tetrachloroethylene	Cyclohexane	+	+	-	Uniflow	Density analysis
Hilbert	1991	31	N_2 + biacetyl + seeds	N_2	+	-	-	Loop	Particles illum. by laser
Ikeda	1991	32	Air + seeds	Air	-	-	-	Loop	LDV
Fansler	1992	33	Air + seeds	Air	-	-	-	Loop	LDV
Ghandhi	1992	34	Air + seeds	Air	-	-	-	Loop	LDV
Nino	1992	35	Air + seeds	Air	-	-	-	Uniflow	Two-color LDV
Guibert	1993	36	Air + seeds	Air	-	-	-	Loop	PIV
Lee	1994	37	N_2, fuel	N_2	-	-	-	Loop	LIF

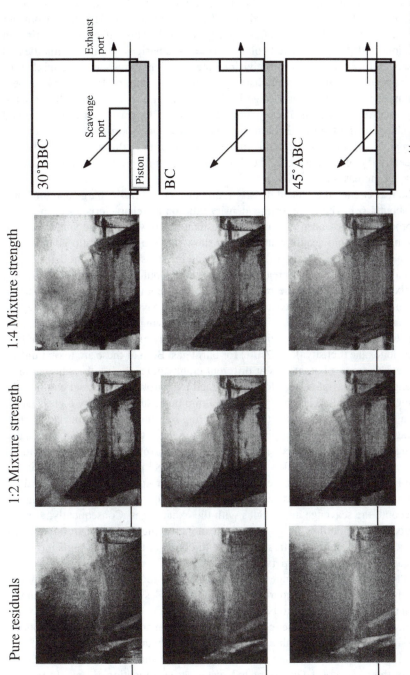

Figure 3-4 Interface profiles between the fresh charge and the burnt gas in a Schnürle-scavenged motored model engine.[14] The interface was made visible by using a dye (top set of photos) or by using a color-change pH indicator (middle and bottom sets of photos). See text for details.

(NaOH) of about 1% concentration, and acid ions were added to the water that initially occupied the cylinder, to form a dilute solution of sulfuric acid (H_2SO_4). The sulfuric acid concentration was adjusted so that a prescribed portion of the alkali solution would be completely neutralized by one portion of the acid solution. A pH indicator powder (phenolphthalein) was added to the fresh-charge reservoir. This indicator is colorless in acid solutions and changes to a red color at pH = 9—a basic solution. The middle and bottom sets of three pictures show the visible interface that corresponds to a mixture of acid (burnt gases) and basic (fresh charge) solutions at a ratio of 1:2 and 1:4, respectively. It appears that, at an early stage in the scavenging process (30° BBC), a large portion of the fresh charge is already diluted. At 45° ABC, about half of the retained cylinder charge is diluted to 1:2 (burnt:fresh) mixture ratio. Short-circuiting is already occurring at BC (if not earlier); at this time about half of the exhaust duct is darkened. The concentration of burnt gases in the short-circuited fluid, however, is quite low (less than 1:2), but increases at 45° ABC.

In the real engine, the typical density ratio between the fresh charge and the burnt gases is about 1.6:1. The effect of this density ratio on the characteristics of the scavenging process has not been thoroughly investigated. However, basic studies on the penetration of a jet into a quiescent medium show that the density ratio is an important parameter.[38] Figure 3-5 shows that increasing the density ratio from 0.48 to 2.08 results in a decrease in the jet spreading angle, which would have a significant impact on the mixing process inside the cylinder, while the depth penetration is lengthened from 10 cm to 13 cm. For this reason, some researchers have designed their model engine scavenging facilities for use with a pair of fluids having the required density ratio. Dedeoglu et al.[18,30] used a pair of liquids: tetrachloroethylene (density 1.62 g/cm^3) for air and cyclohexane (density 0.78 g/cm^3) for burnt gas. Both of these are toxic fluids.

Figure 3-6 shows photographs at different crank angles during the simulation of the scavenging process of a large loop-scavenged two-stroke cycle diesel engine. The scavenging model used a glass cylinder of 160-mm bore, surrounded by a glass jacket filled with the mean mixture of the two fluids during scavenging to minimize distortion of the image. A linkage produces piston motion as in the actual engine, and an isolated scavenging cycle can be obtained. The photos, from the front and side of the cylinder, show the scavenging process takes place in two principal phases. First, fresh air jets penetrate into the in-cylinder burned gas and displace it upward and then toward the exhaust ports. Short-circuiting losses due to "damming-up" of the air occur during this phase. Second, the scavenge air mingles with the exhaust gas. Consequently, as scavenging proceeds, the outflowing gases contain an increasing amount of fresh air. Since the "damming-up" of inflowing fresh air in the direction of the exhaust ports increases, short-circuiting losses probably persist as well.[18]

An alternative way to change the fluid density is to add sugar to the entering charge[17] (for a density ratio of up to 1:1.2). If the two gases involved in the scavenging process in the actual engine are to be simulated by another pair of gases in the scavenging model, a larger variety of gases can be employed: for example, air and an appropriate mixture of air and hydrogen[26]; air and an appropriate mixture of air and CO_2.[27,28]

The progress of the scavenging process can also be evaluated by analyzing the composition (or the purity) of the emerging exhaust gases as a function of time. In fact one

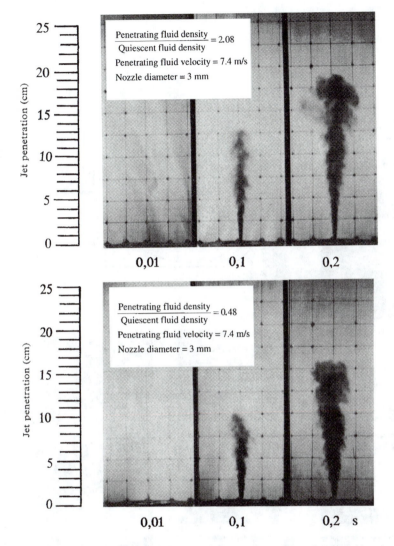

Figure 3-5   The effect of the density ratio between the penetrating and the quiescent fluid on the penetration characteristics of a circular jet.[38] Courtesy Wärtsilä NSD Switzerland Ltd.

of the more accurate methods of evaluating the charging efficiency in either a model or fired two-stroke engine is to locate a sampling valve in the exhaust port, and measure the exhaust gas composition with time. The exhaust gases can either be analyzed continuously or be sampled at discrete time intervals. For modeled engines with dynamic tests, CO_2,[27,28] H_2,[26] NH_3,[15,23,25] Freon 22,[24] acidity,[14] and ion concentration,[19] are examples of possible tracers. Figure 3-7 shows the exhaust gas purity variation with crank angle in a typical Schnürle-scavenged engine model, measured using the pH technique.[14] The exhaust gas purity exhibits a sigmoid-type curve indicating that the

Figure 3-6 Photographs of a denser fluid (dark) scavenging a lighter fluid (clear) in a 160-mm-bore model cylinder simulating a large two-stroke cycle loop-scavenged diesel: (top two rows) view perpendicular to the scavenging loop; (bottom row) orthogonal view. Crank angles after TC are indicated.[18]

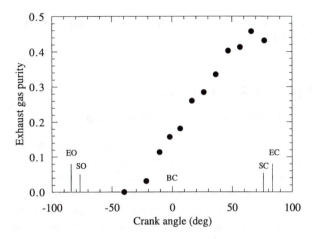

Figure 3-7 The fresh charge content, as a mass fraction, in the gas passing through the exhaust port (the exhaust gas purity) vs. crank angle, as measured in a Schnürle loop-scavenged motored model engine.[14]

short-circuiting losses are associated with a continuous mixing process between the two media, and, therefore, short-circuiting should not be considered as a separate phase in the scavenging process.

When employing two different fluids for the incoming charge and the simulated burnt gas, or when using any kind of tracer to distinguish one fluid from another, the overall performance of the scavenging process can be evaluated by analyzing the composition of either the cylinder contents at the end of the process (in a single-cycle apparatus) or the collected emerging fluid.

The effect of the blowdown period on the scavenging process has not been thoroughly investigated, either. It is expected, however, that its effect would be less significant at low engine speeds and at unthrottled conditions, compared with higher engine speeds and throttled intake conditions. While at low engine speeds, the pressure inside the cylinder has enough time to drop down to the exhaust manifold pressure before the exposure of the scavenge ports; at higher engine speeds a backflow from the cylinder to the crankcase (or to the intake system) through the scavenge ports would be expected. A backflow is also expected when the exhaust pressure is higher than the crankcase (or the intake) pressure, a situation that occurs normally under throttled conditions. The effect of the blowdown period on the fluid motion inside the cylinder prior to the admission of the fresh charge may also be important to the progress of the scavenging process. However, in the absence of relevant experimental results, this effect cannot be estimated in quantitative terms. In some of the experimental model apparatus used,[21,25,27,28] the blowdown process has been included, but its effect has not been evaluated. (See, however, Section 3-3-1 and Fig. 3-21.)

Much effort is now being invested in the development of reliable nonintrusive methods for quantifying the three-dimensional flow field inside the cylinder during the scavenging period. These methods allow in-depth study of the scavenging process under different operation conditions and are starting to be used to optimize the geometry of the cylinder, piston crown, and ports. A summary of these methods and examples of the results they produce can be found in Section 3-3.

3-2 MEASUREMENTS IN FIRED ENGINES

3-2-1 Trapping and Scavenging Efficiency Determination

The characteristics of the scavenging process under real operating conditions can only be examined in full-size engine tests. Table 3-4 summarizes some examples of such scavenging tests. Visualization of the actual process in the real engine, if feasible, would be a valuable tool for guiding the engine designer toward an optimal design. Except for one attempt in the mid-1950s, visualization methods have not been applied successfully for this purpose. This is mainly because the complexity of the 3-D flow inside the cylinder does not allow a simple side-to-side observation. Boyer et al.[39] attempted to apply the Schlieren optical method, which responds to density gradients in the flow field, to visualize the flow inside a large-bore two-stroke engine. The setup provided three views into the cylinder through the head and two views through the cylinder wall. The pictures obtained proved difficult to interpret and analyze in quantitative terms.

One attractive approach for obtaining information about the progress of the scavenging process inside the cylinder as time advances is continuous analysis of the exhausted gas for a particular tracer.[13,40–42] The tracer should be selected so that it accurately represents the instantaneous concentration of either the fresh charge or the burnt gas in the exhaust port and can easily be detected by an available continuous on-line gas-analysis method. When the tracer species represents the burnt gas, the tracer species should be formed during the combustion period, should be distributed homogeneously inside the cylinder before the exhaust ports open, and must be stable after it emerges through the exhaust port. Because of their stability at typical exhaust temperatures (up to 800 K), CO and CO_2[3,9,43,44] are good candidates as tracers for spark-ignition engines. For compression-ignition engines, however, the local air/fuel ratio significantly affects the completeness of CO combustion, and the cylinder contents after combustion are not homogeneous. Nevertheless, some researchers have employed CO[40,42] and CO_2[9,45,46] as burnt gas tracers, for compression-ignition engines, especially at full load but also at other conditions. However, because the cylinder contents in compression-ignition engines operating at part loads are highly nonuniform, a tracer species selected to represent the fresh charge rather than the burnt gas is a better choice.

Where the tracer species represents the fresh charge, that species should burn completely during the combustion process inside the cylinder, and yet also be a stable component at exhaust gas temperatures. Monomethylamine (CH_3NH_2)[47–50] and n-butane (C_4H_{10})[41] have been considered for compression-ignition engine applications, and oxygen (O_2)[13,41,44,51,52] has been used for stoichiometric or rich operating spark-ignition engines. Under lean conditions, unconsumed oxygen and short-circuiting oxygen cannot be separated. When a continuous gas analyzer is not available, sampling of the exhaust gases at specified time intervals and analyzing the samples for a particular tracer or constituent may also be feasible.[9,10,44,47–51]

Table 3-4 Summary of examples of fired engine scavenging tests

Performed by	Year	Ref. no.	Fresh charge tracer	Engine type	Scavenging system	Method of measurement
Schweitzer and DeLuca	1942	47	CH_3NH_2	CI	—	Sampling the exhaust gases and analyzing for the tracer
Boyer et al.	1954	39		CI	—	Schlieren photography
Taylor and Rogowski	1954	48	CH_3NH_2	CI	—	Sampling the exhaust gases and analyzing for the tracer
Hori	1962 1963	45 46	—	CI	Loop	Sampling the cylinder contents before and after the scavenging process and analyzing for CO_2
Bazika and Rodig	1963	40	CO	CI	—	Continuous analysis of the exhaust gas for CO
Miyabe and Shimomura	1963	51	—	SI	Schnürle	Sampling the exhaust gases and analyzing for O_2 for rich mixtures or CO_2 for lean mixtures
Isigami et al.	1962 1969	49 50	CH_3NH_2	CI	Loop	Sampling the exhaust gases and analyzing for the tracer
Booy	1967	57	—	SI	Cross	Skipping a cycle and evaluating its effect on imep
Oggero	1968	9	—	CI + fuel injected SI	Cross	Sampling the exhaust gases during blowdown and analyzing for CO_2
Ohigashi and Hamamoto	1971	52	—	SI	Loop	Sampling the cylinder contents before and after the scavenging process and analyzing for O_2
Huber	1971	41	C_4H_{10}	CI	—	Continuous analysis of the exhaust gas for the tracer
			—	SI	—	Continuous analysis of the exhaust gas for O_2
Wallace and Cave	1971	42	—	CI	Opposed	Continuous analysis of the exhaust gas for CO

Table 3-4 (Continued).

Performed by	Year	Ref. no.	Fresh charge tracer	Engine type	Scavenging system	Method of measurement
Blair et al.	1976 1980	43 3	—	SI	Loop	Sampling the cylinder contents and analyzing for CO_2
Hashimoto et al.	1985	44	—	SI	Schnürle	Sampling the exhaust gases and analyzing for O_2 for rich mixtures or CO_2 for lean mixtures
Nuti and Martorano	1985	10	—	Fuel injected SI	Loop	Sampling both the cylinder contents and the exhaust gases and analyzing for gas composition
Sata et al.	1992	13	—	SI	U	Continuous analysis of the exhaust gas for O_2
Tobis et al.	1994	54	CO_2	SI	Schnürle	Sampling the cylinder contents and analyzing for CO_2
Murayama et al.	1996	53	HC	SI	—	Sampling the exhaust gases and analyzing for HC

When only the overall performance of the scavenging process is of interest, rather than the details of how scavenging progresses, cumulative sampling of the exhaust gases over a large number of cycles may be employed. The collected gases are then analyzed for a specific tracer or gas constituent. This method, however, by its nature averages the scavenging efficiencies of a large number of cycles, and cycle-by-cycle variations cannot be evaluated.

One well-established method of this type is determining trapping and scavenging efficiencies from measurements of oxygen in the exhaust gas. When the engine is operated rich, with a premixed fuel–air charge, there is negligible oxygen in the burnt gases so oxygen in the exhaust gas stream results from scavenging air that has passed directly through the engine. Assuming that the short-circuiting air and burnt gas mix uniformly in the exhaust system, measurement of O_2 in the exhaust gases, $[O_2]_{exh}$, directly gives the engine's trapping efficiency via

$$\eta_{tr} = 1 - [O_2]_{exh}/[O_2]_{atm} \tag{3-8}$$

where $[O_2]_{atm}$ is the oxygen concentration in air, 21%.[44] The sampling position in the exhaust must be chosen to ensure adequate mixing has occurred. Figure 3-8 shows an example of such measurements as a function of distance from the exhaust port.[44] The data indicate that sufficient mixing has occurred by some 0.5 m downstream of the

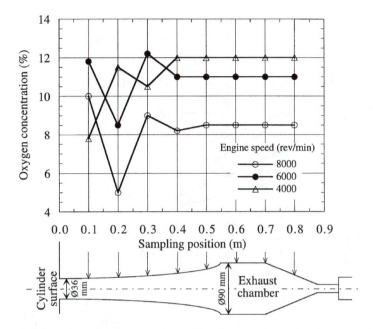

Figure 3-8 The effect of the gas sampling position along an exhaust pipe of a two-stroke spark-ignition engine on the concentration of O_2 in the exhaust gas.[44]

cylinder exit plane. The oxygen concentration also depends weakly on the engine's relative air/fuel ratio and spark timing via the effect of these variables on η_{tr}, and possibly through reaction between the rich exhaust gases and the short-circuiting air. Measurements of O_2 should therefore be taken at values of A/F and relative spark timing that give the maximum bmep, the normal engine operating condition.

To estimate the scavenging efficiency, the relative charge, $c_r = m/m_0$, where m is the mass in the cylinder and m_0 is the reference mass $V_d \rho_{a,0}$ (see Section 2-1), must be determined. If the gas pressure and temperature within the cylinder after combustion, p_b and T_b, when the exhaust port is closed are known (measured and/or estimated) and the cylinder volume is V_b then, from the ideal gas law,

$$c_r = \frac{m}{m_0} = \frac{p_b}{p_0} \cdot \frac{V_b}{V_d} \cdot \frac{T_0}{T_b} \tag{3-9}$$

where p_0 and T_0 are ambient pressure and temperature. Then, from Eqs. (2-3), (2-5), and (2-8),

$$\eta_{sc} = \eta_{ch}/c_r = \Lambda \eta_{tr}/c_r \tag{3-10}$$

where Λ is the delivery ratio. Figure 3-9 shows trapping and scavenging efficiencies determined in this manner, for a single-cylinder high-speed high-performance motorcycle engine.[44]

A different method for determining the scavenging efficiency of a particular engine design, while still allowing study of the cyclic variability, is to extract two samples of

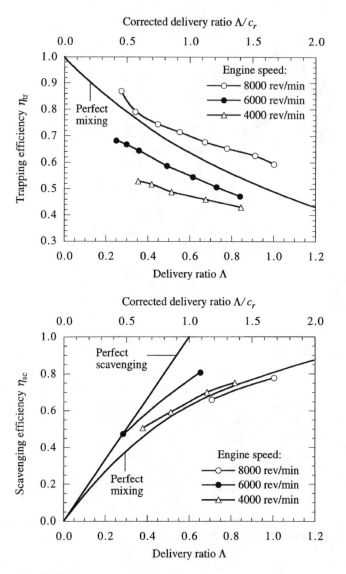

Figure 3-9 Trapping and scavenging efficiencies determined from exhaust gas sampling and oxygen concentration measurement, in a high-speed motorcycle two-stroke spark-ignition engine.[44]

gas from the cylinder: one at the time of exhaust port opening and the other at exhaust port closing.[3,43,45,46,52] The samples are then analyzed for their chemical compositions (CO, CO_2,[3,43,45,46] O_2,[52] HC,[53] or any other tracer).

The following analysis illustrates how in-cylinder CO_2 measurements can be used to determine η_{sc}, η_{tr}, and the fraction of the air that is short-circuited, $x_{a,short}$.[54] Figure 3-10 shows, schematically, the combustion products and air proportions of the in-cylinder gases (precombustion and postcombustion), and the tailpipe gases. The pairs

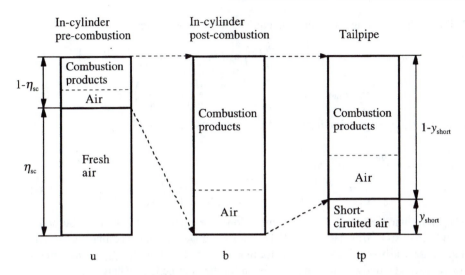

Figure 3-10 Composition of in-cylinder gases, in terms of combustion products and fresh air, before combustion (u) and after combustion (b), and exhaust gases in the tailpipe (tp). Pairs of dotted arrows indicate where gas composition remains the same.[54]

of dotted arrows define those fractions of the cylinder and tailpipe contents whose gas composition is the same. The engine used direct injection of fuel into the cylinder, so fuel is added between stages u and b. The concentration of CO_2 in the precombustion gas sample is the sum of that from the residual gases and the ambient CO_2 in the fresh air (a small component). Neglecting the small difference between the molecular weights of residual, air, and their mixture,

$$[CO_2]_u = [CO_2]_b(1 - \eta_{sc}) + [CO_2]_a \eta_{sc} \tag{3-11}$$

where u and b denote precombustion and postcombustion conditions, respectively, and a in ambient air. Solving for η_{sc},

$$\eta_{sc} = \frac{[CO_2]_b - [CO_2]_u}{[CO_2]_b - [CO_2]_a} \tag{3-12}$$

From the definition of trapping efficiency,

$$\eta_{tr} = \frac{\dot{m}_{tr}}{\dot{m}_{del}} = \frac{\dot{m}_{del} - \dot{m}_{short}}{\dot{m}_{del}} \tag{3-13}$$

To obtain an expression for the short-circuited air, consider the exhaust to be made up of two parts: short-circuited air; and burnt gas that includes any excess air above that needed for combustion. This can be written in the form of mole fractions y, where

$$1 = y_{short} + y_b$$

A mass balance between the exhaust and inputs gives

$$\dot{m}_{short} + \dot{m}_b = \dot{m}_{del} + \dot{m}_f$$

Utilizing this equation, the following equivalency can be written:

$$\dot{m}_{short} = \dot{m}_{short}\left(\frac{\dot{m}_{del} + \dot{m}_f}{\dot{m}_{short} + \dot{m}_b}\right) = (\dot{m}_{del} + \dot{m}_f)\left(\frac{\dot{m}_{short}}{\dot{m}_{short} + \dot{m}_b}\right)$$

$$= (\dot{m}_{del} + \dot{m}_f)\left(\frac{y_{short}M_a}{y_{short}M_a + (1 - y_{short})M_b}\right) \tag{3-14}$$

Substituting the expression for $\dot{m}_{short}$ from Eq. (3-14) into Eq. (3-13) gives

$$\eta_{tr} = 1 - \frac{\dot{m}_{short}}{\dot{m}_{del}}$$

$$= 1 - \left(\frac{\dot{m}_{del} + \dot{m}_f}{\dot{m}_{del}}\right)\left(\frac{y_{short}M_a}{y_{short}M_a + (1 - y_{short})M_b}\right) \tag{3-15}$$

The molecular weight of the postcombustion gases can be estimated via burned gas sample analysis or from Heywood.[55] The molecular weights of the combustion products and air are nearly equal, so the last factor in Eq. (3-15) simplifies to y_{short}. An expression for y_{short} can be derived from an expression for the tailpipe CO_2 where

$$[CO_2]_{tp} = y_{short}[CO_2]_a + (1 - y_{short})[CO_2]_b \tag{3-16}$$

Solving for y_{short},

$$y_{short} = \frac{[CO_2]_b - [CO_2]_{tp}}{[CO_2]_b - [CO_2]_a} \tag{3-17}$$

Substituting Eq. (3-17) into Eq. (3-15) yields the desired relationship for η_{tr} as a function of measured CO_2 concentrations (referenced to wet exhaust gas conditions, that is, exhaust gas with water vapor still present). Because of the nonhomogeneity of the cylinder contents at the time of sampling, the quantity extracted should be large enough to ensure a reliable representation of the average composition of the gases in the cylinder. However, because it is not possible to obtain two representative samples from the same cycle, the two samples should be separated by several undisturbed cycles.

A different method[57] for evaluating the scavenging efficiency is based on the assumption that when a small number of nonfired (skipped) cycles are introduced between normally fired cycles, the scavenging efficiency in each skipped cycle is equal to that of a fired cycle. Then, if $(1 - \eta_{sc})$ is the residual mass fraction for a normal cycle, $(1 - \eta_{sc})^2$ is the residual mass fraction of a fired cycle if the engine fires only every other cycle, $(1 - \eta_{sc})^3$ if it is fired every third cycle, and so on. The corresponding fresh charge mass fractions are $\eta_{sc}, 1 - (1 - \eta_{sc})^2, 1 - (1 - \eta_{sc})^3$ and should be in the same proportion as the corresponding indicated mean effective pressures. If the indicated mean effective pressure is measured for two different cases of spark interruption, the ratio of the two indicated mean effective pressures represents the ratio between the two fresh charge mass fractions: that is,

$$\frac{imep_n}{imep} = \frac{1 - x_r^{n+1}}{1 - x_r} = \frac{1 - (1 - \eta_{sc})^{n+1}}{1 - (1 - \eta_{sc})} \tag{3-18}$$

where x_r is the residual gas mass fraction and n is the number of skipped cycles; $imep_n$ with n skipped cycles is compared with the imep of a normal cycle. Equation (3-18)

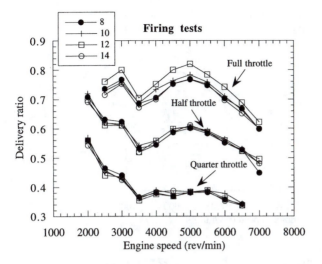

Figure 3-11 The effect of engine speed and throttle position on the delivery ratio for four different designs of cylinders (8, 10, 12, and 14).[3] The engine is a single-cylinder, crankcase-compression, loop-scavenged spark-ignition engine. (See Figs. 3-2 and 3-3 for the steady flow test results of cylinder design 12.)

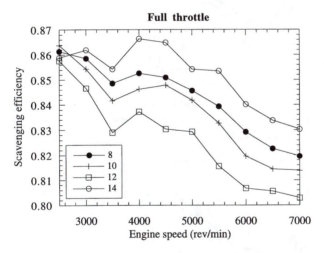

Figure 3-12 The effect of engine speed on the scavenging efficiency at wide-open throttle, for the same engine as in Fig. 3-11.[3]

is then solved for η_{sc}. The skip-cycle method ignores the effect of the density ratio between the incoming charge and the cylinder contents, which may significantly affect the scavenging process (see Section 3-1-2).

Figures 3-11–3-15 illustrate how complex the relationships are between the delivery ratio, the scavenging efficiency, the trapping efficiency, bmep and bsfc, and throttle position and engine speed. These measurements were made on a single-cylinder, crankcase-blown, loop-scavenged spark-ignition engine.[3] The scavenging and the trap-

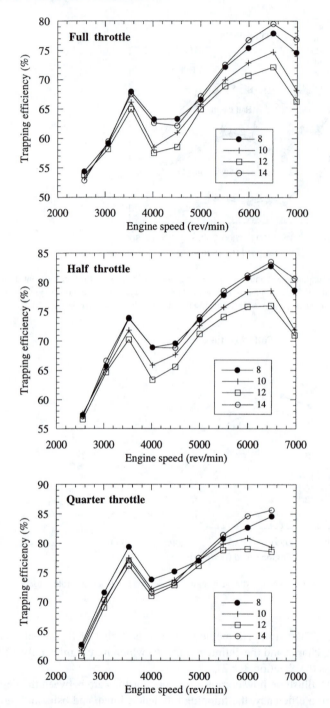

Figure 3-13 The effect of engine speed and throttle position on the trapping efficiency, for the same engine as in Fig. 3-11.[3]

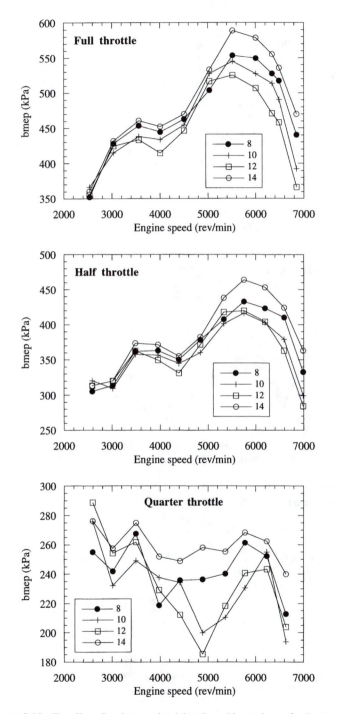

Figure 3-14   The effect of engine speed and throttle position on bmep, for the same engine as in Fig. 3-11.[3]

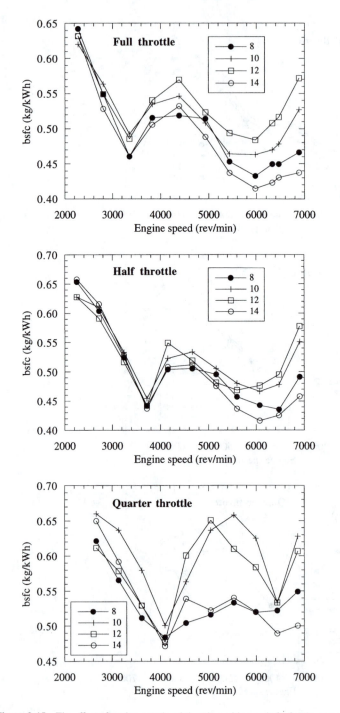

Figure 3-15 The effect of engine speed and throttle position on bsfc, for the same engine as in Fig. 3-11.[3]

ping efficiencies were determined by analyzing samples taken from the in-cylinder gases for O_2. The samples were extracted through a special port in the cylinder head. Figures 3-11–3-13 show the effect of the engine speed on the delivery ratio, the scavenging efficiency, and the trapping efficiency for four different designs of cylinder. For cylinder design 12 (whose steady-flow characteristics are shown in Figs. 3-2 and 3-3), while the delivery ratio decreases from 0.75 to 0.62 when the engine speed increases from 2500 to 7000 rev/min, and the scavenging efficiency drops from 0.86 to 0.81, the trapping efficiency increases from 0.55 to 0.65. At full throttle, while the delivery ratio peaks at 3000 and 5000 rev/min, the scavenging efficiency decreases essentially monotonically and the trapping efficiency peaks at higher speeds, 3500 and 6500 rev/min. The bmep (Fig. 3-14) and bsfc (Fig. 3-15), however, correlate well with each other as would be expected, and with the trapping efficiency at full throttle. At low loads (when the throttle is barely open), the bsfc of the engine increases substantially at mid engine speeds, an effect most likely caused by combustion problems resulting from the high amount of residual burnt gas remaining inside the cylinder under these operating conditions.

3-2-2 Air/Fuel Ratio Determination

Determination of the overall air/fuel ratio from measurements of the species in the engine exhaust gases has long been a standard methodology for four-stroke cycle engines.[55] This approach eliminates the need for direct measurement of engine air flow rate, which, especially in two-stroke engines, may modify the engine's air flow. In two-stroke engines, high concentrations of oxygen (and fuel in homogeneous-charge engines) are present in the exhaust due to the short-circuiting of some of the scavenging flow, introducing additional uncertainties. In engines with direct-injection of fuel into the cylinder after the exhaust ports close, where air short-circuits but fuel does not, the trapped gas air/fuel ratio and the exhaust gas air/fuel ratio are different. Thus use of exhaust gas composition analysis in two-stroke cycle engines introduces some new issues.

One important issue is the high level of hydrocarbons (HC) in the two-stroke exhaust, which must be accurately measured. Flame ionization detector (FID) HC analyzers accurately determine total HC emissions because the FID is effectively a carbon atom counter. However, nondispersive infrared (NDIR) analyzers compare the IR absorption of n-hexane with that of the exhaust HC, which, in four-stroke engines, underestimates the HC by about a factor of 2.[55] At low HC concentrations this correction factor for NDIR HC measurements is not that critical; for homogeneous-charge two-stroke spark-ignition engines it becomes important. The limited data available suggests that this same factor of 2 is still appropriate.[56]

An extension of the analysis first proposed by Spindt[58] provides an accurate determination of overall air/fuel ratio. The overall chemical element balance for fuel–air mixture reacting to produce burnt (and some unburnt) gas can be written as[59]

$$CH_y + A(O_2 + 3.727N_2 + 0.044Ar) + BH_2O \rightarrow$$
$$aCO + bCO_2 + cO_2 + dH_2O + eCH_y$$
$$+ fH_2 + gN_2 + hNO + jAr \qquad (3-19)$$

where

$$B = 4.77A[p_w/(p_0 - p_w)] \tag{3-20}$$

and p_w is the partial pressure of water vapor in the atmosphere and p_a is atmospheric pressure. The air/fuel ratio can then be calculated as follows:

$$A/F = \frac{138.18A}{(12.011 + 1.008y)} = K_f A \tag{3-21}$$

where K_f is a constant that depends only on the fuel composition. The value of A is obtained from the balance equations for C, H, O, and N in Eq. (3-19), the measured concentrations of CO_2, CO, O_2, HC, and NO, and using one of the following assumptions to determine the H_2 concentration (which is not usually measured):

1. The value of water–gas reaction equilibrium constant K is specified, where

$$K = \frac{[H_2][CO]}{[H_2][CO_2]} \tag{3-22}$$

A value of 3.5 is usually used for four-stroke cycle engines. (The exact value is not critical, and the same assumption is appropriate in two-stroke engines.)

2. The H_2 concentration is a fixed fraction of the CO concentration: about one-half (see data in Heywood[55]). †

   With the first assumption (the Spindt method modified for humidity and NO, usually small corrections), the moles of H_2O (d) is given by

$$d = \left[\frac{K[CO_2]}{[CO] + K[CO_2]} \right] \left[\frac{y}{2}(a + b) + B \right] \tag{3-23}$$

Then, solving for A gives

$$A = K_h \left[\frac{a}{2} + b + c + \frac{K_w}{4} y(a + b) + \frac{h}{2} \right] \tag{3-24}$$

where the humidity correction K_h is given by

$$K_h = \left[1 + 2.385(1 - K_w) \left(\frac{p_w}{p_0 - p_w} \right) \right]^{-1} \tag{3-25}$$

and

$$K_w = \frac{K[CO_2]}{[CO] + K[CO_2]} \tag{3-26}$$

†This approach was developed by D.R. Blackmore and K.C. Crawford in "The Measurement of Gasoline Engine Air–Fuel Ratios," Report MOR 641F, Shell Int. Petroleum Co., Ltd., 1973.

a, b, and so on are obtained from the measured mole fractions in the exhaust gases and the total number of moles n_t on the right-hand side (the products) of Eq. (3-19). A carbon balance on Eq. (3-19) gives

$$n_t([CO] + [CO_2] + [HC]) = 1$$

where [ ] denotes the measured mole fractions, and [HC] is expressed as "methane" or C_1. Then, for example,

$$a = n_t[CO] = [CO/([CO] + [CO_2] + [HC]) \qquad (3\text{-}27)$$

and so on. The equation for A/F becomes[59]

$$\begin{aligned} A/F = K_f K_h \{ &(1/2)[CO] + [CO_2] + [O_2] \\ &+ (1/4)K_w y([CO] + [CO_2]) + (1/2)[NO] \} \\ /([CO] &+ [CO_2] + [HC]) \end{aligned} \qquad (3\text{-}28)$$

where

$$K_f = 138.18/(12.011 + 1.008y)$$

and the concentrations are mole fractions in the exhaust.

The second assumption, that $[H_2] = 0.5[CO]$, gives equivalent accuracy (mean error 2–3%) with simpler algebra. d and A are given by

$$d = \frac{y}{2}(a + b) - \frac{a}{2} + B$$

and

$$A = \frac{a}{4} + b + c + \frac{y}{4}(a + b) + \frac{h}{2}$$

and the A/F equation becomes

$$\begin{aligned} A/F = K_f \{ &(1/4)[CO] + [CO_2] + [O_2] \\ &+ (1/4)y([CO] + [CO_2]) + (1/2)[NO] \} \\ /([CO] &+ [CO_2] + [HC]) \end{aligned} \qquad (3\text{-}29)$$

The trapped or burnt zone A/F can also be determined from exhaust gas composition measurements. The trapping efficiency of air and of fuel must be estimated; for homogeneous-charge two-stroke cycle engines these are essentially identical but for direct-injection engines they are different. A simple equation for estimating the air trapping efficiency is based on the exhaust oxygen concentration, Eq. (3-8). A more accurate equation that takes account of the presence of fuel is

$$\eta_{tr,air} = 1 - \frac{[1 + (A/F)[O_2]_e]}{(A/F)[O_2]_{atm}} \qquad (3\text{-}30)$$

Full consideration of the exhaust composition given by Eq. (3-19) leads to the following relationship:

$$\begin{aligned} \eta_{tr,air} = \{ &(1/2)[CO] + [CO_2] + (1/4)K_w y([CO] + [CO_2]) + (1/2)[NO] \} \\ /\{ &(1/2)[CO] + [CO_2] + [O_2] + (1/4)K_w y([CO] \\ &+ [CO_2]) + (1/2)[NO] \} \end{aligned} \qquad (3\text{-}31)$$

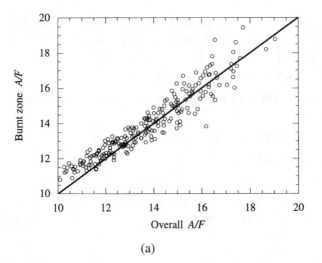

(a)

Figure 3-16 Burnt zone A/F estimates as a function of A/F overall: (a) homogeneous-charge two-stroke spark-ignition engine; (b) direct-injection two-stroke spark-ignition engine with different injection timings.[59]

The fuel trapping efficiency (based on the composition of the in-cylinder burnt gases) is given by

$$\eta_{tr,fuel} = \frac{[CO] + [CO_2]}{[CO] + [CO_2] + [HC]} \tag{3-32}$$

Then the A/F of the burnt gas mixture trapped inside the cylinder is

$$(A/F)_{burnt\ zone} = (A/F)_{overall}(\eta_{tr,air}/\eta_{tr,fuel})$$

Combining Eqs. (3-29), (3-31) and (3-32) yields

$$(A/F)_{burnt\ zone} = K_f K_h \{(1/2)[CO] + [CO_2]$$
$$+ (1/4)K_w y[CO] + [CO_2] + (1/2)[NO]\}$$
$$/([CO] + [CO_2]) \tag{3-33}$$

When the burnt zone is fuel lean, some of the oxygen in the exhaust comes from the lean burnt gases, and Eq. (3-33) will always yield the stoichiometric A/F. To avoid this problem, air trapping efficiencies determined under rich mixture conditions should be used, even at lean operating conditions: that is,[59]

$$(A/F)_{burnt\ zone} = (A/F)_{overall}(\eta_{tr,air(rich)}/\eta_{tr,fuel}) \tag{3-34}$$

The overall A/F calculated from Eq. (3-28) or (3-29) correlates well (mean error or standard deviation 2–3%) with A/F obtained directly from air flow and fuel flow measurements.[54] Burnt zone A/F estimates as a function of overall A/F for a homogeneous charge two-stroke spark-ignition engine and for a direct-injection (stratified charge) two-stroke spark-ignition engine are shown in Fig. 3-16. For the homogeneous charge engine, the burnt zone A/F correlates well with A/F overall. In the stratified charge two-stroke engine, the burnt zone A/F is substantially richer than the overall

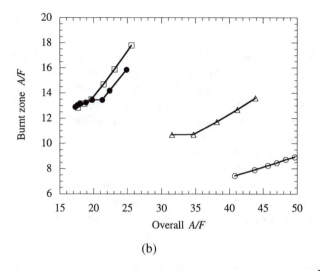

(b)

Figure 3-16   (Continued).

A/F. The different curves are for different injection timings (all direct injection into the cylinder) with the richer burnt zone A/F curves corresponding to injection later in the compression stroke.[59]

3-3 MEASUREMENTS USING LASER DIAGNOSTICS

3-3-1 Gas Velocity Measurements

Laser doppler velocimetry (LDV) has been used to study in-cylinder flows in internal combustion engines since the late 1970s. With this velocity measurement technique, interference fringes are produced within the small volume created by the intersection of two laser beams in the flow field. When a small particle passes through this volume, it scatters light at a frequency proportional to the particle velocity. By seeding the flow with particles small enough to be carried without slip by the flow and collecting the scattered light, the local flow velocity can be determined. Standardized LDV systems and methodologies for their use have now been developed (see, e.g., Miles et al. [60]). LDV can be applied to motored and firing engines, requires optical access to the measurement location, and has become a well-established diagnostic. It is, however, a point measurement, and obtaining the data to quantify complex flow fields such as the two-stroke engine scavenging flow requires measurements at a substantial number of individual locations. Note that each measurement, which results from a small particle being carried by the flow through the small measurement volume, gives the instantaneous local flow velocity at that point in time. Mean velocities and root mean square (rms) fluctuations about the mean are then determined from the ensuing data ensemble of velocity–crank-angle pairs obtained over many engine cycles.[55] Some examples of LDV measurements of the two-stroke engine scavenging process will now be reviewed. The insight that detailed flow velocity measurements provide about aspects of the scavenging flow is substantial.

Fansler and French[33] carried out extensive LDV measurements on a three-cylinder crankcase-compression loop-scavenged two-stroke engine designed for marine applications. One cylinder was modified by adding a cylinder head with a flat quartz window, and a flat-topped (instead of domed) piston. The LDV measurements were done under motored conditions. The exhaust and scavenge port layout of the engine is shown in Fig. 3-17. The piston uncovers the (single) exhaust port at 90° ATC and, of course, closes it at 90° ABC. The main (B,G) and auxiliary (C,F) transfer ports are uncovered at 120° ATC; the boost transfer ports (D,E) are uncovered 4° later. The engine was motored, not fired: extensive firing operation, with substantial-size windows, creates significant engine cooling issues. Figure 3-18 shows measured velocity vectors within the cylinder during a period (130–150° ATC) when the scavenging air flows strongly into the cylinder. The efflux from each port is roughly a plug flow, directed more toward the axis of the cylinder than the port shape suggests. The in-cylinder flow vectors show the developing strong upward flow along the liner wall opposite the exhaust port that the Jante methodology demands (see Section 3-1-1), but also shows that despite the nominal symmetry of the exhaust and transfer port geometries there is swirl and some flow asymmetry. These authors also generated the Jante scavenging diagrams for this engine under motored conditions, and under steady-flow conditions, with the piston at BC. These are shown in Fig. 3-19 and both diagrams exhibit a region of upward flow on the boost port side of the cylinder occupying slightly less than half the cylinder cross section. The zero velocity contour is nearly flat across most of the cylinder. However, side-wall "tongues" are apparent (an undesirable feature because these increase the surface area across which mixing can occur), and the slight asymmetry about the exhaust–boost port plane of symmetry apparent in the LDV data is also present. Both the tongues and asymmetry will increase the short-circuiting of fresh air through the exhaust port. This extensive study illustrates the value of detailed flow field measurements.

A substantial complexity in the two-stroke engine scavenging process is the wide range of speeds and loads over which the engine must operate satisfactorily. In homogeneous charge two-stroke spark-ignition engines, a throttle plate or valve is used to control the amount of air drawn into the crankcase. As speed and throttle position are varied, so the pressure difference between the crankcase and cylinder driving the scavenging flow varies, the velocity in the transfer ports varies, and the amount of air blown into the cylinder each cycle varies. LDV velocity data provide a means of determining the volumetric flow of air into the cylinder each cycle, by integrating the measured transfer port velocities over the cross-sectional area of the port. Figure 3-20 shows examples of mean transfer port velocity, pressure difference (crankcase to exhaust), and reed valve lift through the total scavenging period, for throttle open area ratios of 100 (WOT), 50, and 20%, for a motored engine.[32] Note that although most of the transfer port flow is positive, a negative flow occurs after BC just before the transfer port closes. The positive, negative, and net air flow per cycle as a function of speed and throttle opening are shown in Fig. 3-20, also. One would expect the scavenging flow to vary substantially over this operating map.

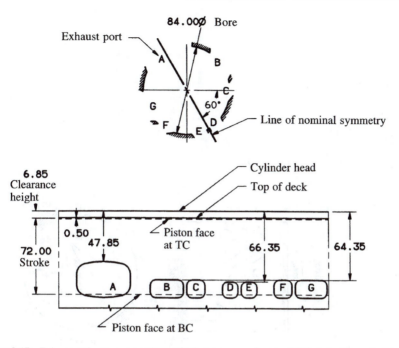

Figure 3-17 Exhaust and scavenge port layout for loop-scavenged engine LDV study. Dimensions in mm.[33]

Detailed velocity data can be used to address the question of how well motored engine scavenging measurements match the fired engine scavenging process. Because the motored engine has no combustion energy release, the cylinder pressure at exhaust port opening is close to or below atmospheric pressure. There is therefore no blowdown process in a motored engine, and the velocity fields, during this part of the scavenging process at least, would be expected to be different. Figure 3-21 shows that the mean radial velocities at different locations along the cylinder axis of a Schnürle-type loop-scavenged window-in-head two-stroke homogeneous-charge spark-ignition engine with flat piston crown, during blowdown for the fired engine and the motored engine are substantially different. During the actual scavenging process (early, middle, and late periods) the velocity data become much more similar, although significant differences do still exist. Thus, although during scavenging, and after the ports have been closed, the general features of the flow under motored conditions approximately match those of the fired engine they are not quantitatively representative of the actual scavenging process. A major advantage of flow field measurement techniques like LDV is that they can be used on firing engines.

An example of such firing engine velocity data is shown in Fig. 3-22.[62] Measurements were made in a loop-scavenged motorcycle engine of 98 cm^3 cylinder displacement, in the exhaust port, in one of the transfer ports, in the combustion chamber, and within the swept cylinder volume, as indicated. The operating conditions, 1500 rev/min and 10% throttle opening, correspond to engine idle. Prior to scavenge port opening, during blowdown (112°), the combustion chamber and swept volume flows are almost

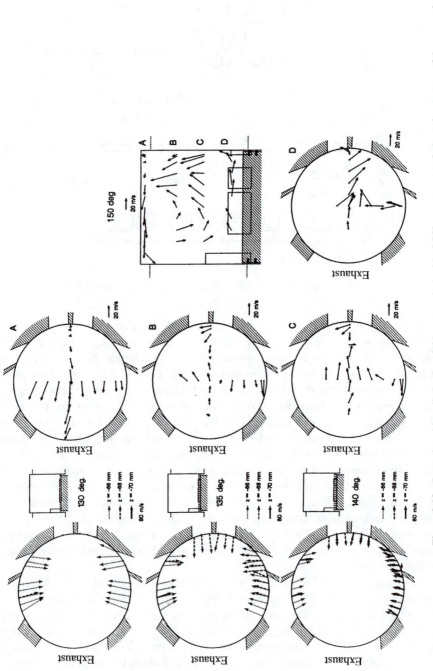

Figure 3-18 Port exit and in-cylinder flow field vectors obtained by LDV measurements during primary stage of the scavenging process.[33]

106

Motored

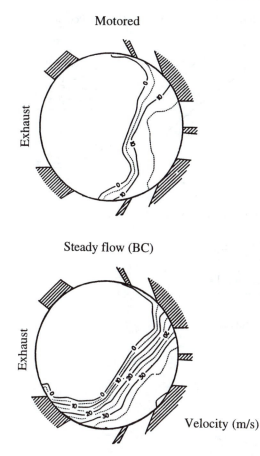

Steady flow (BC)

Velocity (m/s)

Figure 3-19 Jante scavenging diagrams for the engine in Fig. 3-17 under motored (upper) and steady flow (lower) conditions. For the steady flow test the piston was fixed at BC.[33]

uniform toward the exhaust port. At 130°, the transfer ports are already opening, the in-cylinder velocities are irregular, and the combustion chamber flows are lower. At 150° the flow from the two boost or transfer ports (opposite the exhaust port) has set up a strong flow across the combustion chamber. By about 180°, the scavenge flow and exhaust flow velocities have decreased significantly, and the circulating flow across the combustion chamber is starting to decay. One interesting conclusion from this detailed scavenging flow study is that while the exhaust flow and transfer port flow directly follow the pressure difference between the transfer ports and exhaust pipe, the in-cylinder and combustion chamber flows were not strongly influenced by this (varying) pressure difference. Figure 3-23 illustrates this difference in behavior of the in-cylinder and passageway flows.[62]

LDV measurements on the cylinder axis have been used to examine the differences between the scavenging flows with a crankcase-scavenged and external-blower Schnürle loop-scavenged two-stroke engine.[60] The exhaust blowdown is not significantly af-

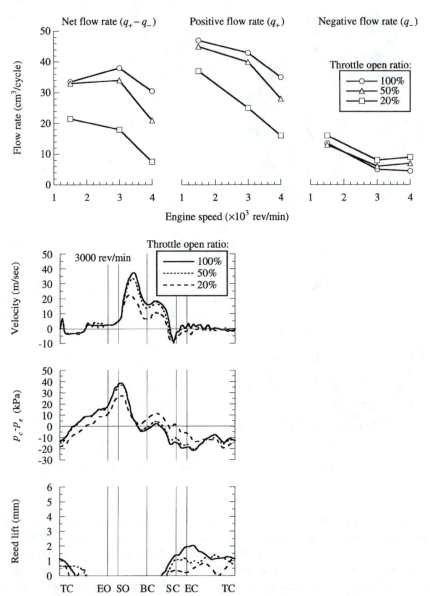

Figure 3-20 Flow rate through the scavenge ports (air volume per cycle) as a function of engine speed and throttle open area ratio.[32] Positive flow is into the cylinder, negative flow out. Also shown are mean port velocity, pressure difference across the cylinder, and crankcase reed valve lift or opening.

fected. During the early stages of scavenging (120–140° ATC), both scavenging methods develop the expected tumbling (looping) in-cylinder flow but the crankcase-driven flow develops a strong almost uniform flow across the cylinder toward the exhaust port. The blower-driven flow builds up more slowly to a similar across-the-cylinder flow, with

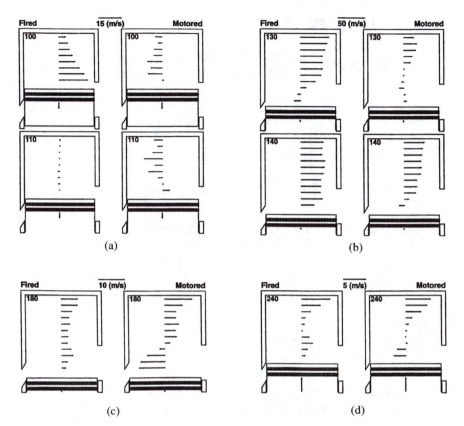

Figure 3-21 Comparison of mean radial velocity profiles along the cylinder axis for motored and fired engine operation, at various crank angles,[61] during (a) cylinder blowdown, (b) early scavenging, (c) mid-scavenging, and (d) late scavenging.

maximum velocities later (150–165° rather than about 140°, at a delivery ratio of 0.6). Late in the scavenging process, the blower-driven flow is stronger as is the tumble pattern across the top of the cylinder. This tumbling flow persists after scavenging, during compression, with the blower-driven flow the stronger. These scavenging-produced in-cylinder flow patterns did not vary greatly as the delivery ratio was varied. Clearly the type of compressor and ducting geometry affect, to a degree, the details of the scavenging flow.

The temporal evolution of the scavenging loop during compression determines the flow pattern and turbulence within the combustion chamber during combustion, and is therefore important. Figure 3-24 shows how the velocity vectors set up during the high entry-flow phase of scavenging (around 140° ATC) persist into compression; this flow then decays and breaks up into turbulence. In this loop-scavenged engine (see Fig. 3-17 for details of port geometry), the scavenge-flow-produced vortex first speeds up, then breaks down near TC, converting an appreciable portion of the mean flow kinetic energy into turbulence.[33]

Firing 1500 r/min Throttle 10%

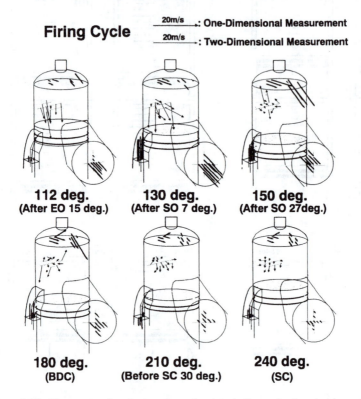

Figure 3-22 Flow patterns in exhaust and transfer ports, in the combustion chamber and in the swept volume, during the scavenging process in a loop-scavenged motorcycle engine.[62]

Because LDV measures gas velocities at a specific point in the flow, a large number of measurements are required to define the full flow field. Two-dimensional optical flow field measurement techniques are being developed that have the potential of overcoming this drawback. One of these, particle image velocimetry, is now being used to define in-cylinder flows. A thin sheet of light illuminates seed particles in the flow at a given time; a short time later, a second image in the same plane is obtained either with a second laser pulse or by a special camera. These two images are then processed to determine the two-dimensional velocity distribution, where each velocity vector is obtained from the pair of images from each particle observed. Preliminary results for two-stroke cycle engines can be found in Nino et al.[35] and Guibert et al.[36]

3-3-2 Fuel Distribution Measurements

This chapter is focused on experimental methods that help define the two-stroke engine scavenging process. This scavenging process is so important to two-stroke engine operation because the amount of fuel that can be burned in each cycle in each cylinder is

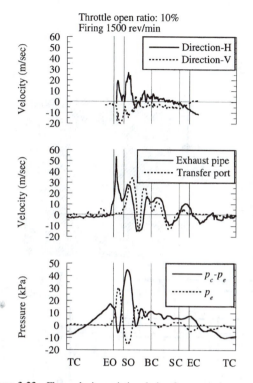

Figure 3-23 Flow velocity variation during the scavenging process (in the cylinder, in the exhaust pipe, and in the transfer port) and the exhaust port pressure and the crankcase-to-exhaust pressure difference,[62] for the same engine as in Fig. 3-22.

limited by the amount of air that is trapped within the cylinder; the scavenging process determines the amount of trapped air. Thus the amount of fuel–air mixture that can be burned each cycle and the process by which that mixture is produced connect strongly with the scavenging process.

In homogeneous charge spark-ignition engines where the gasoline fuel, mostly in vapor form, enters the cylinder with the scavenging air, the mixing of fuel with the burnt gases and the distribution of fuel within the cylinder closely follow the mixing and distribution of air. The mixture of fuel, air, and residual burnt gas at time of ignition is certainly not uniform, but is usually well enough mixed for a successful combustion process to occur. With direct-injection two-stroke engines, the injection, vaporization, and mixing of fuel with air and hot residual gas occurs after the exhaust port closes and the scavenging process has ended. As described in the previous section, however, the scavenging flow persists through much of the compression process and it therefore influences the in-cylinder mixture preparation processes. This is especially the case with two-stroke direct-injection spark-ignition engines where fuel injection usually commences soon after exhaust port closing to allow time to achieve adequate mixing of fuel vapor with air. This mixing process has a major impact on ignition and combustion, yet following its progress experimentally or with models is a challenging task. A num-

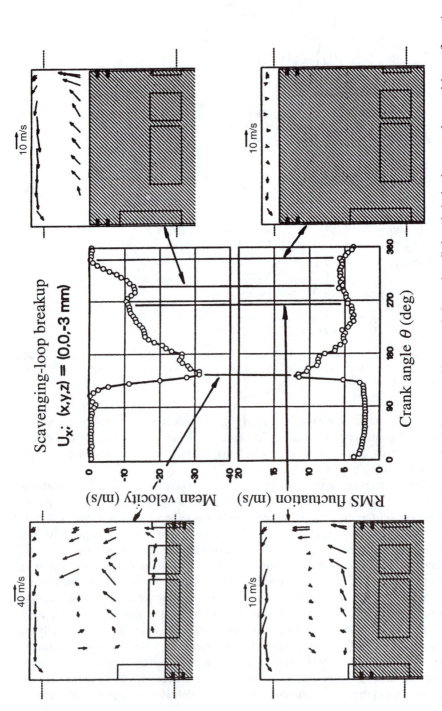

Figure 3-24 Scavenging flow evolution and subsequent breakdown: ensemble mean velocity on the cylinder axis, in the clearance volume, and the rms fluctuation, with in-cylinder flow vectors.[33]

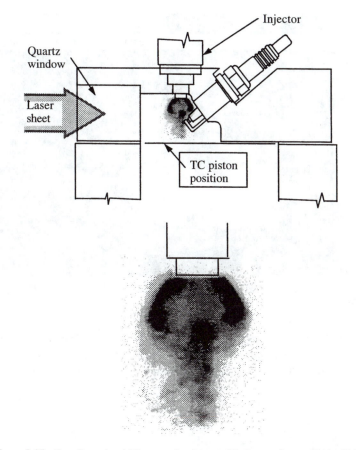

Figure 3-25 Two-dimensional Mie-scattering image of fuel spray from a digitized high-speed film superimposed on combustion chamber cross section for two-stroke direct-injection spark-ignition engine.[63]

ber of laser-based diagnostics have been developed for visualizing and quantifying fuel behavior in engine cylinders over the past decade, and have been used to examine the fuel behavior in direct-injection two-stroke spark-ignition engines. Some of the insights these new diagnostic methods are providing will now be reviewed.

Two-dimensional imaging of the fuel distribution can be accomplished by passing a thin sheet of laser light through the cylinder and using light scattering or fluorescence techniques to produce images dependent on local fuel concentration. Mie scattering of the laser light from liquid fuel droplets can readily be used in motored and fired engines to define the behavior of the liquid fuel in the spray. Figure 3-25 shows a spray image, digitized from a high-speed film, of the gasoline spray in a direct-injection two-stroke spark-ignition engine. Injection duration was from 83 to 37° BTC at these conditions, 1500 rev/min and 180 kPa imep. The image shows that the spray, which initially leaves the injector as a hollow cone, as expected from the injector poppet geometry, subsequently collapses and the core fills in with liquid. The spray penetrates to the bottom of

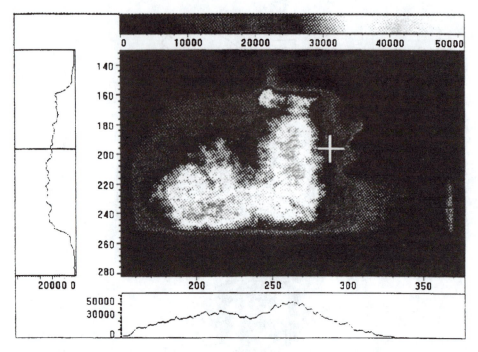

Figure 3-26 Fuel vapor produced LIF intensity distribution along combustion chamber center plane, 1° before ignition in the engine of Fig. 3-25. Vertical and horizontal intensity profiles through the cursor location, which is at the spark plug gap, are shown.[63]

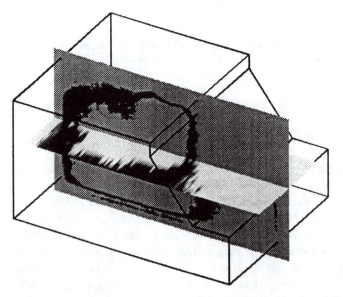

Figure 3-27 Ignitable mixture region ($\phi \approx 0.6$–1.6) 1° before ignition from LIF image analysis in the engine of Fig. 3-25. Dark regions show the relatively thin zone of ignitable mixture in each cutting plane.[63]

Liquid	Vapor		Liquid	Vapor

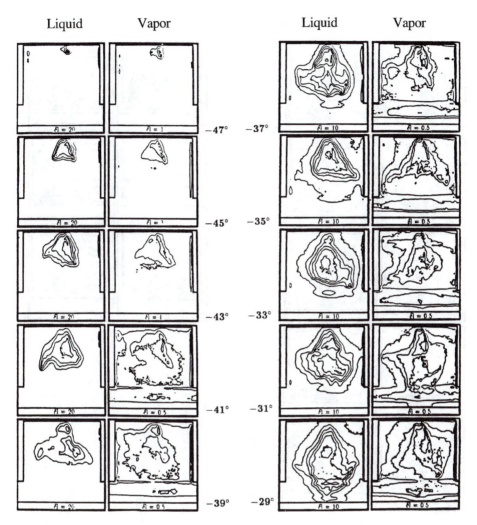

Figure 3-28 Exciplex fluorescence images of liquid and vapor fuel on spray symmetry plane, through the fuel injector axis as a function of crank angle. Equal intensity contours are shown. Two-stroke direct-injection spark-ignition engine with hollow cone injected spray at 1600 rev/min.[37] (Continued on p. 116)

the field of view by about 65–70° BTC. The Mie-scattering intensity then diminishes, and by about 50° BTC the liquid spray extends only as far as the spark plug, as shown.[63] Such studies show that essentially almost all the fuel has vaporized by time of ignition.

Laser induced fluorescence (LIF) is a useful compliment to Mie-scattering images, which respond only to liquid fuel droplets. Also, LIF can be made quantitative because the fluorescence intensity is proportional to the mass of fuel present in the laser sheet. Gasolines contain compounds that fluoresce. The fuel can also be doped with a ketone or acetaldehyde to increase the fluorescence intensity: however, the image is then controlled by the dopant distribution. Figure 3-26 shows the LIF intensity distribution

Liquid   Vapor   Liquid   Vapor

Figure 3-28 (Continued).

from the fuel cloud across the combustion chamber center plane, 1° before ignition.[63] Because almost all the fuel has vaporized this will be the gasoline vapor distribution. Fluorescence intensity profiles are shown. Measurements with multiple planes for the laser sheet allow a three-dimensional picture of the fuel distribution to be built up. Figure 3-27 shows the region of ignitable mixture that injection, vaporization, and mixing have produced just before ignition.[63]

Exciplex fluorescence has been applied to DI two-stroke engines to quantify both the liquid fuel and fuel vapor distributions. The simpler LIF technique summarized in the previous paragraph does not distinguish between the two, but liquid fluorescence dominates when both liquid and vapor are present due to the much higher liquid-phase fuel concentrations. However, exciplex imaging requires the addition of several dopants to the fuel (and hence measures dopant distribution) and its use is limited to motored engines with nitrogen instead of air, because the fluorescence is strongly quenched by oxygen. Figure 3-28 shows constant fluorescence intensity contour images of liquid and vapor fuel on the symmetry plane of an initially hollow-cone spray through the injection period in a DI two-stroke spark-ignition engine. In the early part of injection, the spray cone angle is about equal to the injection angle of 53°. As the cylinder pressure rises, the spray increasingly collapses and the core becomes filled with liquid. Penetration continues rapidly axially and radially to fill the combustion chamber, with vapor fuel spreading more rapidly than liquid. Finally, at time of ignition, the fuel vapor distribution is seemingly uniform.[37] Note that the fuel was primarily decane, the scavenging medium was nitrogen, and the engine was motored.

These laser sheet diagnostics that produce two-dimensional, and with multisheets three-dimensional, images of the fuel distribution within the cylinder offer great promise for connecting mixture preparation with scavenging in a detailed quantitative way.

REFERENCES

1. Jante, A., "Scavenging and Other Problems of Two-Stroke Cycle Spark-Ignition Engines," SAE Paper 680468, *SAE Trans.*, vol. 77, 1968.

2. Blair, G.P., "Studying Scavenge Flow in a Two-Stroke Cycle Engine," SAE Paper 750752, 1975.

3. Blair, G.P., and Kenny, R.G., "Further Developments in Scavenging Analysis for Two-Cycle Engines," SAE Paper 800038, *SAE Trans.*, vol. 89, 1980.

4. Smyth, J.G., Kenny, R.G., and Blair, G.P., "Steady Flow Analysis of the Scavenging Process in a Loop Scavenged Two-Stroke Cycle Engine—A Theoretical and Experimental Study," SAE Paper 881267, 1988.

5. Ishihara, S., Murakami, Y., and Ishikawa, K., "Improvement of Pilot Tube Set for Obtaining Scavenging Pictures of Two-Stroke Cycle Engines," SAE Paper 880171, 1988.

6. Sung, N.W., and Patterson, D.J., "Air Motion in a Two-Stroke Engine Cylinder—The Effects of Exhaust Geometry," SAE Paper 820751, *SAE Trans.*, vol. 91, 1982.

7. Sher, E., Hossain, I., Zhang, Q., and Winterbone, D.E., "Calculations and Measurements in the Cylinder of a Two-Stroke Uniflow-Scavenged Engine under Steady Flow Conditions," *Exp. Ther. Fluid Sci.*, vol. 4, pp. 418–431, 1991.

8. Percival, W.H., "Method of Scavenging Analysis for 2-Stroke Cycle Diesel Cylinder," *SAE Trans.*, vol. 63, pp. 737–751, 1955.

9. Oggero, M., "FIAT Research Methods for the Experimental Study of the Two-Stroke Large Bore Diesel Engines," *FIAT Tech. Bull.*, vol. 21, no. 3, pp. 86–92, 1968.

10. Nuti, M., and Martorano, L., "Short-Circuit Ratio Evaluation in the Scavenging of Two-Stroke S.I. Engines," SAE Paper 850177, 1985.

11. Ferro, A., "Investigation Means of Models into the Scavenging of Two-Stroke Internal Combustion Engines," *The Engineers' Digest*, vol. 19, pp. 512–522, Dec. 1958.

12. Oka, T., and Ishihara, S., "Relation between Scavenging Flow and Its Efficiency of a Two-Stroke Cycle Engine," *Bull. JSME*, vol. 14, no. 69, pp. 257–267, 1971.

13. Sata, K., Ukawa, H., and Nakano, M., "A Two-Stroke Cycle Gasoline Engine with Poppet Valves in the Cylinder Head—Part II," SAE Paper 920780, 1992.

14. Sher, E., "Investigating the Gas Exchange Process of a Two-Stroke Cycle Engine with a Flow Visualization Rig," *Israel J. Tech.*, vol. 20, pp. 127–136, 1982.

15. Phatak, R.G., "A New Method of Analyzing Two-Stroke Cycle Engine Gas Flow Patterns," SAE Paper 790487, *SAE Trans.*, vol. 88, 1979.

16. Smyth, J.G., Kenny, R.G., and Blair, G.P., "Motored and Steady Flow Boundary Conditions Applied to the Prediction of Scavenging Flow in a Loop Scavenged Two-Stroke Cycle Engine," SAE Paper 900800, *SAE Trans.*, vol. 99, 1990.

17. Sanborn, D.S., Blair, G.P., Kenny, R.G., and Kingsbury, A.H., "Experimental Assessment of Scavenging Efficiency of Two-Stroke Cycle Engines," SAE Paper 800975, *SAE Trans.*, vol. 89, 1980.

18. Dedeoglu, N., "Scavenging Model Solves Problems in Gas Burning Engine," SAE Paper 710579, *SAE Trans.*, vol. 80, 1971.

19. Sanborn, D.S., and Roeder, W.M., "Single Cycle Simulation Simplifies Scavenging Study," SAE Paper 850175, 1985.

20. Mirko, C., and Radislav, P., "A Model Method for the Evaluation of the Scavenging System in a Two-Stroke Engine," SAE Paper 850176, 1985.

21. Sammons, H., "A Single-Cycle Test Apparatus for Studying 'Loop Scavenging' in a Two-Stroke Engine," *Proc. Inst. Mech. Eng.*, vol. 161, pp. 233–246, 1949.

22. Rizk, W., "Experimental Studies of the Mixing Processes and Flow Configurations in Two-Cycle Engine Scavenging," *Proc. Inst. Mech. Eng.*, vol. 172, no. 1, pp. 417–437, 1958.

23. Ohigashi, S., Kashiwada, Y., and Achiwa, J., "Scavenging the 2-Stroke Diesel Engine," *Bull. JSME*, vol. 3, no. 9, pp. 130–136, 1960.

24. Ohigashi, S., and Kashiwada, Y., "On the Scavenging of Two-Stroke Diesel Engine," *Bull. JSME*, vol. 4, no. 16, pp. 669–705, 1961.

25. Ohigashi, S., and Kashiwada, Y., "A Study on the Scavenging Air Flow through the Scavenging Ports," *Bull. JSME*, vol. 9, no. 36, pp. 777–784, 1966.

26. Martini, R., and Oggero, M., "Experimental Methods for the Study of Two-Stroke Engine Scavenging," SAE Paper 710145, 1971.

27. Kannapan, A., "Cumulative Sampling Technique for Investigating the Average Process in Two-Stroke Engine," ASME Paper 74-DGP-11, 1974.

28. Sweeney, M.E.G., Kenny, R.G., Swann, G.B.G., and Blair, G.P., "Single Cycle Gas Testing Method for Two-Stroke Engine Scavenging," SAE Paper 850178, *SAE Trans.*, vol. 94, 1985.

29. Reddy, K.V., Ganesan, V., and Gopalakrishnan, K.V., "Under the Roof of the Cylinder Head—An Experimental Study of the In-Cylinder Air Movement in a Two-Stroke Spark Ignition Engine," SAE Paper 860166, *SAE Trans.*, vol. 95, 1986.

30. Sanborn, D.S., and Dedeoglu, N., "Investigation on Scavenging of Two-Stroke Engines," SAE Paper 881264, 1988.

31. Hilbert, H.S., and Falco, R.E., "Measurements of Flows During Scavenging in a Two-Stroke Engine," SAE Paper 910671, *SAE Trans.*, vol. 100, 1991.

32. Ikeda, Y., Hikosaka, M., and Nakajima, T., "Scavenging Flow Measurements in a Motored Two-Stroke Engine by Fiber LDV," SAE Paper 910669, *SAE Trans.*, vol. 100, 1991.

33. Fansler, T.D., and French, D.T., "The Scavenging Flow Field in a Crankcase-Compression Two-Stroke Engine—A Three-Dimensional Laser Velocimetry Survey," SAE Paper 920417, *SAE Trans.*, vol. 101, 1992.

34. Ghandhi, J.B., and Martin, J.K., "Velocity Field Characteristics in Motored Two-Stroke Ported Engines," SAE Paper 920419, *SAE Trans.*, vol. 101, 1992.

35. Nino, E., Gajdeczko, B.F., and Felton, P.G., "Two-Color Particle Image Velocimetry Applied to a Single Cylinder Two-Stroke Engine," SAE Paper 922309, 1992.

36. Guibert, P., Murat, M., Hanet, B., and Keribin, P., "Particle Image Velocimetry Measurements: Application to In-Cylinder Flow for a Two-Stroke Engine," SAE Paper 932647, 1993.

37. Lee, C.F., and Bracco, F.V., "Initial Comparisons of Computed and Measured Hollow-Cone Sprays in an Engine," SAE Paper 940398, *SAE Trans.*, vol. 103, 1994.

38. Lustgarten, G., "Model Investigations of the Mixing and Combustion Process in the Diesel Engine," *Sulzer Tech. Review*, pp. 6–18, 1974.

39. Boyer, R.L., Craig, D.R., and Miller, C.D., "A Photographic Study of Events in a 14-In. Two-Cycle Gas Engine Cylinder," *Trans. ASME*, vol. 76, pp. 97–108, 1954.

40. Bazika, V., and Rodig, J., "A New Method of Determining the Scavenging Efficiency of Oil Engine Cylinders," *The Engineers' Digest*, vol. 24, no. 3, 1963.

41. Huber, E.W., "Measuring the Trapping Efficiency of Internal Combustion Engines through Continuous Exhaust Gas Sampling," SAE Paper 710144, *SAE Trans.*, vol. 80, 1971.
42. Wallace, E.J., and Cave, P.R., "Experimental and Analytical Scavenging Studies on a Two-Stroke Opposed Piston Diesel Engine," SAE Paper 710175, 1971.
43. Blair, G.P., and Ashe, M.C., "The Unsteady Gas Exchange Characteristics of a Two-Cycle Engine," SAE Paper 760644, *SAE Trans.*, vol. 85, 1976.
44. Hashimoto, E., Tottori, T., and Terata, S., "Scavenging Performance Measurements of High Speed Two-Stroke Engines," SAE Paper 850182, *SAE Trans.*, vol. 94, 1985.
45. Hori, K., "A Method of Measuring Scavenging Efficiency and Its Application," *Bull. JSME*, vol. 5, no. 18, pp. 327–334, 1962.
46. Hori, K., "The Influence of Volume and Period of Gas-Sampling upon the Accuracy of the Measurements of Scavenging Efficiency," *Bull. JSME*, vol. 6, no. 22, pp. 289–296, 1963.
47. Schweitzer, P.H., and DeLuca, F., "The Tracer Gas Method of Determining the Charging Efficiency of 2-Stroke Cycle Engine," NACA Tech. Note 838, Jan. 1942.
48. Taylor, C.F., and Rogowski, A.R., "Scavenging the 2-Stroke Engine," *SAE Trans.*, vol. 62, pp. 487–502, 1954.
49. Isigami, S., "On Scavenging of the Crank Compression Type Two-Stroke Cycle Diesel Engine," *Bull. JSME*, vol. 6, no. 23, pp. 531–539, 1963.
50. Isigami, S., Tanaka, Y., and Tamari, M., "The Trapping Efficiency Measurement of Two-Stroke Cycle Diesel Engine by Tracer Gas Method," *Bull. JSME*, vol. 6, no. 23, pp. 524–539, 1969.
51. Miyabe, H., and Shimomura, T., "Measurement of Trapping Efficiencies of a Two-Stroke Engine," *Bull. JSME*, vol. 6, no. 22, pp. 297–307, 1963.
52. Ohigashi, S., and Hamamoto, Y., "Cylinder Gas Composition of Small 2-Stroke Gasoline Engine," SAE Paper 710143, *SAE Trans.*, vol. 80, 1971.
53. Murayama, T., Sekiya, Y., Sugiarto, B., and Chikahisa, T., "Study on Exhaust Control Valves and Direct Air–Fuel Injection for Improving Scavenging Process in Two-Stroke Gasoline Engine," SAE Paper 960367, 1996.
54. Tobis, B.J., Meyer, R., Yang, J., Brehob, D.D., and Anderson, R.W., "Scavenging of a Firing Two-Stroke Spark-Ignition Engine," SAE Paper 940393, *SAE Trans.*, vol. 103, *J. Engines*, Sect. 3, pp. 535–546, 1994.
55. Heywood, J.B., *Internal Combustion Engine Fundamentals*, McGraw-Hill, New York, 1988.
56. Eltinge, L., "Fuel/Air Ratio and Distribution from Exhaust Gas Composition," SAE Paper 680114, 1968.
57. Booy, R.R., "Evaluating Scavenging Efficiency of Two-Stroke Cycle Engines," SAE Paper 670029, 1967.
58. Spindt, R.S., "Air/Fuel Ratios from Exhaust Gas Analysis," SAE Paper 650507, *SAE Trans.*, vol. 74, 1965.
59. Douglas, R., "AFR Calculations for Two-Stroke Cycle Engines," SAE Paper 901599, *SAE Trans.*, vol. 99, 1990.
60. Miles, P.C., Green, R.M., and Witze, P.O., "In-Cylinder Gas Velocity Measurements Comparing Crankcase and Blower Scavenging in a Fired Two-Stroke Cycle Engine," SAE Paper 940401, *SAE Trans.*, vol. 101, 1994.
61. Miles, P.C., Green, R.M., and Witze, P.O., "Comparison of In-Cylinder Scavenging Flows in a Two-Stroke Cycle Engine under Motored and Firing Conditions," Proceedings of the International Symposium on Applications of Laser Techniques to Fluid Mechanics, Lisbon, Portugal, July 11–14, 1994.
62. Ohira, T., Ikeda, Y., Kakemizu, K., and Nakajima, T., "In-Cylinder Flow Measurement and Its Application for Cyclic Variation Analysis in a Two-Stroke Engine," SAE Paper 950224, *SAE Trans.*, vol. 104, 1995.
63. Fansler, T.D., French, D.T., and Drake, M.C., "Fuel Distributions in a Firing Direct-Injection Spark-Ignition Engine Using Laser-Induced Fluorescence Imaging," SAE Paper 950110, *SAE Trans.*, vol. 104, 1995.

FOUR

MATHEMATICAL MODELS OF THE SCAVENGING PROCESS

Reliable models of the gas exchange process in two-stroke cycle engines are a valuable tool to the engine designer. Such models have been developed for a variety of purposes and the user will obviously choose the most appropriate model for his or her particular purpose. For example, optimizing the engine's structure requires a model that connects the geometry of the cylinder and ports to the details of the gas flows inside the ports and cylinder. Such models, however, require a large and complex computer program, which is not a practical tool when the task is to predict the operating characteristics of a given engine under many different operating conditions. For this latter purpose, a simpler model that relates the scavenging and trapping efficiencies, and the delivery ratio (or other parameters), is more appropriate.

Because there is no one class of model that is capable of predicting the scavenging process in any type of scavenging system for the full range of cylinder geometries and operating parameters of interest, this chapter will review the available scavenging models and illustrate their application. It is convenient to classify these models into three main categories[1,2]: simple single-phase one- or two-zone models, which are useful for overall engine cycle simulations; multizone multiphase models, which are useful for evaluating the effects of design and operating parameters on the performance characteristics of the engine; and multidimensional models (or computational fluid dynamic models), which are useful for examining phenomena that depend on the detailed flow fields in the ports and engine cylinder, such as the combustion process or the geometric design of the ports, piston crown, and cylinder head. Here *phase* refers to a time period: the scavenging process is assumed to occur as a sequence of one, two, or three distinct phases. *Zone* is used to denote a part of the cylinder volume. Commonly used zones are

Table 4-1 Classification of scavenging models and their applications

Category	Applications
Single-phase, one- or two-zone models	Simple engine cycle simulations to study approximate engine behavior trends
Multizone, single-phase or multiphase models	Engine performance simulations where phenomenological models of scavenging are sufficiently accurate for study of issues associated with:
	Engine speed and load
	Ambient conditions
	Exhaust tuning
	Air pollutant emissions and noise
	Spark and fuel-injection timing
	Heat transfer to the cylinder walls
	Supercharging and turbocharging
	Friction and lubrication
Multidimensional or computational fluid dynamic models	Study of issues where the flow processes in the exhaust and transfer ducts and ports, and the cylinder dominate:
	Engine design (detailed geometry of cylinder and port assemblies)
	Port and valve timings
	Mixture preparation
	Ignition and combustion processes
	Air pollutant formation mechanisms
	Cyclic variability

fresh charge zone, burnt gas (or residual gas) zone, mixing zone, and short-circuiting zone. Table 4-1 summarizes these three categories of models and their applications.

4-1 SIMPLE SINGLE-PHASE MODELS

The two classic single-phase, one- or two-zone models, the pure-displacement and the perfect-mixing models, are simple and easy to use. They do not, however, predict realistic values for the charging efficiency and the scavenging efficiency. The pure-displacement model overestimates the charging efficiency, the perfect-mixing model underestimates it. The two models are still important, however, because they provide the upper and lower bounds for the engine's scavenging performance. Also, these two models introduce two important conceptual concepts: displacement and mixing. The perfect-mixing model is sometimes used in simple cycle simulations when the user is interested only in an approximate or qualitative assessment of the engine's behavior.

4-1-1 The Pure Displacement Model

The pure-displacement model, the upper bound for the charging efficiency, is considered to be the ideal scavenging process. It is a single-phase two-zone model (see Fig. 4-1a).

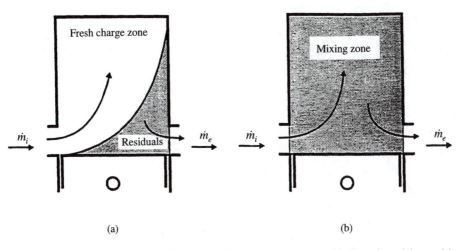

(a)   (b)

Figure 4-1   (a) Perfect-displacement model; (b) perfect-mixing models.

In this model the following are assumed:

a. The entering fresh charge pushes the burnt gas out of the cylinder by a pure displacement mechanism.
b. The process occurs at a constant cylinder volume and pressure.
c. No mass or heat is allowed to cross the interface between the fresh charge and burnt gas.
d. The cylinder walls are adiabatic.

The isothermal pure-displacement model was introduced in Section 2-1. Here we develop the nonisothermal version of this model. From the definitions of delivery ratio Λ, and charging efficiency η_{ch}, Eqs. (2-1) and (2-2), it follows that

$$\eta_{ch} = \begin{cases} \Lambda & \text{for } \Lambda < \rho_i/\rho_0 \\ \rho_i/\rho_0 & \text{for } \Lambda \geq \rho_i/\rho_0 \end{cases} \tag{4-1}$$

where ρ_i is the density of the fresh charge at the cylinder pressure and ρ_0 is the ambient gas density. If we further assume that the characteristic cylinder volume during the scavenging process equals the cylinder displaced volume, then

$$V_d = V_i + V_b \tag{4-2}$$

Also, $m_b = \rho_b V_b$, where ρ_b is the burnt gas density, $m_i = \rho_i V_i$, and $m_0 = \rho_0 V_d$; therefore, for $m_i \leq \rho_i V_d$,

$$\eta_{sc} = \frac{\rho_i V_i}{\rho_i V_i + \rho_b V_b} = \rho_i V_i \left[\rho_i V_i + \rho_b \left(\frac{m_0}{\rho_0} - V_i \right) \right]^{-1} \tag{4-3}$$

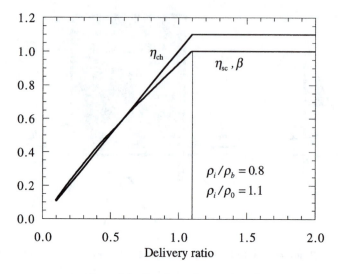

Figure 4-2 Charging and scavenging efficiency results of the perfect-displacement model.

Rearranging terms results in[1,2]:

$$\eta_{sc} = \begin{cases} \left[1 + \dfrac{\rho_b}{\rho_i}\left(\dfrac{1}{\Lambda}\dfrac{\rho_i}{\rho_0} - 1\right)\right]^{-1} & \text{for } \Lambda < \dfrac{\rho_i}{\rho_0} \\ 1.0 & \text{for } \Lambda \geq \dfrac{\rho_i}{\rho_0} \end{cases}$$

(4-4)

The exhaust gas purity β, which is the instantaneous mass fraction of the fresh charge in the emerging exhaust gas, is given by

$$\beta = \begin{cases} 0 & \text{for } \Lambda < \rho_i/\rho_0 \\ 1.0 & \text{for } \Lambda \geq \rho_i/\rho_0 \end{cases}$$

(4-5)

Note that if we further assume that $\rho_b = \rho_a = \rho_0$ (i.e., this is an isothermal process) and $M_0 = M_i = M_b$, where M is the molecular weight, then Eqs. (4-1), (4-4), and (4-5) reduce to the well-known form of Eq. (2-9). The pure-displacement model is the upper bound for the real scavenging process, which in practice is never as efficient. Figure 4-2 shows the results predicted by this model.

4-1-2 The perfect mixing model

The perfect-mixing model gives the lower bound for the charging efficiency. It is a single-zone single-phase model and is illustrated in Fig. 4-1b. In this model the following are assumed:

a. The scavenging process is one where mixing is perfect. As fresh charge enters the cylinder, it mixes instantaneously with the cylinder contents to form a homogeneous mixture. The instantaneous composition of the mixture of fresh charge and burnt gas leaving the cylinder through the exhaust port is, therefore, the composition of this homogeneous in-cylinder mixture (i.e., $\beta = \eta_{sc}$).

b. The process occurs at a constant cylinder volume and pressure.

c. The cylinder walls are adiabatic.

d. The two gases involved follow the ideal gas law and have the same molecular weights with an identical and constant specific heat.

The charging efficiency, the scavenging efficiency, and the exhaust gas purity relationships for this model may be evaluated as follows.[1,2] The energy balance for an open system boundary filling the cylinder is

$$\frac{d(mc_v T)}{dt} = \dot{m}_i c_p T_i - \dot{m}_e c_p T_e \tag{4-6}$$

Since pV is constant, the ideal gas law $pV = mRT$, where m is the mass in the cylinder, shows the left-hand side is zero. Thus

$$\dot{m}_e = \dot{m}_i (T_i / T_e) \tag{4-7}$$

The ideal gas law also yields

$$\dot{m}T = -\dot{T}m \tag{4-8}$$

The mass conservation law, with Eq. (4-7), gives

$$\dot{m} = \dot{m}_i - \dot{m}_e = \dot{m}_i [1 - (T_i / T_e)] \tag{4-9}$$

and the perfect mixing assumption requires that $T_e = T$. Substitution of Eq. (4-9) into Eq. (4-8) yields

$$\dot{m}_i = -\frac{\dot{T}(mT)}{T(T - T_i)} \tag{4-10}$$

The difference between inlet and ambient conditions is usually small. If this is neglected, Eq. (4-10) may be integrated to give

$$\int_0^{\Lambda m_0} dm_i = -(m_1 T_1) \int_{T_4}^{T_1} \frac{dT}{T(T - T_i)} \tag{4-11}$$

where T_4 and T_1 are the temperatures of the cylinder contents at the end of the blowdown period and at the end of the scavenging period, respectively. Completing the integration yields

$$T_1 = \frac{T_i}{1 - (1 - T_i / T_4) \exp(-\Lambda)} \tag{4-12}$$

The energy of the mixture of fresh charge and burnt gas in the cylinder at the end of the scavenging process is the sum of the energy of the mixture components. This gives

$$T_1 = \eta_{sc} T_i + (1 - \eta_{sc}) T_4 \tag{4-13}$$

The scavenging efficiency is obtained from Eqs. (4-12) and (4-13):

$$\eta_{sc} = 1 - (T_1 / T_4) \exp(-\Lambda) \tag{4-14}$$

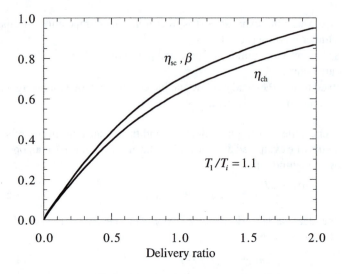

Figure 4-3 Charging and scavenging efficiency results of the perfect-mixing model.

or, alternatively,

$$\eta_{sc} = (T_1/T_i)[1 - \exp(-\Lambda)] \tag{4-15}$$

By definition, the perfect mixing process gives

$$\beta = \eta_{sc} \tag{4-16}$$

The charging efficiency is related to the scavenging efficiency by

$$\eta_{ch} = (T_i/T_1)\eta_{sc} \tag{4-17}$$

So,

$$\eta_{ch} = 1 - \exp(-\Lambda) \tag{4-18}$$

Thus, for the perfect-mixing model, the charging efficiency is solely dependent on the delivery ratio, and independent of the temperature ratio between the burnt gas and the fresh charge.

In modern engines, the gas exchange performance is significantly better than the perfect-mixing model predicts. However, this model is still useful as a reference for comparison, and the shapes of the η_{sc} and η_{ch} versus Λ curves are similar to those of measured values. For an isothermal perfect-mixing process, Eq. (4-15) reduces to the simple form of Eq. (2-12), first proposed in 1914 by Hopkinson.[3] Figure 4-3 shows results obtained from this model.

4-2 MULTIZONE MULTIPHASE MODELS

In analyzing the gas exchange process while keeping the mathematical treatment straightforward, it is convenient to divide the cylinder domain into two or more distinct zones, and to consider scavenging as proceeding in several distinct phases rather than as a continuous process.[4–8] This approach stems from interpretion of visual observations of the scavenging process in motored engines. Such approaches are useful for understanding the progress of the process and are still simple enough to be used in overall engine cycle simulations. These models are to some extent arbitrary, however, and often require several empirical constants that must then be determined from experimental measurements. These constants may depend not only on the scavenging method and the geometry of the cylinder and ports assembly, but also on the specific operating conditions of the particular engine.

In this class of models, the cylinder is subdivided into regions or zones that contain fresh charge, products of combustion, and a mixture of fresh charge and burnt gas. In each zone the temperature may be different but is assumed uniform. Further, the in-cylinder pressure is uniform, and any heat transfer between the different zones is neglected. The scavenging process is assumed to proceed in two or three principal phases; displacement, mixing, and short-circuiting.

4-2-1 The Isothermal Three-Zone Model of Maekawa

The first model of this kind, an isothermal three-zone model incorporating mixing, was proposed by Maekawa[4] in 1957. Maekawa assumed that the scavenging process occurs in a single phase at constant cylinder volume, pressure, and temperature and that the two gases (fresh charge and burnt gas) have identical density and specific heat. The fresh charge entering the cylinder is split into three streams, and the cylinder volume contains three corresponding zones; see Fig. 4-4. A fraction of the fresh charge, $(1 - K)\dot{m}_i$, flows directly to the exhaust and is lost (the short-circuiting fraction). A second fraction, $q K \dot{m}_i$, mixes with the burnt gases in a mixing zone. The third fraction, $(1 - q)K\dot{m}_i$, forms a fresh charge zone that over time displaces the mixing zone into the exhaust. If

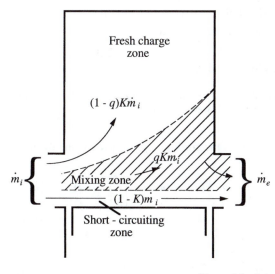

Figure 4-4 Schematic of the three-zone model of Maekawa.[4]

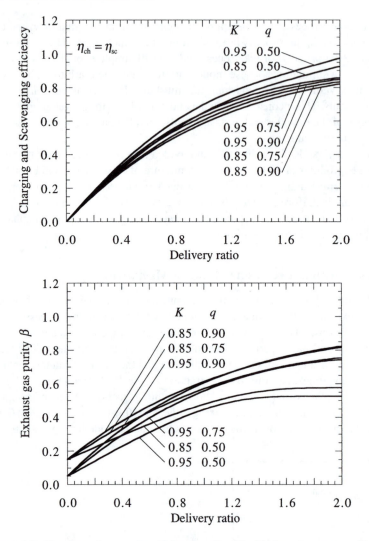

Figure 4-5 Charging and scavenging efficiency results of the Maekawa two-zone model. Reprinted from Sher,[1] *Prog. Energy Combust. Sci.*, with permission from Elsevier Science.

the mixing zone is fully exhausted, gas from the fresh charge zone exits the cylinder. The results of this analysis are[1,2,4]

$$\eta_{ch} = \begin{cases} K(1-q)\Lambda + [1 - K(1-q)\Lambda]\eta_m & \text{for } \Lambda < [K(1-q)]^{-1} \\ 1.0 & \text{for } \Lambda \geq [K(1-q)]^{-1} \end{cases} \quad (4\text{-}19)$$

$$\eta_{sc} = \eta_{ch} \quad (4\text{-}20)$$

and

$$\beta = \begin{cases} (1-K) + \eta_m K & \text{for } \Lambda < [K(1-q)]^{-1} \\ 1.0 & \text{for } \Lambda \geq [K(1-q)]^{-1} \end{cases} \quad (4\text{-}21)$$

where η_m is the mass fraction of the charge in the mixing zone and is defined by

$$\eta_m = q\left\{1 - \exp\left[\frac{-K\Lambda}{1 - K(1-q)\Lambda}\right]\right\} \tag{4-22}$$

Note that for $K = 1$ (which means that no short circuiting occurs) and $q = 1$ (which means that no fresh charge zone exists), Eqs. (4-19)–(4-21) reduce to the form of the isothermal perfect-mixing model, Eq. (2-12). Typical values for K and q are 0.90 and 0.75, respectively.[4] Figure 4-5 shows some results from this model. The model is capable of predicting realistic values for both the charging and the scavenging efficiencies, which are between the upper (pure-displacement) and the lower (perfect-mixing) bounds.[1,2] However, the exhaust gas purity predictions do not follow, even qualitatively, the actual behavior. This subject will be discussed in more detail in Section 4-3. Nevertheless, a modified version of the Maekawa model, in which no direct short-circuiting was allowed but heat transfer between the two zones was permitted and the fresh charge zone was replaced by a stratified zone, was applied successfully by Sato and Kido[9] to a small crankcase-scavenged engine.

4-2-2 The Isothermal Three-Zone Model of Benson and Brandham

To adapt the Maekawa model to their observations in a motored engine, Benson and Brandham[5] suggested that it would be more realistic to assume that instead of the fresh charge zone, the cylinder contains a burnt gas zone. This burnt gas zone is always adjacent to the exhaust port, and the mixing zone is always adjacent to the scavenge port (Fig. 4-6). They assumed that, initially, the mixing zone contains a mass of $(1 - x)m_0$ and the burnt gas zone contains a mass of xm_0. The incoming fresh charge splits into two substreams: $\dot{y}m_i$ is short-circuited to the exhaust port, and the other, $(1 - y)\dot{m}_i$, mixes in the mixing zone. The exhaust gas zone exhausts first. It follows that[5]

$$\eta_{ch} = \begin{cases} (1-y)\Lambda & \text{for } \Lambda \leq x/(1-y) \\ 1 - (1-x)\exp[x - (1-y)\Lambda] & \text{for } \Lambda > x/(1-y) \end{cases} \tag{4-23}$$

$$\eta_{sc} = \eta_{ch} \tag{4-24}$$

$$\beta = \begin{cases} y & \text{for } \Lambda \leq x/(1-y) \\ 1 - (1-y)(1-x)\exp[x - (1-y)\Lambda] & \text{for } \Lambda > x/(1-y) \end{cases} \tag{4-25}$$

Figure 4-7 shows some results from the Benson–Brandham model. For this model, as for the Maekawa model, two empirical parameters have to be specified. The important improvement of the Benson–Brandham model is that it predicts a more realistic profile for the exhaust gas purity; it thus approaches real engine behavior more closely (see Section 4-3).

A version of the Brandham–Benson model modified to include pressure and volume variations has been applied to an opposed-piston engine by Wallace and Cave.[7] In the uniflow-scavenged opposed-piston engine, the combustion process demands a high swirl rate. In such an engine, Wallace and Cave postulated a scavenging process where the incoming air adheres to the cylinder walls (held there by centrifugal forces), passes along the cylinder in an annular space, and leaves a central core unscavenged (Fig. 4-8). The annular space was assumed to be perfectly scavenged with a high scavenge ratio,

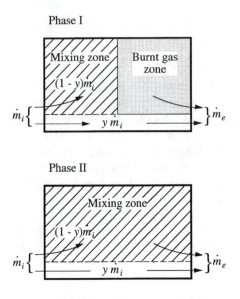

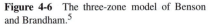

Figure 4-6 The three-zone model of Benson and Brandham.[5]

while the central core was treated as unscavenged or perfectly mixed with a low scavenge ratio. Because cylinder pressure and volume variations were allowed, the model equations had to be solved on a step-by-step basis.

Another version of the Brandham–Benson model has been applied by Streit and Borman[8] to a special design of opposed-piston engine. Here again the scavenging process was treated on a step-by-step basis. During the scavenging period, the cylinder was assumed to consist of two thermodynamic subsystems (Fig. 4-9). System I is always connected to the scavenge port, and system II to the exhaust port. It was assumed that in each system a uniform mixture and temperature exist and that both systems are at the same pressure. Heat transfer from or to the walls in each system was handled separately, but it was assumed that there is no heat transfer between the two systems. The mass exchange from system I into system II and vice versa was specified separately as a function of inlet port mass flow rate and delivery ratio, as follows:

$$\dot{m}_{21} = \begin{cases} 0 & \text{for } \Lambda \le 0.40 \\ 2.5 \Lambda \dot{m}_i & \text{for } \Lambda > 0.40 \end{cases} \tag{4-26}$$

and

$$\dot{m}_{12} = \begin{cases} 0 & \text{for } \Lambda \le 0.40 \\ 0.8 \Lambda \dot{m}_i & \text{for } \Lambda > 0.20 \end{cases} \tag{4-27}$$

When $m_2 \le 0.01(m_1 + m_2)$, m_1 and m_2 were mixed instantaneously. For this model, four calibration parameters are needed. Figure 4-10 shows the results obtained. The exhaust gas purity function exhibits a sigmoid curve. This type of curve for β occurs with all types of scavenging systems, and the Streit–Borman model was the first to predict it. The success of this model results from considering the scavenging process as a continuous process that involves a high degree of mixing, instead of a process that proceeds in distinct phases.

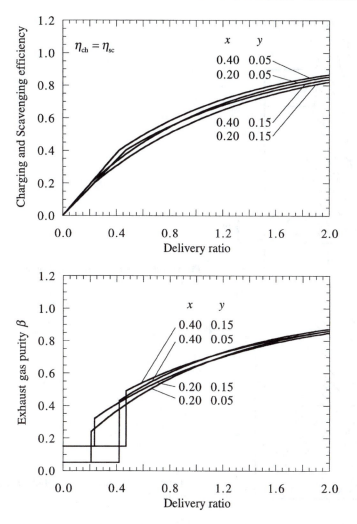

Figure 4-7 Charging and scavenging efficiency results of the Benson–Brandham three-zone model. Reprinted from Sher,[1] *Prog. Energy Combust. Sci.*, with permission from Elsevier Science.

Sanborn[10] suggested a modification to the model of Streit and Borman that correlated his test results better. He assumed that each inflow increment occupies a separate portion of the cylinder volume prior to its mixing. The mixing of each mass increment is assumed to take place after it has traversed a prescribed path length within the cylinder. Selection of the path length before mixing provides a means for model calibration. This modification resulted in a more flexible model that fit Sanborn's experimental observations, but this was obtained at the expense of introducing two more calibration constants.

Baudequin and Rochelle[11] faced several difficulties as they tried to correlate their experimental data with the predictions of the Benson–Brandham model. They found

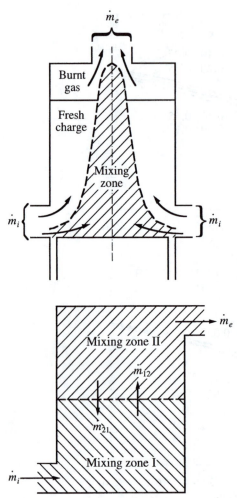

Figure 4-8 The three-zone model of Wallace and Cave.[7]

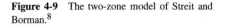

Figure 4-9 The two-zone model of Streit and Borman.[8]

that for small delivery ratios the charging efficiency appears to be smaller than that resulting from perfect mixing, whereas for high delivery ratios the charging efficiency is higher. The Maekawa model does predict this transition, but does not consider an initial displacement-scavenging phase. To overcome these difficulties, Baudequin and Rochelle proposed a synthesis of these two models. The resulting model required a large number of calibration constants, which detracts significantly from understanding the physics of the process.

4-2-3 The Three-Zone Model of Benson

Based on careful observation of photographs from flow visualization rigs of the scavenging process in loop- and cross-scavenged engines,[12] Benson[6] suggested that the gas exchange process takes place in three phases (Fig. 4-11). In phase I, the fresh charge

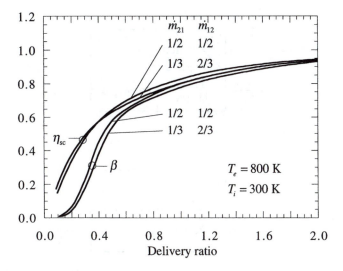

Figure 4-10 Scavenging efficiency and exhaust gas purity results of the Streit–Borman two-zone model. Reprinted from Sher,[1] *Prog. Energy Combust. Sci.*, with permission from Elsevier Science.

jets enter the cylinder and displace some of the cylinder contents. Mixing occurs at the boundaries of the jets. In phase II some of the fresh charge entering the cylinder passes straight through the exhaust ports. Only fresh charge is passing through the exhaust ports; mixing, however, continues to take place at the boundaries of the jets. In phase III, a homogeneous mixture of fresh charge and burnt gas leaves the cylinder; the composition of this mixture varies with time. Benson[6] assumed that the fresh charge entering the mixing zone at any time is a fixed proportion of the instantaneous fresh charge flow rate entering the cylinder, say $b_1 \dot{m}_1$. Further, he assumed that at any time the proportions of burnt gas and fresh charge entering the mixing zone are constant (i.e., he assumed a fixed gas entrainment ratio). Alternatively, Benson has suggested (version II) that the volume of the burnt gas zone is reduced linearly with time.

Under these conditions, by applying the first law of thermodynamics to the cylinder contents while assuming that the specific heat ratio γ, as well as the molecular weights of the two gases involved are the same, the cylinder pressure can be calculated from the following relation:

$$\frac{dp_c}{dt} = \frac{(\gamma - 1)\dot{Q}_c - \dot{m}_e \gamma RT_e + \dot{m}_i \gamma RT_i}{V_c} - \frac{\gamma p_c \dot{V}_c}{V_c} \tag{4-28}$$

where subscripts c, e, and i denote cylinder, exhaust, and fresh charge (intake) values. The temperature of the fresh charge zone, in any phase, may be calculated by

$$\frac{dT_{fcz}}{dt} = \frac{\gamma - 1}{\gamma} \left[\frac{\dot{Q}_{fcz}}{m_{fcz} R} + \frac{\gamma}{(\gamma - 1)} \frac{\dot{m}_i}{m_{fcz}} (T_i - T_{fcz}) + \frac{T_{fcz}}{p_c} \frac{dp_c}{dt} \right] \tag{4-29}$$

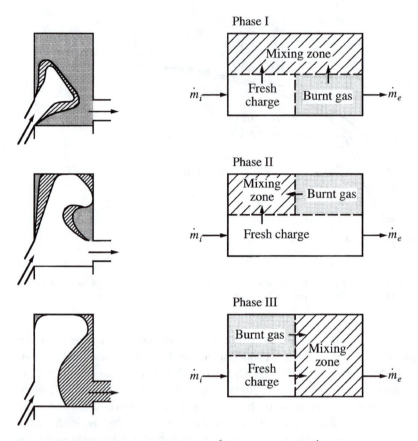

Figure 4-11 The three-zone model of Benson.[6] Reprinted from Sher,[1] *Prog. Energy Combust. Sci.*, with permission from Elsevier Science.

and the temperature of the burnt gas zone, in any phase, by

$$\frac{dT_b}{dt} = \frac{\gamma - 1}{\gamma} \left[\frac{\dot{Q}_b}{m_b R} + \frac{T_b}{p_c} \frac{dp_c}{dt} \right] \tag{4-30}$$

By integrating these equations through the scavenging process, the cylinder pressure, temperature, and volume of each zone are determined. The duration of the scavenging phase periods and the coefficients b_1 and b_2 (where $b_2 \dot{m}_i$ is the instantaneous burnt gas flow rate entering the mixing zone) recommended by Benson are as follows:

Phase I $95° \le \theta < 155°$ ATC

Phase II $155° \le \theta < 200°$ ATC

Phase III $200° \le \theta < 265°$ ATC

while $b_1 = 0.2$ and $b_2 = 0.1$. The scavenge and exhaust mass flow rates as well as the heat transfer from each zone to the cylinder walls are calculated separately by using the scavenge and exhaust manifold conditions, and an empirical correlation for the heat transfer. For version II of the model, Benson recommended $b_1 = 0.1$, and the final proportion of the cylinder volume occupied by the burnt gas zone be 0.1 at low speed, and 0.3 at high speed.

As the three-zone model of Benson[6] requires the solution of four differential equations at each time step and introduces as many as five calibration constants, it cannot be considered a practical model. Nevertheless, this model was the first to suggest that the scavenging process proceeds in several sequential phases, a concept later adopted by many investigators[13-15] in attempts to understand better their experimental data.

4-2-4 The "S"-Shape Model

A semiempirical model to simulate the scavenging process in cross-, loop- or uniflow-scavenged engines has been proposed by Sher.[16] The model is based on the experimental finding that, in all engine configurations, the variation of the exhaust gas purity β with crank angle is well approximated by a sigmoid ("S"-type) curve. Thus, an empirical correlation for β has been suggested:

$$\beta = 1 - \exp(-c\Lambda\tau^b) \tag{4-31}$$

where b and c are the form and shape factors, respectively, and τ is defined as

$$\tau = (\theta - \theta_{so})/(\theta_{sc} - \theta_{so}) \tag{4-32}$$

Based on this assumption, the appropriate expressions for the charging and scavenging efficiencies may be derived as follows. The mass of delivered charge retained at any time is

$$m_{rt} = \int_0^t \dot{m}_i \, dt - \int_0^t \beta\dot{m}_e \, dt \tag{4-33}$$

where

$$\dot{m}_i = m_0 \frac{d\Lambda}{dt} \tag{4-34}$$

and m_0 is the ideal charging mass. If we assume a constant-volume, constant-pressure scavenging process and that the gases follow the ideal gas law with a constant specific heat, then

$$pV = mRT = \text{constant} \tag{4-35}$$

and, from the energy balance,

$$\dot{m}_e = \dot{m}_i(T_i/T_e) \tag{4-36}$$

where

$$T_e = \beta T_i + (1 - \beta)T_b \tag{4-37}$$

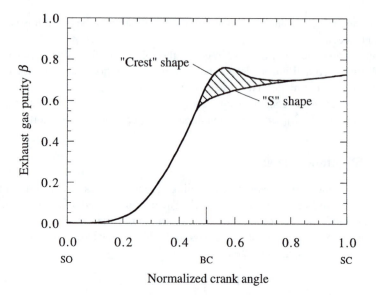

Figure 4-12 The exhaust gas purity profiles from the "S" model[16] and the "crest" model.[17] The hump may be interpreted as the occurrence of backflow and extensive short-circuiting. (Reproduced from the Proceedings of the Institution of Mechanical Engineers by permission of the Council of the Institution of Mechanical Engineers.)

Here T_b is the temperature of the unmixed burnt gas. Substitution of these results into the definition of the charging efficiency yields

$$\eta_{ch} = \frac{m_{rt}}{m_0} = \Lambda - \int_0^\tau \frac{\beta(d\Lambda/d\tau)}{\beta + (1 - \beta)(T_b/T_i)} \, d\tau \qquad (4\text{-}38)$$

The scavenging efficiency is related to the charging efficiency by

$$\eta_{sc} = (T_1/T_i)\eta_{ch} \qquad (4\text{-}39)$$

This model suggests that the scavenging process may be interpreted as a combination of pure-displacement scavenging [the first term in Eq. (4-38)], and charging losses (the second term). The character of the process is determined by the form of Eq. (4-31) and the shape factors b and c: a pure displacement process is represented by $c = 0$, an isothermal perfect-mixing process by $c = 1$ and $b = 0$, and a pure short-circuiting process by $c \to \infty$. An appropriate selection of b and c can produce any curve between the perfect mixing and the pure displacement process. Sher[16] recommended a value of 2.0 for the form factor b, and 1.7 for the shape factor c.

This "S"-shape model prescribes the exhaust gas composition; it is not related to the instantaneous cylinder contents. Because these are not independent, this can lead to anomalies. In addition, careful examination of the composition of exhaust gases shows that it does not precisely follow the prescribed "S" shape, but exibits a "crest" profile: a protuberance usually appears on the latter portion of the curve. A modification to improve the fitting of the "S"-shape curve to the experimental observations, and to

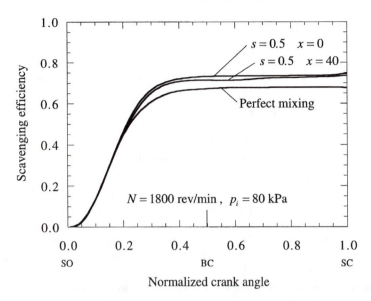

Figure 4-13 Scavenging efficiency results of the "crest" model.[17] (Reproduced from the Proceedings of the Institution of Mechanical Engineers by permission of the Council of the Institution of Mechanical Engineers.)

take into account the effects of varying engine speed, has been suggested by Sher and Harari.[17] The modified model provides a means of including the occurrence of back-flow as well as extensive short-circuiting during the scavenging process. Under these conditions, a rounded hump may appear on the latter portion of the sigmoid curve as shown in Fig. 4-12, forming a "crest" rather than an "S" profile. An alternative function for β, to the exponential of Eq. (4-31), has been suggested:

$$\beta = \eta_{sc}(1 - \varepsilon)/x \qquad (4\text{-}40)$$

where the parameters x and ε reflect the influence of mixing and short-circuiting, respectively. Thus the exhaust composition depends strongly on the composition of the cylinder contents. The short-circuiting degree ε is evaluated from

$$\varepsilon = s\tau'(d\Lambda/d\tau') \qquad (4\text{-}41)$$

where s is a calibration parameter and τ' is the dimensionless scavenging time, defined in terms of engine speed N as

$$\tau' = (\theta_{sc} - \theta_{so})/(2\pi N) \qquad (4\text{-}42)$$

It was found that both x and ε are inherent properties of the engine design and are independent of engine speed and load. For a typical Schnürle-type scavenging system, a value of 0.57 was recommended for x and a value of 0.81 for s/s_{max}. The charging and scavenging efficiencies are evaluated using Eqs. (4-38) and (4-39). Figure 4-13 compares results obtained from the "crest" model with those obtained by assuming a perfect mixing process. In both cases, the delivery ratio was derived from an arbitrary function of the form

$$dΛ/dt = 1 - \cos 2\pi\tau \qquad (4\text{-}43)$$

A second function for the delivery ratio can be used at low engine speed, when exhaust backflow occurs:

$$dΛ/dt = 2(1 - \tau - \cos 2\pi\tau) \qquad (4\text{-}44)$$

Improvements to the model have been developed that account for variations in pressure and mass transfer rates through the ports, heat transfer between the cylinder contents and the cylinder wall, and mass fraction of burnt gases in the mixture. The model is usually calibrated by choosing the mixing factor x to fit the exhaust gas purity curve at the end of the scavenging period, and chosing the short-circuiting factor s to fit the exhaust gas purity curve at the beginning of the scavenging period.[17,2]

It should be noted that a careful examination of the scavenging process in flow visualization rigs[12,18] reveals that distinct phases are not apparent during the process; rather a complex combination of the three phases—displacement, mixing, and short-circuiting—occurs. Thus the preceding approach, which considers a pure displacement process while simultaneously accounting for charging losses, appears to be more realistic than any model based on distinct phases.

4-3 ASSESSMENT: THE "β TEST"

The scavenging process in a two-stroke cycle engine is a complex process involving several important coupled phenomena, such as momentum, mass, and heat transfer, in a turbulent flow. The success of the scavenging process is a function of the engine's geometry, the method of scavenging, and the operating conditions. The engine performance, however, is not only a function of the scavenging efficiency but also of the success of the ignition and combustion processes, which depend strongly on the preparation of the mixture prior to ignition (especially mixture homogeneity and turbulence intensity).

This chapter has discussed the different types of phenomenological models that have been developed to analyze engine scavenging. These were classified as single-phase, one- or two-zone models, and multizone multiphase models. Their different underlying assumptions, organization, and computational cost make them suitable for analyzing different aspects of how the scavenging process affects engine operation. The engine designer–researcher should select the best model for his or her particular purpose. Optimizing engine structural design, which includes port and valve geometry, plug location, and combustion chamber design, would require a detailed multidimensional model (see Section 4-4). However, such models require large computer programs, which are costly and time-consuming to run. The perfect-mixing model is very simple, but its use is appropriate when only the trend behavior of an engine is required. In a complete cycle simulation for studying problems associated with engine speed and load, ambient conditions, valve, spark, and fuel-injection timing, heat transfer, supercharging and turbocharging, lubrication, and noise, a realistic yet simple-to-use scavenging correlation may be preferable.

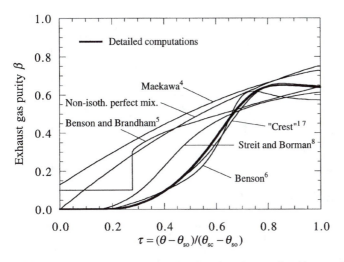

Figure 4-14 Exhaust gas purity versus normalized crank angle as predicted by several different scavenging models for loop-scavenged (Schnürle-type) engines. For the same set of operating conditions, each model has been calibrated to yield an overall scavenging efficiency of 87% and to fit the exhaust gas purity curve of the multidimensional model as closely as possible.[24] (Reproduced from the Proceedings of the Institution of Mechanical Engineers by permission of the Council of the Institution of Mechanical Engineers.)

How can the accuracy of these models be assessed? Because the charging and the scavenging efficiencies are essentially integrated parameters, short time duration phenomena such as short-circuiting and backflows are smoothed out. The most sensitive parameter for evaluating any scavenging model is the variation of the exhaust gas purity β, with the crank angle: the "β test." Experimental observations in real[19] and motored[20] engines and in flow visualization rigs[18] have revealed that, for all engine configurations, speeds, and loads, the exhaust gas purity exhibits a sigmoid-type curve. This phenomenon has also been predicted by detailed computations of the in-cylinder flow field[21–23]. Figure 4-14 shows the exhaust gas purity curve as predicted by a multidimensional model for a loop-scavenged (Schnürle-type) engine. Also shown are predictions calculated by several different single-phase and multiphase multizone models.[24]

The failure of the single-phase single-zone models is pronounced during the first half of the scavenging process, where the perfect-mixing assumption is far from reality. Among the multiphase multizone models, only three models (and their derivative models) are capable of producing a sigmoid shape for the β parameter: the Streit–Borman model,[8] the Benson model (version II),[6] and the "S" and "crest" models.[16,17] The precise shape of the curve may be correlated with experimental results by means of each model's calibration parameters. This procedure of calibrating the model by fitting the exhaust gas purity curve is practical and convenient, because one of the most accurate methods of measuring the charging efficiency in a fired two-stroke engine is to locate a sampling valve in the exhaust pipe just outside the exhaust ports and trace the exhaust gas composition over time.[20] Note that only two of the models are capable of predicting the occurrence of extensive short-circuiting (the cause of the rounded protuberance that appears on the "S" shape's back). These are the Benson[6] and the "crest"[17] models.

Table 4-2 Comparison of single-phase and multiphase multizone models

Model or Author	Year	Ref.	Model essentials			Calculation method			Applications[†]		Test[‡]
			No. of zones	No. of phases	No. of algebraic eqs.	No. of differential eqs.	Order of differential eqs.	No. of empirical parameters	Developed for	Applied to	β test
Perfect displacement	—	—	2	1	1	—	—	—	Any	—	F
Hopkinson (perfect mixing)	1914	3	1	1	1	—	—	—	Any	Any	F
Maekawa	1957	4	2	1	1	—	—	2	L	(L,C)[9,11,25]	F
Benson and Brandham	1969	5	2	2	1	—	—	2	L	(L,C)[10,11] U[7]	F
Streit and Borman	1971	8	2	1	—	2	1	4	U	U[8]	OK
Benson	1977	6	3	3	—	4	1	5	L	(L,C)[14,15] U[13,14]	OK
"S" shape	1985	16	1	1	1	1	1	2	L	(L,C)[26–28]	OK
"Crest"	1991	17	1	1	1	1	1	2	L	(L,C)[28]	OK

[†] L, C, and U are abbreviations for loop, cross, and uniflow scavenging, respectively.

[‡] F and OK stand for "fails" and "satisfies", respectively.

As has already been stated, there is no strict transition from one phase to another: the charging process advances continuously. Also, the charging efficiency does not necessarily increase at a decreasing rate. Rather it may follow the pure-displacement curve for a period, coincide with the perfect-mixing curve for another period, or even intersect it. A realistic model should have the ability to follow the real progress of the charging process. Each of the multiphase multizone models described here has been developed for a specific method of scavenging. The sequence of events as observed for a loop- or cross-scavenged engine may not be the same as in a uniflow-scavenged engine. Even for a particular method of scavenging, say a uniflow-scavenging system, a different port inclination (which may affect the degree of swirl, for example) may result in a different sequence of events. For low swirl designs, the model of Streit and Borman,[8] Kyrtatos and Koumbarelis,[13] or Kishan et al.[14] may represent the scavenging progress well. For high swirl designs, the incoming air adheres to the cylinder wall, held there by centrifugal effects, and passes along the cylinder in an annulus leaving the central core unscavenged. For this case, the model of Wallace and Cave[7] is more realistic. Table 4-2 summarizes these singlephase and multiphase multizone models, and the experience acquired with them.

4-4 MULTIDIMENSIONAL MODELS

A full description of the two-stroke engine gas exchange process requires the solution of the complete set of the differential equations that govern the process, for the time-varying profiles of the flow field, mixture composition, and temperature. These equations include conservation laws of momentum in one, two, or three dimensions, mass of each species, and energy, and models for transport coefficients, turbulent flow phenomena, and the boundary conditions. These fluid-mechanics-based multidimensional models can provide a clearer picture of the in-cylinder events and can identify problems such as fresh mixture short-circuiting and poorly scavenged regions, for a given cylinder and port assembly geometry. Multidimensional model analysis shows promise of defining the flow field details with sufficient precision to guide inlet and exhaust port design. Although these models are not yet practical for a complete cycle simulation that predicts overall engine performance, their application to two-stroke engines has led to significant progress in our understanding of the gas exchange process. An introduction to this topic, and comprehensive reviews on fluid motion within the cylinder of internal combustion engines, may be found elsewhere.[29–31] This section reviews the application of multidimensional computational fluid dynamic (CFD) models to two-stroke cycle engines and explains their predictive capabilities. Examples of the application of multidimensional models to various designs of two-stroke engines are summarized in Table 4-3.

These fluid-dynamics-based engine process computational codes solve the partial differential equations for the conservation of mass, momentum, energy, and species concentrations. These differential equations are transformed into finite-difference equations, which are then solved numerically. The computational grid arrangement that defines the number and positions of the locations at which the flow parameters are to be calculated, the discretization practices used to transform the differential equations of the

Table 4-3 Examples of applications of multidimensional models to two-stroke cycle engines

Author	Year	Ref.	Scavenge flow	Dimensions	Code used
Ramos et al.	1979	32	Uniflow	Axisymmetric	Own
Sher	1980	22	Loop	Quasi-3	Own, based on Gosman and Johns[33]
Sher	1985	34	Loop	Quasi-3	Own, based on Gosman and Johns[33]
Swann et al.	1985	35	Loop	3	PHOENICS[36]
Diwakar	1987	37	Uniflow	3	KIVA[38]
Carapanayotis and Salcudean	1988	39	Uniflow	3	Own, based on Harlow and Amsden[40]
Sher et al.	1988	41	Uniflow	Axisymmetric	DICE[42]
Uzkan	1988	43	Uniflow	3	CONCHAS (KIVA[38])
Ahmadi-Befrui et al.	1989	44	Loop	3	FMCS
Fabvre and Ferreira	1990	45	Loop	3	KIVA[38]
Smyth et al.	1990	46	Loop	3	Own
Epstein et al.	1991	47	Cross	3	KIVA[38]
Amsden et al.	1992	48	Cross	3	KIVA[38]
Kuo and Reitz	1992	49	Cross	3	KIVA[38]
Ravi and Marathe	1992	50	Uniflow	3	Own—CARE
Lai et al.	1993	51	Cross	3	Own
Huh et al.	1993	52	U (4 valves)	3	KIVA[38]
Das and Dent	1993	53	U (4 valves)	3	KIVA[38]
Tamamidis and Assanis	1993	54	Uniflow	3	Own—ARIS
Haworth et al.	1993	55	Loop	3	Own
McKinley et al.	1994	56	Loop	3	PHOENICS
Leep et al.	1994	57	Loop	3	FIRE
Corcione et al.	1997	58	Loop	3	KIVA-3
Yu et al.	1997	59	Loop	3	STAR-CD

mathematical model into algebraic equations, and the solution algorithms employed to obtain the flow parameters from the discrete equations are the three important numerical features of the multidimensional method.

To apply a computer code to the solution of a continuum problem, the continuum must be represented by a finite number of discrete elements. The most common method of discretization is to divide the region of interest into a number of small zones or cells. These cells form a grid or mesh that serves as a framework for constructing finite volume approximations to the governing partial differential equations. The computing mesh must match the topography of the combustion chamber, scavenge ports, and scavenge ducts, including moving boundaries such as the piston top. Three-dimensional formulations of the finite-difference equations are required for practical engine calculations. Two-dimensional (or axisymmetric) formulations can be useful, however, under simpler flow situations. Although the dynamic characteristics of intake and exhaust flows can usually be studied with one-dimensional unsteady fluid-dynamic calculations, flows within the cylinder, and the scavenge and exhaust ports, are inherently unsteady and three dimensional. Flows in the scavenge ducts strongly affect the scavenging process and cannot, in general, be supplied merely as boundary conditions to the in-cylinder flows. Furthermore, the computing mesh should allow control of local resolution to obtain the maximum accuracy with a given number of grid points; thus, it is desirable that the mesh allows concentration of grid points in regions where steep gradients exist such as jets and boundary layers. Figure 4-15 shows an example of a three-dimensional nonorthogonal lagrangian–eulerian mesh. The mesh does not correspond to a cylindrical coordinate system, but is instead derived from a nonperiodic mapping of a single block of cells. Also, because it avoids the singularity associated with having a mesh axis, use of such a mesh requires fewer cells and allows larger time steps when there are strong flows across the cylinder. The ports attached to this cylinder were created by processing coordinate data digitized from the actual asymmetric ports. As the piston moves, cells are added or subtracted to the mesh as indicated. The entire mesh has 20,729 grid points.[48]

The multidimensional CFD engine flow models are time-marching programs that solve finite-difference approximations to the governing differential equations. The time variable is similarly discretized into a sequence of small time intervals (time steps), and the transient solution is marched out in time; the solution at time t_{i+1} is calculated from the known solution at time t_i. In general, numerical calculations of compressible flows are inefficient at low Mach numbers because of the wide disparity between the time scales associated with convection and with the propagation of sound waves. Although all methods use first-order temporal discretization and are therefore of comparable accuracy, they differ in whether forward or backward differencing is employed in the transport equations leading to implicit or explicit discrete equations, respectively. In explicit schemes, this inefficiency occurs because the time steps needed to satisfy the sound speed stability condition are much smaller than those needed to satisfy the convective stability condition alone. In implicit schemes, the inefficiency manifests itself in the additional computational labor needed to solve the implicit (simultaneous) system of equations at each time step. This solution is usually performed using iterative techniques. The computing time requirements of these two approaches scale with the number of equations n, and the number of mesh points m, as follows. For explicit methods, computing time scales as nm, but the time step is limited by the stability condition. For implicit methods, computing time scales as $n^3 m$, and Δt is limited only by accuracy considerations. The implicit equations that result from forward differencing consist of simultaneous sets for all variables and thus require more elaborate methods of solution. However, they contain no intristic stability constraints.

The engine flow processes are turbulent. In the engine cylinder, the flow involves a complicated combination of turbulent shear layers, recirculating regions, and wall

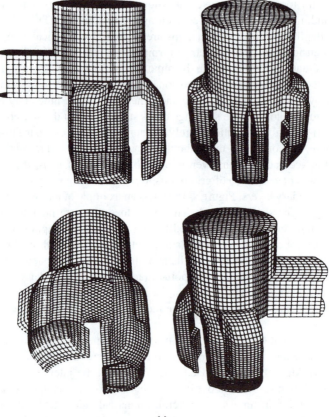

(a)

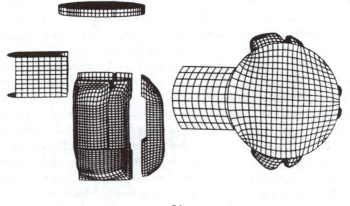

(b)

Figure 4-15 An advanced computing mesh arrangement for engine cylinder, ports, and ducts assembly, with the piston at (a) BC, and (b) TC.[48]

boundary layers. The flow is unsteady and may exhibit substantial cycle-to-cycle fluctuations. Both large-scale and small-scale turbulent motions are important factors governing the overall behavior of the flow. An important characteristic of a turbulent flow
is its irregularity or randomness, and statistical methods are necessary to define the flow
field. The quantities normally used are the mean velocity, the fluctuating velocity about
the mean (characterized by its root-mean-square value, the turbulence intensity), and
several length and time scales. In engines, the application of these definitions is complicated by the fact that the flow pattern changes during the engine cycle. Also, although
the overall features of the flow repeat each cycle, the details do not because the mean
flow can vary significantly from one engine cycle to the next. There are both cycle-
to-cycle variations in the mean or bulk flow at any point in the cycle, and turbulent
fluctuations about that specific cycle's mean flow. The fact that turbulent flows exhibit
important spatial and temporal variations over a range of scales makes direct numerical
solution of these governing equations impractical for flows of engine complexity. Re-
course must therefore be made to some form of averaging or filtering that removes the
need for direct calculation of the small-scale motions. Two approaches have been developed for dealing with this turbulence modeling problem: full-field modeling (FFM),
which is sometimes called statistical flux modeling, and large-eddy simulation (LES), or
subgrid-scale simulation. In FFM, one works with the partial differential equations de-
scribing suitably averaged quantities, using the same equations everywhere in the flow.
LES is an approach in which one actually calculates the large-scale three-dimensional
time-dependent turbulence structure in a single realization of the flow. Thus, only the
small-scale turbulence need be modelled. An important difference between the two approaches is their definition of "turbulence." In FFM the turbulence is the deviation of the
flow at any instant from the average over many cycles of the flow at the same point in
space and oscillation phase, and thus it contains some contributions from cycle-by-cycle
flow variations. LES defines turbulence in terms of variations about a local average; thus
in LES turbulence is related to events in the current cycle.[29]

To illustrate the potential for multidimensional modeling of two-stroke cycle engine flows, examples of the output from such calculations will now be presented. A
large amount of information on many fluid flow and state variables is generated with
each calculation, and the processing, organization, and presentation of this information
are tasks of comparable scope to its generation. Flow field results are usually presented
in terms of the gas velocity vectors at each grid point of the mesh in appropriately selected planes. Arrows are used to indicate the direction and magnitude (by length) of
each vector. Figure 4-16 shows an example of measured[60] and predicted[55] mean velocity vectors within the cylinder of a motored loop-scavenged two-stroke cycle engine.
The scavenge port layout is shown in Fig. 3-17. The data shown correspond to a crank
angle of 150° ATC, well after the normal fired-engine blowdown process (not present
in the motored engine) would be over. The measured data are the top set of figures. The
scavenge-flow-driven "tumble" loop is by now well established. Two computational
cases are shown. The base case (BASE) employs the nominal port geometry. The port-
shifted case (PSHIFT) explores the effect of asymmetry in the opening of the transfer
ports. The two main ports B and C on one side of the cylinder (see Fig. 3-17) were
shifted 1 mm up the cylinder liner and the measured swirl level in the residual burnt

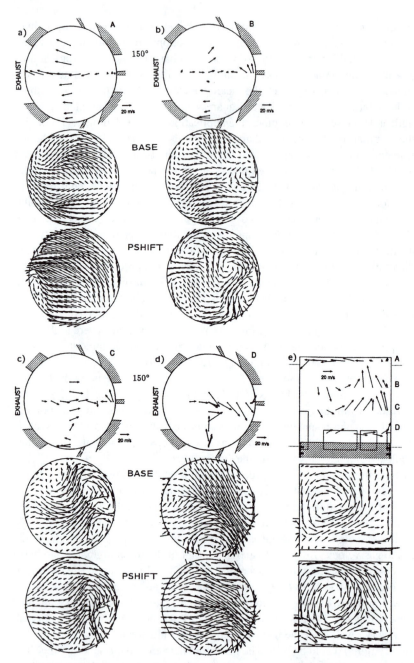

Figure 4-16 Velocities at 150° ATC in a loop-scavenged engine: measured (top)[60] and calculated (middle, bottom)[55]. Port geometry is shown in Fig. 3-17. The vertical viewing plane is through the nominal symmetry axis of the cylinder. The horizontal planes are identified in the right-hand top cross section. BASE is computation with nominal port geometry. PSHIFT is computation with main scavenge ports on one side of the engine raised 1 mm to produce asymmetry in scavenging flow comparable to that measured.

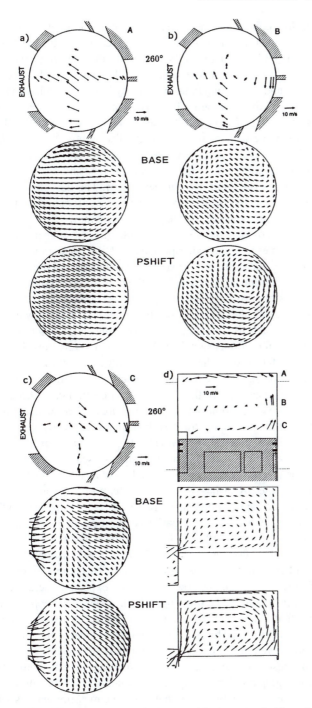

Figure 4-17 Measured (top) and calculated (middle, bottom) velocities at 260° crank angle ATC. The vertical viewing plane is through the nominal symmetry axis of the cylinder. Same details as Fig. 4-16.[55,60]

gas was used as an initial flow condition. These changes explore the likely causes of the asymmetry apparent in the measured flow data. The gross features of the measured flow are well predicted by the CFD calculations. With the introduced port asymmetry (PSHIFT calculation) the finer details are also reasonably well matched (with some localized exceptions). Planes A and D (top and bottom of the cylinder) match best; these flows are most directly driven by the scavenging flow. The matching in the intermediate planes C and D is less complete. The port shift does increase the in-cylinder flow asymmetry and bring it more in line with the measurements. The authors[60] suggest that a combination of different pressures in the scavenge ports on either side of the engine and a slightly tilted piston would contribute to this flow asymmetry; and differences in ring and ring gap location could also result in asymmetric flow. The flow field inside the cylinder is sensitive to small changes in effective port geometry, probably due to the steep velocity gradients the two-stroke gas exchange process generates.

Figure 4-17 shows a similar comparison at 260° ATC, after the scavenge ports have closed. Again reasonable agreement is obtained, especially with the calculations where asymmetries were introduced via the shifting of two of the scavenge ports 1 mm axially. Note how the exhaust process continues, driven by both the upward piston motion and scavenging flow, after the transfer ports have closed.

An important issue with these calculations is the specification of boundary conditions: the flow velocity distribution across the transfer port openings. Figure 4-18 shows the streamlines predicted with a multidimensional CFD code, for a steady flow through the cylinder from the two transfer ports to the exhaust port, for a plug flow (uniform) input velocity boundary condition and for the measured velocity distribution across the port.[56] The differences in the streamlines for these two cases are significant. Accurate specification of the flow boundary conditions at the ports requires either detailed measurements or CFD modeling of the flow in the transfer ducts that feed these ports.

Multidimensional model calculations also provide local gas composition information during (and after) the scavenging process. Figure 4-19 shows a scalar concentration map, for a cross section in the plane of symmetry through the exhaust port (on the left), for a loop-scavenged engine.[57] The engine has two main transfer ports, one on each side of the exhaust port, two auxiliary transfer ports, and a rear transfer port opposite to the exhaust. The shading level varies from white "old gas" (the burnt gas in the cylinder at exhaust port opening), to black "new gas" (the fresh scavenging flow). Between about 225° and 240° ATC an exhaust backflow driven by oscillations in exhaust pressure resulting from the interaction between the exhaust flows from the different cylinders of the engine is apparent. The fresh scavenging flow follows the expected overall loop-scavenged pattern. Some short-circuiting and trapping of burnt gases in the combustion chamber are evident. The extent of mixing between fresh and burnt gases is also apparent. The nonuniform nature of the residual burnt gas distribution in the in-cylinder gases after the scavenging process is over, in this particular example, is significant.

These CFD calculations also provide the information required for calculating the charging, trapping, and scavenging efficiencies, by suitable integration. Since these are "average" parameters, good agreement between computations and measurements is usually obtained because any local differences between the computed and measured flow fields cancel out.

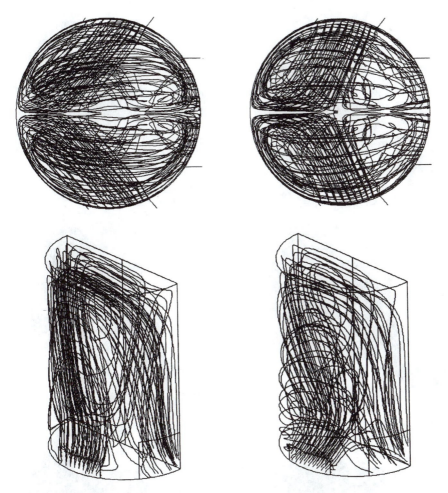

Figure 4-18 Two CFD predictions of scavenging flow streamlines for steady flow through a loop-scavenged engine, with different transfer port velocity boundary conditions: (left) using plug flow boundary condition (uniform velocity through port); (right) using laser doppler velocimetry measured velocity distribution as boundary condition.[56]

With direct-injection engines, these codes, with a suitable fuel spray model, can predict the development of the fuel vapor distribution inside the cylinder following start of injection. This is an important process that, with direct-injection gasoline engines, determines the composition of the mixture at the spark plug location at time of spark (and hence the quality of ignition) as well as the extent and composition of the fuel-vapor–air mixture region through which a flame must rapidly propagate. Figure 4-20 shows predictions of the in-cylinder flow field in a loop-scavenged direct-injection engine, the liquid fuel spray characteristics (the injector is at the top of the cylinder head), and the fuel vapor fraction distribution that results.[59] The darkness of the shaded areas indicates the local value of fuel vapor mass fraction. The tumbling loop-scavenging-

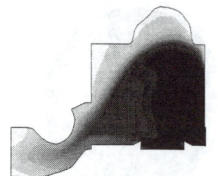

CA = 194

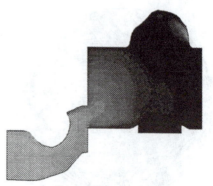

CA = 242

CA = 210

CA = 254

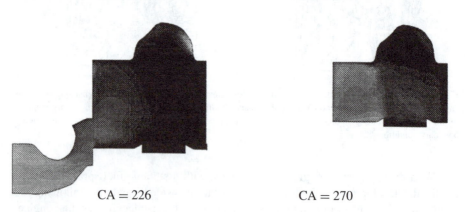

CA = 226

CA = 270

Figure 4-19 Computed passive scalar maps showing distribution of burnt and fresh charge within the cylinder of a loop-scavenged engine, during the scavenging process. Plane shown is plane through the cylinder axis bisecting the exhaust port on the left. Main scavenge ports adjacent to exhaust, auxiliary ports next (dark section at bottom of cylinder), rear scavenge port on right. "Old gas", burnt gas at exhaust port opening, is white; "new gas", fresh charge, is black.[57]

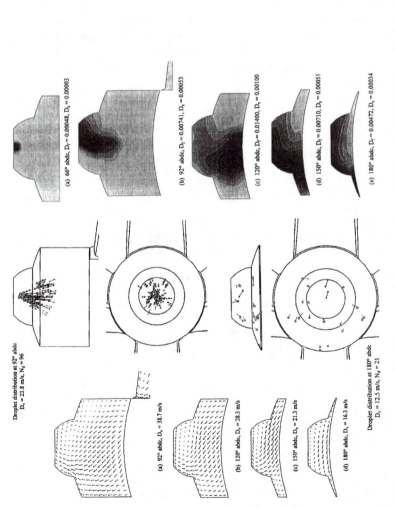

Figure 4-20 Computed in-cylinder gas flow pattern, liquid fuel spray droplet distribution, and fuel vapor distribution, in a loop-scavenged direct-injection spark-ignition two-stroke engine. (Left) Flow velocity distribution during compression; D_v is the highest local gas velocity. (Center) Droplet distribution in spray; D_v and N_d are maximum droplet velocity (darker arrows, higher velocity) and number of droplets (representative number only). Note significant droplet vaporization. (Right) Fuel vapor distribution; D_f is maximum fuel vapor concentration and D_s is the difference in fuel vapor concentration between contours (relative values).[59]

In the images, the following labels appear:

Droplet distribution at 92° abdc
$D_v = 23.8$ m/s, $N_d = 96$

Droplet distribution at 180° abdc
$D_v = 12.5$ m/s, $N_d = 21$

(a) 92° abdc, $D_v = 38.7$ m/s
(b) 120° abdc, $D_v = 28.3$ m/s
(c) 150° abdc, $D_v = 21.3$ m/s
(d) 180° abdc, $D_v = 16.3$ m/s

(a) 66° abdc, $D_f = 0.00048$, $D_s = 0.00003$
(b) 92° abdc, $D_f = 0.00741$, $D_s = 0.00053$
(c) 120° abdc, $D_f = 0.01400$, $D_s = 0.00100$
(d) 150° abdc, $D_f = 0.00710$, $D_s = 0.00051$
(e) 180° abdc, $D_f = 0.00472$, $D_s = 0.00034$

generated in-cylinder flow is apparent on the left. The axial conical spray of liquid fuel droplets initially is little affected by the gas flow until the drops have, through vaporization, become small enough to be convected significantly away from the spray and cylinder axis. The fuel vapor distribution corresponds to this penetrating and vaporizing fuel spray, and the convection of fuel vapor by the scavenge-created tumbling flow. In these plots, D_v is the highest local gas velocity, D_f is highest local value of fuel vapor fraction, and D_s is the interval between adjacent contours. The value of these calculations for exploring the effects of injector location, spray geometry, cylinder head, piston crown, and transfer port geometries on fuel vapor distribution, and hence combustion, is clear.

Analyzing the scavenging flows of an operating two-stroke engine is a particularly challenging area in which experiment and modelling usually play complementary roles. Such calculations give valuable information on scavenging that cannot easily be obtained experimentally. However, due to the early stage of development of both diagnostic and computational tools, high-quality measurements inside the cylinder in fired engines and reliable calculations of the flow field (especially during combustion) are still not readily available.

REFERENCES

1. Sher, E., "Scavenging the Two-Stroke Engine," *Prog. Energy Combust. Sci.*, vol. 16, pp. 95–124, 1990.
2. Merker, G.P., and Gerstle, M., "Evaluation on Two Stroke Engine Scavenging Models," SAE Paper 970358, 1997.
3. Hopkinson, B., "The Charging of Two-Cycle Internal Combustion Engines," *Trans. NE Coast Inst. Engineers Shipbuilders*, vol. 30, pp. 433–462, 1914.
4. Maekawa, M., "Text of Course," JSME No. G36, p. 23, 1957.
5. Benson, R.S., and Brandham, P.T., "A Method for Obtaining a Quantitative Assessment of the Influence of Charging Efficiency on Two-Stroke Engine Performance," *Int. J. Mech. Sci.*, vol. 11, pp. 303–312, 1969.
6. Benson, R.S., "A New Gas Dynamic Model for the Gas Exchange Process in Two-Stroke Loop and Cross Scavenged Engines," *Int. J. Mech. Sci.*, vol. 19, pp. 693–711, 1977.
7. Wallace, E.J., and Cave, P.R., "Experimental and Analytical Scavenging Studie on a Two-Stroke Opposed Piston Diesel Engine," SAE Paper 710175, 1971.
8. Streit, E.E., and Borman, G.L., "Mathematical Simulation of a Large Turbocharged Two-Stroke Diesel Engine,", SAE Paper 710176, *SAE Trans.*, vol. 80, 1971.
9. Sato, K., and Kido, K., "Simulation of the Gas Exchange Process in a Small Two-Stroke Cycle Engine," *Bull. JSME*, vol. 26, pp. 1178–1187, 1983.
10. Sanborn, D.S., "Paper Powerplants Promote Performance Progress," SAE Paper 750016, *SAE Trans.*, vol. 84, 1975.
11. Baudequin, F., and Rochelle, P., "Some Scavenging Models for Two-Stroke Engines," *Proc. Inst. Mech. Eng.*, vol. 194, pp. 203–210, 1980.
12. Dedeoglu, N., "Scavenging Model Solves Problems in Gas Burning Engine," SAE Paper 710579, *SAE Trans.*, vol. 80, 1971.
13. Kyrtatos, N.P., and Koumbarelis, I., "A Three-Zone Scavenging Model for Two-Stroke Uniflow Engines," *J. Eng. Gas Turbines Power*, vol. 110, p. 531, 1988.
14. Kishan, S., Bell, S.R., and Caton, J.A., "Numerical Simulation of Two-Stroke Cycle Engines Using Coal Fuels," *J. Eng. Gas Turbines Power*, vol. 108, pp. 661–668, 1986.
15. Changyou, C., and Wallace, F.J., "A General Isobaric and Isochoric Thermodynamic Scavenging Model," SAE Paper 871657, *SAE Trans.*, vol. 96, 1987.

16. Sher, E., "A New Practical Model for the Scavenging Process in a Two-Stroke Cycle Engine," SAE Paper 850085, *SAE Trans.*, vol. 94, pp. 487–495, 1985.

17. Sher, E., and Harari, R., "A Simple and Realistic Model for the Scavenging Process in a Crankcase-Scavenged Two-Stroke Engine," *J. Power Energy*, vol. 205, pp. 129–137, 1991.

18. Sher, E., "Investigating the Gas Exchange Process of a Two-Stroke Cycle Engine with a Flow Visualization Rig," *Israel J. Tech.*, vol. 20, pp. 127–136, 1982.

19. Blair, G.P., *The Basic Design of Two-Stroke Engines*, SAE, Warrendale, PA, 1990.

20. Kannapan, A., "Cumulative Sampling Technique for Investigating the Average Process in Two-Stroke Engine," ASME Paper 74-DGP-11, 1974.

21. Kenny, R.G., Smyth, J.G., Fleck, R., and Blair, G.P., "The Scavenging Process in the Two-Stroke Engine Cylinder," Motorcycle Symposium, Gratz, April, 1989.

22. Sher, E., "An Improved Gas Dynamic Model Simulating the Scavenging Process in a Two-Stroke Cycle Engine," SAE Paper 800037, 1980.

23. Sweeney, M.E.G., Swann, G.B.G., Kenny, R.G., and Blair, G.P., "Computational Fluid Dynamics Applied to Two-Stroke Engine Scavenging," SAE Paper 851519, *SAE Trans.*, vol. 94, 1985.

24. Sher, E., "Scavenging Models for Numerical Simulations of Two-Stroke Engines: A Comparative Study," Proceedings of the 2nd International Conference on Computers in Engine Technology, pp. 135–139, IMechE, Cambridge, Sept. 1991.

25. Oka, T., and Ishihara, S., "Relation between Scavenging Flow and Its Efficiency of a Two-Stroke Cycle Engine," *Bull. JSME*, vol. 14, pp. 257–267, 1971.

26. Malik, K.A., Mehta, P.S., and Gupta, C.P., "Phenomenological Scavenge- Combustion Model for Two Stroke Cycle SI Engine," SAE Paper 881268, 1988.

27. Sher, E., "The Effect of Atmospheric Conditions on the Performance of an Air-Borne Two-Stroke Spark-Ignition Engine," *J. Power Energy (Proc. Inst. Mech. Eng.)*, vol. 198D, pp. 239–251, 1984.

28. Harari, R., and Sher, E., "The Effect of Ambient Pressure on the Performance Map of a Two-Stroke SI Engine," SAE Paper 930503, 1993.

29. Heywood, J.B., *Internal Combustion Engine Fundamentals*, McGraw-Hill, New York, 1988.

30. Heywood, J.B., "Fluid Motion within the Cylinder of Internal Combustion Engines," *J. Fluids Eng.*, vol. 109, pp. 4–35, 1987.

31. Arcoumanis, C., and Whitelaw, J.H., "Fluid Mechanics of Internal Combustion Engines—A Review," *Proc. Inst. Mech. Eng.*, vol. 201, no. C1, pp. 57–74, 1987.

32. Ramos, J.I., Humphrey, J.A.C., and Sirignano, W.A., "Numerical Prediction of Axisymmetric Laminar and Turbulent Flows in Motored, Reciprocating Internal Combustion Engines," SAE Paper 790356, *SAE Trans.*, vol. 88, 1979.

33. Gosman, A.D., and Johns, R.J.R., "Development of a Predictive Tool for In-Cylinder Gas Motion in Engines," SAE Paper 780315, 1978.

34. Sher, E., "Prediction of the Gas Exchange Performance in a Two-Stroke Cycle Engine," SAE Paper 850086, 1985.

35. Swann, B.G., Kenny, R.G., and Blair, G.P., "Computational Fluid Dynamics Applied to Two-Stroke Engine Scavenging," SAE Paper 851519, *SAE Trans.*, vol. 94, 1985.

36. Spalding, D.B., "A General Purpose Computer Program for Multi-Dimensional One and Two-Phase Flow," *Mathematics and Computers in Simulation*, vol. 23, pp. 267–276, 1981.

37. Diwakar, R., "Three-Dimensional Modeling of the In-Cylinder Gas Exchange Processes in a Uniflow-Scavenged Two-Stroke Engine," SAE Paper 870596, 1987.

38. Amsden, A.A., Ramshaw, J.D., O'Rourke, P.J., and Dukowicz, J.L., "KIVA: A Computer Program for Two and Three-Dimensional Fluid Flows with Chemical Reactions and Fuel Sprays," Los Alamos National Laboratory Report LA-10245-MS, Feb. 1985.

39. Carapanayotis, A., and Salcudean, M., "Mathematical Modeling of the Scavenging Process in a Two-Stroke Diesel Engine," *Basic Processes in Internal Combustion Engines*, ASME—Eleventh Annual Energy-Sources Technology Conference and Exhibition, New Orleans, Jan. 1988.

40. Harlow, F.H., and Amsden, A.A., "A Numerical Fluid Dynamics Method for All Flow Speeds," *J. Comput. Phys.*, vol. 8, pp. 197–213, 1971.

41. Sher, E., Hossain, I., Zhang, Q., and Winterbone, D.E., "Calculations and Measurements in the Cylinder of a Two-Stroke Uniflow-Scavenged Engine under Steady Flow Conditions," *Exp. Ther. Fluid Sci.*, vol. 4, pp. 418–431, 1991.

42. Ahmadi-Befrui, B., Gosman, A.D., Jahanbakhsh, A., and Watkins, A.P., "The DICE Computer Program Codes for Prediction of Laminar and Turbulent Flow and Heat Transfer in Idealized Motored Diesel Engine Combustion Chambers," internal report, Imperial College of Science and Technology, Department of Mechanical Engineering, Nov. 1980.

43. Uzkan, T., "Analytically Predicted Improvements in the Scavenging and Trapping Efficiency of Two-Cycle Engines," SAE Paper 880108, 1988.

44. Ahmadi-Befrui, B., Brandstatter, W., and Kratochwill, H., "Multidimensional Calculation of the Flow Process in a Loop-Scavenged Two-Stroke Engine," SAE Paper 890841, *SAE Trans.*, vol. 98, 1989.

45. Fabvre, A., and Ferreira, C., "Three Dimensional Modeling of Flow and Mixture Preparation in A Two Stroke Engine," *Proceedings of the International Symposium on Diagnostics and Modeling of Combustion in Internal Combustion Engines*, COMODIA 90, Kyoto, Japan, pp. 475–480, 1990.

46. Smyth, J.G., Kenny, R.G., and Blair, G.P., "Motored and Steady Flow Boundary Conditions Applied to the Prediction of Scavenging Flow in a Loop Scavenged Two-Stroke Cycle Engine," SAE Paper 900800, *SAE Trans.*, vol. 99, 1990.

47. Epstein, P.H., Reitz, R.D., and Foster, D.E., "Computations of a Two-Stroke Engine Cylinder and Port Scavenging Flows," SAE Paper 910672, *SAE Trans.*, vol. 100, 1991.

48. Amsden, A.A., O'Rourke, P.J., Butler, D., Meintjes, K., and Fansler, T.D., "Comparisons of Computed and Measured Three-Dimensional Velocity Fields in a Motored Two-Stroke Engine," SAE Paper 920418, *SAE Trans.*, vol. 101, 1992.

49. Kuo, T.W., and Reitz, R.D., "Three-Dimensional Computations of Combustion in Pre-Charge and Direct-Injected Two-Stroke Engines," SAE Paper 920425, *SAE Trans.*, vol. 101, 1992.

50. Ravi, M.R., and Marathe, A.G., "Effect of Port Sizes and Timings on the Scavenging Characteristics of a Uniflow Scavenged Engine," SAE Paper 920782, *SAE Trans.*, vol. 101, 1992.

51. Lai, Y.G., Przekwas, A.J., and Sun, R.L., "Three-Dimensional Computation of the Scavenging Flow Process in a Motored Two-Stroke Engine," SAE Paper 930499, 1993.

52. Huh, K.Y., Kim, K.K., Choi, C.R., Park, S.C., Moon, S., and Lee, K.Y., "Scavenging Flow Simulation of a Four-Poppet-Valved Two-Stroke Engine," SAE Paper 930500, 1993.

53. Das, S., and Dent, J.C., "A CFD Study of a 4-Valved, Fuel Injected Two-Stroke Spark Ignition Engine," SAE Paper 930070, 1993.

54. Tamamidis, P., and Assanis, D.N., "Optimization of Inlet Port Design in a Uniflow-Scavenged Engine Using a 3-D Turbulent Flow Code," SAE Paper 931181, 1993.

55. Haworth, D.C., Huebler, M.S., El Tahry, S.H., and Matthes, W.R., "Multidimensional Calculations for a Two-Stroke-Cycle Engine: A Detailed Scavenging Model Validation," SAE Paper 932712, 1993.

56. McKinley, N.R., Kenny, R.G., and Fleck, R., "CFD Prediction of a Two-Stroke, In-Cylinder Steady Flow Field: An Experimental Validation," SAE Paper 940399, 1994.

57. Leep, L.J., Strumolo, G.S., Gmaznov, V.L., Sengupta, S., Brohmer, A.M., and Meyer, J., "CFD Investigation of the Scavenging Process in a Two-Stroke Engine," SAE Paper 941929, 1994.

58. Corcione, F.E., Rotondi, R., Gentili, R., and Migliaccio, M., "Modeling the Mixture Formation in a Small Direct-Injected Two-Stroke Spark-Ignition Engine," SAE Paper 970364, 1997.

59. Yu, L., Campbell, T., and Pollock, W., "A Simulation Model for Direct-Fuel-Injection of Two-Stroke Gasoline Engines," SAE Paper 970366, 1997.

60. Fansler, T.D., and French, D.T., "The Scavenging Flow Field in a Crankcase Compression Two-Stroke Engine—A Three-Dimensional Laser-Velocimetry Survey," SAE Paper 920417, *SAE Trans.*, vol. 101, 1992.

FIVE

PORT, VALVE, INTAKE, AND EXHAUST SYSTEM DESIGN

5-1 GAS EXCHANGE, SHORT-CIRCUITING, AND BACKFLOW

The purpose of the gas exchange process is to remove the burned gases from the cylinder at the end of the power stroke and admit the fresh charge for the next cycle. The fresh charge is either air or a mixture of air and fuel (and, possibly, recycled exhaust gases for emissions control). The proportions of these components depend on the type of cycle, operating conditions, and the emission control requirements. For direct-injection gasoline and diesel two-stroke engines, the fresh charge is air alone. For carburetted or port-injected engines, it is a mixture of air and fuel vapor.

The maximum indicated power of an internal combustion engine, at a given speed, is proportional to the maximum indicated work per cycle, which is proportional to the mass of air trapped in the cylinder at the start of the compression process. Achieving good mixture preparation is also an important goal of the gas exchange process. Good mixture preparation includes both the appropriate overall mixture composition in each cylinder, and uniform distribution or the desired stratification of air, fuel, and burnt residual within the cylinder. Minimizing pumping power is also important for optimizing fuel consumption and brake power. Minimizing short-circuiting losses is especially important for good fuel consumption and low hydrocarbon emissions. Achieving these objectives depends on the design of critical engine subsystems such as the valves (if any), ports, intake and exhaust manifolds and ducting, mufflers, as well as the engine's operating conditions. This chapter addresses the relevant design issues.

We first discuss the possible port and valve scavenging arrangements. Figure 5-1 shows the port (and valve) open area diagrams of six scavenging arrangements.[1] The

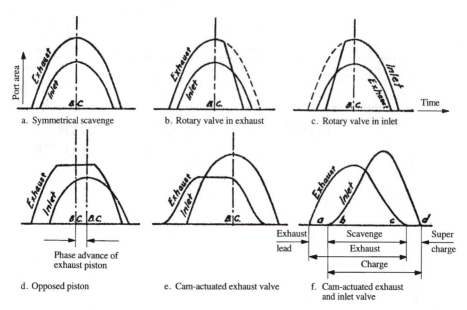

a. Symmetrical scavenge   b. Rotary valve in exhaust   c. Rotary valve in inlet

d. Opposed piston   e. Cam-actuated exhaust valve   f. Cam-actuated exhaust and inlet valve

Figure 5-1 Exhaust and scavenge port or valve open areas as a function of crank angle of various scavenging arrangements. Note that the simplest approach, with ports, is inherently symmetrical about BC.[1]

standard exhaust and scavenging port approach, because it is controlled by piston motion, is symmetrical about bottom center. The selection of the port opening times determines their closing times and port heights. It is advantageous to open the exhaust port(s) sufficiently before scavenging commences to allow the exhaust blowdown process to be essentially complete. It is also attractive to continue the scavenging flow after the exhaust closes. Meeting both these objectives requires nonsymmetric timing diagrams, and Fig. 5-1 shows various ways that such timing diagrams can be achieved. Exhaust and/or scavenging valves are required. This chapter will discuss the relevant prior experience with these arrangements, with primary focus on the simplest approach: the use of scavenge and exhaust ports.

In a simple two-stroke engine, the opening and closing times of the ports are fixed and are symmetrical about the BC crank position. The port timings are optimized at well-defined conditions of operation where excellent performance is most important. Excellent performance means high trapping, and usually scavenging, efficiency with the minimum feasible delivery ratio, without sacrificing too much of the effective stroke of the piston during compression and expansion. At off-design engine speeds and loads (or throttle positions), both the delivery ratio and charging efficiency can deteriorate significantly due to the short circuiting of fresh charge through the exhaust port, and due to backflow through the inlet and scavenge ports. These unwanted flows are a major cause of the convex shape of the typical torque versus engine speed curve, high hydrocarbon emission levels, poor fuel economy, and substantial cycle-to-cycle combustion (and, hence, torque) variability. Due to the gas exchange mechanisms used in simple ported two-stroke engines, it is impossible to prevent short-circuiting of fresh charge to the exhaust port. Under normal operating conditions, a carburetted crankcase-scavenged engine typically loses a considerable portion, of the order of 15–20%, of its fuel due to this short-circuiting of fresh charge to the exhaust. The direct loss of fuel may be reduced with charge stratification (see Section 2-3-6), and completely eliminated when operating with a direct-fuel-injection system.

Backflows occur when the pressure difference that normally drives the forward flow through the inlet ports or valves into the crankcase, and the scavenging flow through the transfer ports into the cylinder, reverses direction. At design conditions, this does not normally occur to any significant degree. However, at substantially different off-design engine speeds, backflow can become important. It is useful to distinguish between three types of backflows:

Type I—At low engine speeds, there is enough time to fill the crankcase volume each cycle with fresh charge during the upward stroke of the piston. When the piston reverses direction and the crankcase volume decreases, backflow of fresh charge through the intake system may occur before the intake valve or port closes (IC).

Type II—Also at low engine speeds, backflow may occur in the scavenge ducts. As the piston ascends, the pressure in the crankcase falls below the cylinder pressure before the scavenge ports close (SC). A backflow of mixture from the cylinder into the crankcase may then occur.

Type III—At high engine speeds, the time available for exhaust blowdown may be too short for the process to be completed before the scavenge ports open. Then, if the pressure in the cylinder as the scavenge ports open (SO) is higher than the crankcase pressure, a backflow of exhaust gases into the crankcase through the scavenge ducts may occur.

As we discuss the geometric design issues of the intake and exhaust systems relevant to optimizing the gas exchange process, we will focus especially on minimizing short-circuiting through the exhaust port, and backflows through the inlet and scavenge ports.

5-2 INLET AND TRANSFER FLOW CONTROL

5-2-1 Reed Valves

A reed valve is constructed of flexible leaf springs, of cantilever design, typically made of thin spring steel (0.1–0.3 mm thick), firmly anchored at one end and free to deflect substantially at the other end under load. The objective of installing a reed valve at the intake port is to improve low-speed engine operation by eliminating backflows of type I. The reed valve allows the intake mixture to enter the crankcase volume through the inlet port when the intake pressure exceeds the crankcase pressure by a small amount, and seals the intake port under all other conditions. Figure 5-2 shows an example of a prismatic reed valve. Figure 5-3 shows the more important parameters that define reed valve geometry. Table 5-1 summaries the critical geometric proportions of a typical reed valve and block assembly.

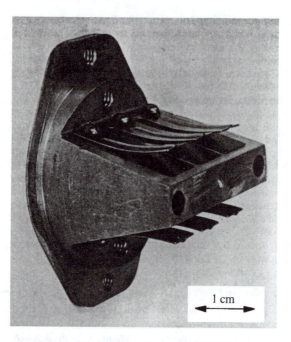

1 cm

Figure 5-2 A prismatic reed valve used in a Perkins 14-kW twin outboard two-stroke engine. Note the six individual reeds (in closed position), and the six curved guard plates with stiffening ribs on each finger which define the fully open reed position. (Courtesy Perkins Engines.)

Table 5-1 Details of typical reed valve and block assembly for a reed port length of L_r

Geometric proportions		Reed petal details	
Total reed length	$1.1–1.2\,L_r$	Reed petal thickness	0.2–0.4 mm
Clamped reed length	$0.1\,L_r$	Reed block angle	30–60°
Reed taper length	$0.2–0.4\,L_r$	Number of reeds	2–8
Reed port width	$0.6–1.0\,L_r$	Reed Young's modulus:	
Reed petal width	$0.6–1.0\,L_r$	Steel	200 GN/m²
Reed width at clamp	$0.2–0.4\,L_r$	Carbon fiber	20 GN/m²
Maximum guard-plate gap	$0.2–0.3\,L_r$	Reed petal density:	
		Steel	7800 kg/m³
		Carbon fiber	1400 kg/m³

Reed petals must be thin and flexible so that they open easily and cause little air flow restriction. At the same time, petals must be thick and stiff enough to avoid any deterioration in crankcase filling at high speeds. At high engine speeds, flexible petals tend to flutter, allowing reverse flow out of the crankcase. They also tend to close and then rebound from their seats due to inertia or resonance in the induction manifold.

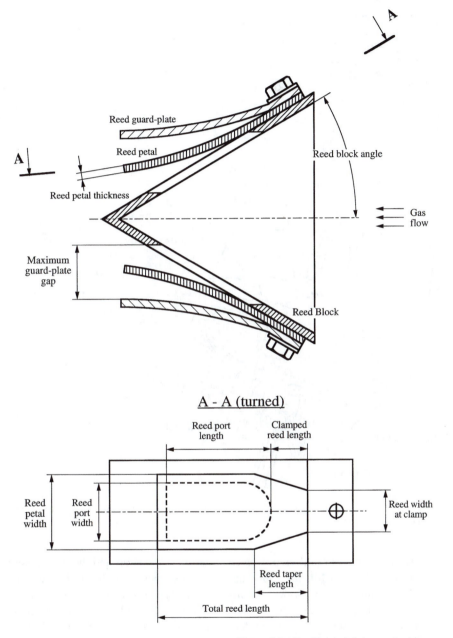

Figure 5-3 Parameters defining reed valve geometry.

Figure 2-12 shows how reed valve petal lift typically depends on engine speed and throttle position. The data show that, at low engine speeds, the reed valve successfully follows the change in the sign of the pressure difference at BC, and opens. At higher

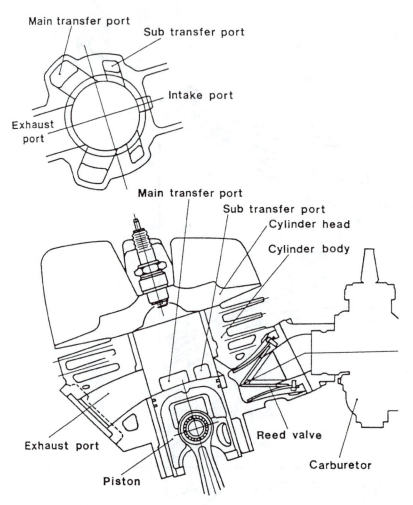

Figure 5-4   A Yamaha two-stroke crankcase-scavenged air-cooled single-cylinder engine with a reed valve installed in the intake port.[2]

engine speeds, this pressure difference reversal period is too short and the reed valve remains closed. The reed lift magnitude apparently depends strongly on the throttle position. As the throttle opens wider and/or the engine speed increases, the mass flow rate of the fresh charge increases, the pressure drop across the reed valve increases, and the reed valve petal experiences a bigger deflection. An example of a reed valve design for a small-bore engine of 199 cm^3 displacement is shown in Fig. 5-4. The engine power improvement when a reed valve is installed at the intake port is shown in Fig. 5-5. Here, the dimensions of the reed valve were selected so that the delivery ratios for both engines were equal at the highest engine speed. The intake pipe system was not tuned.

Employing a dual reed assembly may considerably extend the operating range of the reed valve. A dual reed assembly may be constructed using a thin reed riding on

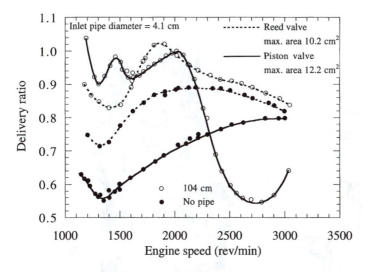

Figure 5-5 The engine power improvement that normally results when a reed valve is installed in the intake port.[3] The delivery ratio of the reed valve engine and the base piston–valve engine is shown, as a function of engine speed. Cylinder bore 80 mm; stroke 90 mm; displacement 452 cm³; crankcase compression ratio 1.49; exhaust pipe diameter 44.5 mm; inlet port open period 109.4°; scavenge port open period 109°; exhaust port open period 148.4°.

top of a thicker reed. The thin reed opens easily under low pressure drops whereas the thicker one takes over at high engine speeds.

It is interesting to note that, although it would be expected that the reed valve geometry details and reed petal material are both important for proper functioning of the reed valve, it has been found that the differences in performance among carbon fiber, glass fiber, reinforced plastics, and steel reeds are small.[4] In practice, stainless steel and carbon fiber reed petals are more popular, whereas fiberglass[5] is sometimes used in outboard engines.

Although reed valves do improve the gas exchange process over a range of operating conditions, they are noisy in operation due to the reeds striking the seats. Also, under some operating conditions, fluttering and occasionally breakage or splitting of the reeds can be serious problems.

5-2-2 Rotary Disc Valves

Another approach by which the desired asymmetrical intake opening and closing can be obtained is to install a rotary disc valve at the intake port. The rotary disc valve was originally developed by Zimmermann in Germany in 1953. A typical modern installation is shown in Fig. 5-6; it consists of a thin disc splined to and able to slide freely along the crankshaft, running in a narrow gap (0.25–0.35 mm) between the wall and the cover. Ports are cut in the cover plate and the crankcase, and the disc is cut away to open the port for the required crank angle period. As the disc valve is free to float laterally, it can take up any required position through use of an auxiliary mechanism. This is done

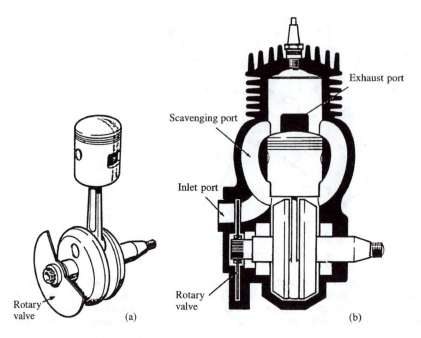

Figure 5-6   (a) A rotary disc valve; (b) its installation in a simple two-stroke engine.[6]

to prevent any backflow through the inlet port over the entire speed range of the engine. (In contrast, note the operating difficulties of the reed valve at high engine speeds as described in Section 5-2-1.)

Okanishi et al.[7] have examined the effects of the size and phasing of the disc opening on the airflow characteristics of a crankcase-scavenged two-stroke engine, and compared the results obtained with an engine using a reed valve, and a four-stroke cycle engine. They found that a volumetric efficiency† higher than 120% could be obtained over a speed range of 3000–6000 rev/min, which was not achievable with the reed valve system. Figure 5-7 shows the volumetric efficiency of the four-stroke engine, and the delivery ratio curves of the original engine installed with a reed valve at the inlet port. Also shown are the delivery ratio characteristics with 110° and 160° rotary disc cutout angles. For each cutout angle, results for three different valve timings, as indicated on the diagrams, are shown. Compared with the delivery ratio curve obtained with a reed valve, the delivery ratio curves obtained with a rotary disc valve are shifted toward higher engine speeds when the inlet valve timing is retarded, and the shift is larger with a larger rotary disc cutout angle. By controlling the cutout angle and the timing of the rotary disc valve, an improved delivery ratio can be obtained over a wider range of engine speeds.

---

†The volumetric efficiency is defined as the volume flow rate of air into the engine at atmospheric conditions, divided by the rate at which the piston displaces volume. It is most commonly used with four-stroke cycle engines.

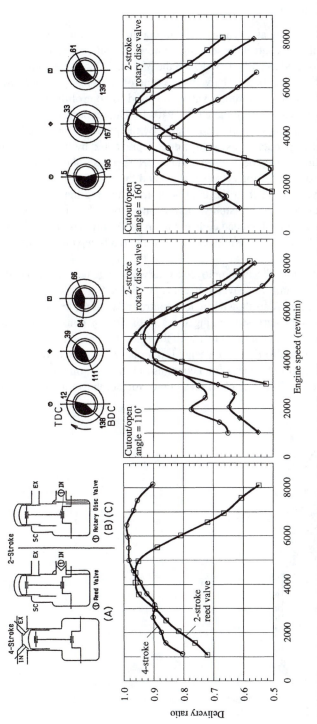

Figure 5-7 The effect of rotary disc inlet valve cutout or open angle (110° or 160°), and cutout phasing, on the delivery ratio as a function of engine speed at wide-open throttle, compared with a reed valve.[7]

5-2-3 Rotary Valves

One method of obtaining an asymmetrical timing diagram with a two-stroke engine is to use a rotary valve to control the scavenging flow. It is interesting to note that a rotary exhaust valve was used as early as 1913, on the 350-cm^3 Quadrant engine to allow earlier opening and closing times relative to the time of scavenge port opening. Another two-stroke engine in which a rotary valve was used to control the scavenge port was produced for racing cars during the early 1940s by Duesenberg (see Fig. 5-8). In this arrangement, the scavenging port is uncovered by the piston before EO, but communication between the cylinder and scavenge duct is prevented by the rotary valve, which is closed at that time. After the exhaust port has been open sufficiently long to allow the cylinder to blow down, the rotary valve opens and the scavenging process begins. The valve remains open longer than the inlet port, and charging therefore continues until the inlet port is closed by the piston some time after the exhaust port has been closed. The Duesenberg design, which had eight cylinders of 79.6-mm bore by 69.8-mm stroke, developed 85.8 kW at 4800 rev/min, corresponding to a bmep (at maximum power) of 386 kPa.

Blundell and Sanford[9] have employed a rotary scavenge port situated in the cylinder head to implement a uniflow scavenging concept. Figure 5-9 shows the rotary scavenging valve construction. Also shown is a flow control valve at the exhaust port (discussed in Section 5-4). The rotary scavenging valve is of tubular construction containing a cylindrical stator, which avoids contact with the rotor through a minimal running clearance. Blundell and Sandford's stator design is internally ported to allow two scavenging streams: the first is air only; the second contains rich fuel–air mixture. The rotary valve is positioned in the cylinder head, to one side of the combustion chamber, allowing a

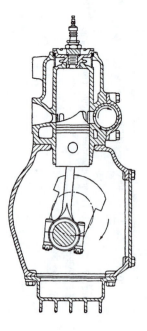

Figure 5-8 Duesenberg engine with rotary valve at scavenge port.[8]

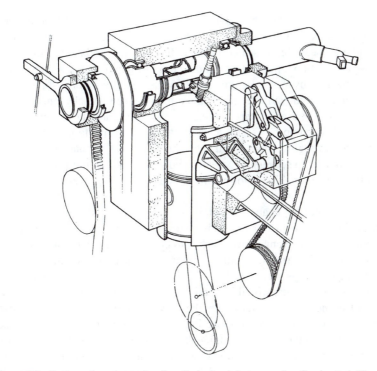

Figure 5-9 Rotary valve construction in cylinder head for scavenging flow control. The scavenge valve, with its two internal streams, enables stratified-charge engine operation.(Note the exhaust flow control valve, also.)[9]

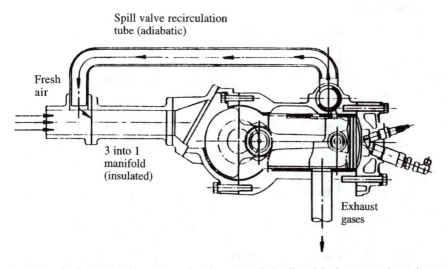

Figure 5-10 Schematic of the rotary spill valve three-cylinder direct-injection two-stroke engine concept developed and tested by Adams et al.[10]

near central spark-plug location. The valve is supported on sealed ball bearings and is driven at engine crankshaft speed by a belt. A radial clearance exists between the valve and the cylinder head that is sealed in the radial direction (relative to the rotary valve axis) using conventional piston rings at the edges of the port, and sealed axially along the sides of the port by Wankel-engine-type apex seals. The stator is supported in bearing caps that attach to the cylinder head, and a minimal running clearance ensures no contact between the stator and rotary valve. Labyrinth seals are placed between the stator and the rotor, backed up by O-ring seals in the bearing caps, to prevent leakage of charge air when scavenging is not occurring. The stator can be rotated to vary the relationship between the two scavenging streams and/or to vary the timing of the rotary valve open or closed position. The stator position is adjusted by a cable-operated servo motor.

In another interesting application, a rotary valve system was employed to spill charge from the cylinder back into the intake duct during the compression stroke. Adams et al.[10] found that the loop-scavenged two-stroke direct-injection engine can achieve adequate stability at idle with stratified combustion at a very lean overall air/fuel ratios, but the exhaust temperature is then very low. To obtain stable engine operation with a low engine delivery ratio, which would increase the exhaust temperature, they opened an additional port in the cylinder liner above the scavenge port and installed a rotary valve to control the flow through the port (see Fig. 5-10). The poor scavenging characteristics of their experimental prototype engine resulted in a higher than desired burnt residual gas content in the spill recirculation flow, which limited the minimum delivery ratio that could be attained. Adams et al. concluded, however, that the concept may be feasible if a high-purity spill recirculation flow can be achieved (i.e., the spill flow is largely air), in conjunction with a high trapping efficiency.

5-2-4 Fluid Diodes

In two-stroke engines with piston-controlled ports, the more desirable asymmetric port opening and closing angles can also be obtained by installing a fluid diode at the port. The fluid diode, an element with no moving parts, allows fluid to pass through it in the forward direction with a relatively low pressure loss, but provides a much higher resistance to flow in the reverse direction (see Fig. 5-11). The higher resistance results from the specific geometry of the device, which permits the development of vortices within the flow. Many types of fluid diodes have been developed and reported in the literature. These include vortex, scroll, tesla, nozzle, turbulence, and cascade diodes. Nagao et al.[12] suggested that a fluid diode can be used to prevent the type I backflow (see Section 5-1) through the inlet port in a crankcase-scavenged two-stroke engine. These authors examined the influence of vortex and scroll diodes experimentally in a motored engine, and found a substantial effect on the delivery ratio at low engine speeds. They also found that the scroll diode is the more effective. Sher,[11] in experiments with a fired engine, showed that engine torque is significantly improved at low engine speeds when a scroll diode is placed at the inlet port. Sher and Yafe[13] have shown that the backflow through the scavenge port (backflow types II and III) in a small crankcase-scavenged engine can be suppressed to a remarkable degree by installing a fluid diode

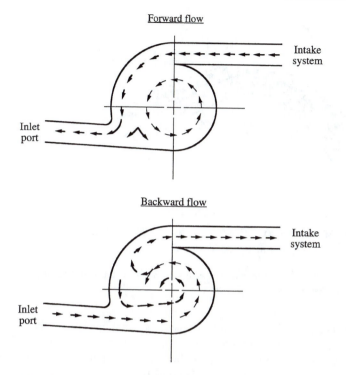

Figure 5-11 The operating principles of the scroll fluid diode.[11] Reproduced by permission of the Council of the Institution of Mechanical Engineers.

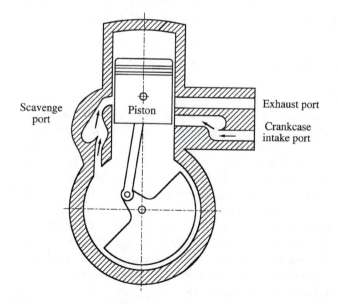

Figure 5-12 A crankcase-scavenged two-stroke engine with fluid diodes installed at both inlet and scavenge ports.[13]

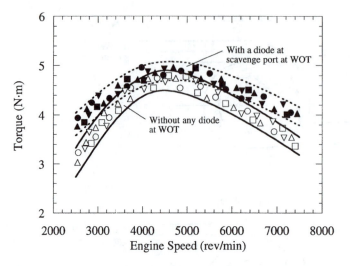

Figure 5-13 Wide-open-throttle engine torque vs. speed with and without a fluid diode installed at the scavenge port.[13] Cylinder displaced volume is 111 cm^3.

at the scavenge port (Fig. 5-12). The engine torque in both configurations, that is, with and without diodes at the scavenge ports, are shown in Fig. 5-13. The torque-versus-speed curve is flatter; the maximum torque is slightly higher than the original value, and is increased by some 20% at low and high engine speeds.

The performance characteristics of these port control devices are compared in Table 5-2. The advantages of the fluid diode are its simplicity and ease of installation. It is very effective at suppressing backflow, although it cannot eliminate it.

5-3 SCAVENGE AND EXHAUST PORT DESIGN

5-3-1 Port Layout

The most commonly used two-stroke engine configuration is fixed scavenge and exhaust ports in the cylinder liner that are opened and closed by the moving piston [arrangement (a) in Fig. 5-1]. The important characteristics of the gas exchange process are the removal of the burned gas from the cylinder at the end of the power stroke, the effectiveness of the charging processes, and the composition of the in-cylinder charge and its spatial distribution. These depend to a large extent on the geometry and layout of the ports, as well as engine operating conditions. A cylinder port system is characterized by the crank angle at which each port opens, the size, geometry, and number of ports, their location around the cylinder liner circumference, and the geometrical shape of the gas passage, which determines the direction and flow pattern of the jet issuing from the port into the cylinder with scavenging ports or out of the cylinder with the exhaust port. Scavenge ports are usually inclined in the axial or in the transverse plane or in both. This is done to direct the flow into the cylinder so as to achieve the desired in-cylinder scavenging process. In the uniflow scavenging system, the scavenge ports are inclined

Table 5-2 Comparison of performance characteristics of port control devices

Performance characteristic	Ideal control	Piston control	Reed valve	Disc or rotary valve	Fluid diode
Timing	Variable	Fixed	Varies	Varied with an auxiliary mechanism	Varies
Period of opening	Variable	Fixed	Varies	Varied with an auxiliary mechanism	Varies
Resistance to back-flow	High	Low	High	High	Medium
Resistance to forward flow	None	Low	Low	Low	Low
Time response	Fast	Fast	Slow	Fast	Fast
Moving parts	No	No	Yes	Yes	No
Noise	No	No	Moderate	No	No
Simple to install at inlet port	Yes	Yes	Yes	No	Yes
Simple to install at scavenge port	Yes	Yes	No	No	Yes

15–60° relative to the radial direction to make the incoming charge swirl around the cylinder axis. Ports are usually rectangular, with rounding of the corners. This shape provides the largest available flow for a given bridge width. The common port shapes are shown in Fig. 5-14.

The parameters that define the geometry of a single scavenge or exhaust port are shown in Fig. 5-15. To provide context, Fig. 5-16 shows two examples of port layout, with dimensions, for small-bore spark-ignition engines. In the Yamaha design, a pair of auxiliary scavenging ports alongside the main ports have been opened to direct the mixture flow toward the rear of the cylinder, and upward. In the Suzuki design, two small-area ports have been added in the rear cylinder wall for the same reason.

Wide ports are cheaper to produce and tend to clog less. Ports that are too wide may, especially with rectangular ports, lead to breakage of the piston rings. In practice, rectangular ports are limited to 15° circumferential width for freely rotating piston rings and to 30° width if the rings are pinned. Both rectangular and rhomboidal ports can cause ring breakage due to impingement of the ring on the port edge, especially when an end of a floating ring is in the port. To prevent such a ring breakage, the upper and the lower edges of the port are frequently rounded. Alternatively, the port ends may be rounded or made with a taper relief (see Fig. 5-14). Also, the radial pressure of the ring is commonly shifted away from the ring ends to form an "apple shaped" pressure pattern (see Section 8-6-2).

In Chapter 3, we reviewed how various experimental methods have been used to quantify aspects of the gas exchange process. Here we summarize the most important methods and the data these produce that are relevant to port design. It is sometimes helpful to examine a single scavenging process rather than a continuously operating engine.

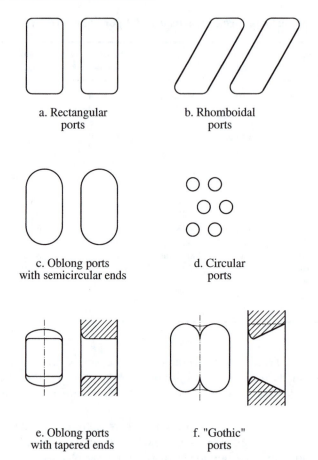

Figure 5-14 Common scavenge and exhaust port shapes[1]: (a) rectangular; (b) rhomboidal; (c) oblong, with semicircular ends; (d) circular; (e) oblong, with tapered ends; (f) "Gothic".

Flow tests on models in which the cylinder is subjected to a steady air motion through its ports while the piston is locked at bottom center are also used. Jante[16] showed that effective scavenging port designs can be developed by testing the engine under steady-state conditions with the cylinder head removed. The fresh charge is introduced steadily into the cylinder and the velocity profile of the flow just above the open cylinder is measured. To minimize short-circuiting, the upward flow from the scavenge ports toward the cylinder head should be concentrated on the wall opposite the exhaust port: "a stable closed rising current is obtained on the wall opposite to the exhaust ports." The zero-velocity interface between upward and downward streams should be located about halfway across the cylinder or slightly farther from the exhaust side. The area of the interface available for turbulent mixing between the two streams is minimized if the zero-velocity contour is a straight line perpendicular to the plan of geometrical symmetry.[16] Figure 3-1 shows results obtained by Jante for four different geometry designs. Figure 5-17 shows examples of two scavenging flows that this measurement technique can

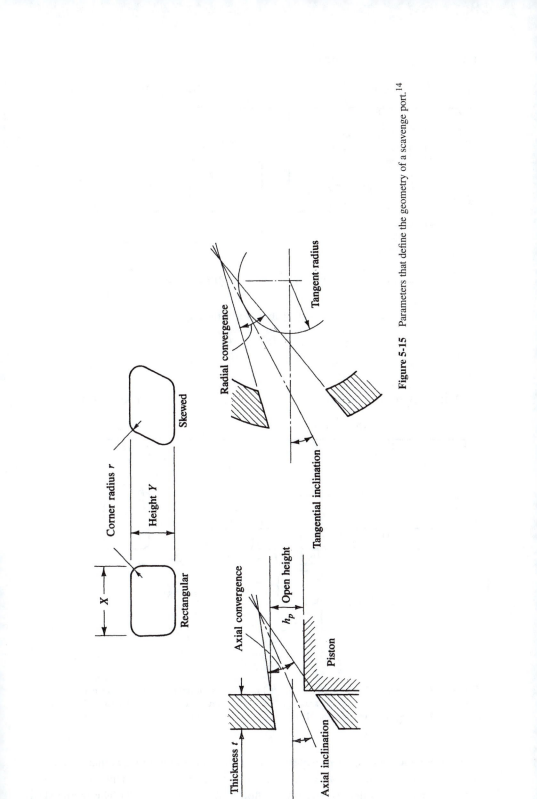

Figure 5-15 Parameters that define the geometry of a scavenge port.[14]

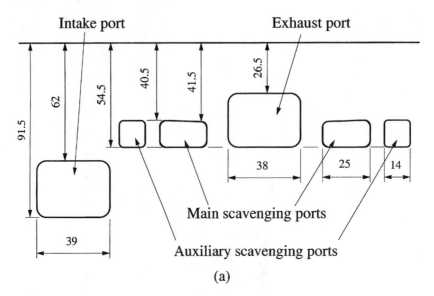

(a)

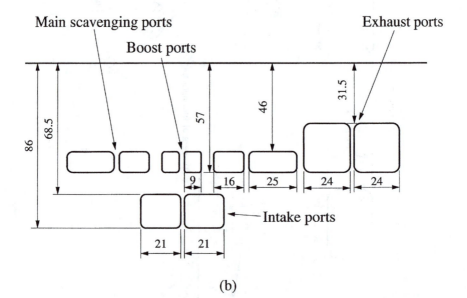

(b)

Figure 5-16 Examples of port system layout and dimensions (in millimeters) for two small-bore spark-ignition engines, illustrating exhaust, scavenge, and intake port arrangement: (a) Yamaha TZ250 engine; (b) Suzuki PE175 engine. Adapted from Two Stroke Performance Tuning by A. Graham Bell.[15]

identify: tongues of fresh mixture penetrating across the middle of the cylinder, or along the walls, circumferentially. The Jante method has been extended to include the effect of piston motion by introducing pulsating air flow (see Figs. 3-2 and 3-3). This method has

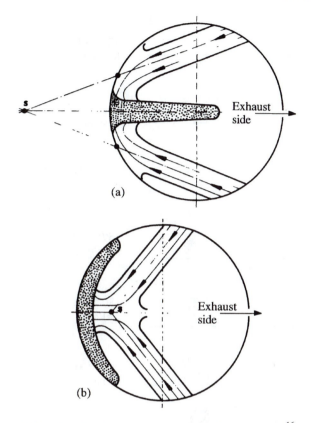

(a)

(b)

Figure 5-17   Schematics of two common scavenging flow deficiencies[16]: (a) tongue of fresh charge penetrating across the middle of the cylinder; (b) tongues penetrating, circumferentially, along the walls.

been further extended by considering the profiles of the three velocity components,[17,18] and the turbulence parameters,[19] at the open end of the cylinder.

Laser doppler velocimetry (LDV), the phase doppler technique, and laser-induced fluorescence (LIF) are examples of the modern optical diagnostics now available for more detailed improvement of the port system[20-27] (see Section 3-3). Figure 5-18 shows how the flow pattern through a piston-controlled scavenge port depends on the piston motion and the port opening fraction. These LDV gas velocity measurements were taken in a motored two-stroke cycle engine. For ports that are more than half open, a plug-like velocity profile, with flow direction varying with piston position, can be identified. Similar behavior was found for a wide range of delivery ratios, with velocity components scaling approximately with delivery ratio. During port opening, the velocity vectors close to the piston crown are inclined relatively steeply as the scavenging flow establishes itself as the port is uncovered. As BC is approached, the velocity vectors still indicate an upward though lower inclination of the flow, and smaller magnitude velocities in the lower section of the port due perhaps to the scavenge-duct flow boundary layer.

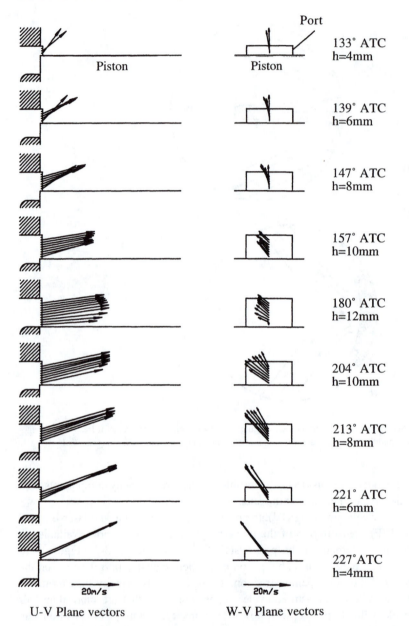

U-V Plane vectors   W-V Plane vectors

Figure 5-18 Flow pattern exiting a scavenge port into the cylinder, as a function of crank angle, as the piston uncovers and then covers the port[29]: engine speed = 600 rev/min and delivery ratio = 1.0.

Figure 5-19 shows how the mean rotation of the emerging jet (as seen from the top) and its elevation angle vary with crank angle and engine speed.[29] The data show that there is little difference between the rotation angle at 225 and 600 rev/min. The variation

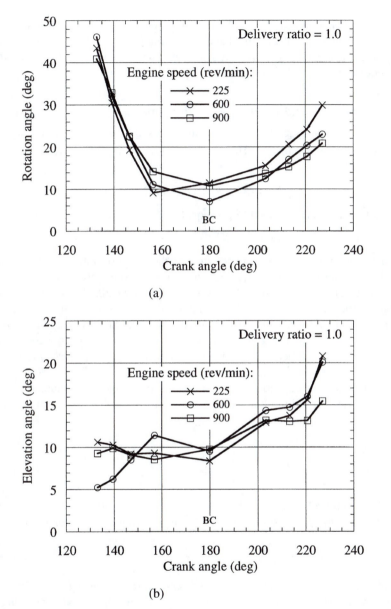

Figure 5-19 Mean scavenging jet efflux angle as a function of crank angle and engine speed,[29] for delivery ratio = 1.0: (a) rotation efflux angle; (b) elevation efflux angle.

of the elevation angle also shows little effect of engine speed over the range examined, and this angle does not change much, especially at the highest speed tested. The general trend is a flow efflux angle that decreases rapidly with port opening, and increases less rapidly with port closing. The higher mean efflux angles during port opening partly result from the steeply inclined vectors close to the piston crown, and from the bulk port flow itself.[29] Ekenberg and Johansson[30] have demonstrated how the LDV technique can be employed to investigate the effect of the scavenging port geometry on the in-cylinder flow field. For this purpose, they used a fired loop-scavenged engine.

5-3-2 Port Geometry and Discharge Coefficients

We now examine how the mass flow rate through a scavenge or exhaust port depends on port geometry and flow parameters. The open area of the partly open port in Fig. 5-15 is given by

$$A_P = Xh_P - r^2(4-\pi)/2 \tag{5-1}$$

When the port is fully open, its height h_P is equal to Y and the full port open area is

$$A_{P,\max} = XY - r^2(4-\pi) \tag{5-2}$$

These are the appropriate areas to use in determining the mass flow rate through the port. The mass flow rate through any flow restriction such as a port or valve, whether into or out of the cylinder, depends on how effectively the actual flow fills the geometric open area available. The equation for calculating the mass flow rate of a compressible gas through a restriction is derived from a one-dimensional isentropic flow analysis, and real flow effects are included by using an experimentally determined discharge coefficient C_D. The product $C_D A_R$, where A_R is an appropriate geometric area characteristic of the port (or valve) open area, is often termed the effective flow area. The gas flow rate $\dot{m}$ is related to the upstream stagnation pressure p_0 and stagnation temperature T_0, static pressure just downstream of the flow restriction (assumed equal to the pressure at the restriction, p_T), and the reference flow area A_R by

$$\dot{m} = \frac{C_D A_R p_0}{(RT_0)^{1/2}}\left(\frac{p_T}{p_0}\right)^{1/\gamma}\left\{\frac{2\gamma}{\gamma-1}\left[1-\left(\frac{p_T}{p_0}\right)^{(\gamma-1)/\gamma}\right]\right\}^{1/2} \tag{5-3}$$

When the flow is choked, that is, the flow velocity at the minimum flow area is sonic, which occurs when $p_T/p_0 \le [2/(\gamma+1)]^{\gamma/(\gamma-1)}$, the appropriate equation for $\dot{m}$ is

$$\dot{m} = \frac{C_D A_R p_0}{(RT_0)^{1/2}}\gamma^{1/2}\left(\frac{2}{\gamma+1}\right)^{(\gamma+1)/2(\gamma-1)} \tag{5-4}$$

For flow into the cylinder through a scavenging port, p_0 is the scavenge flow pressure p_{sc}, and p_T is the cylinder pressure. For flow out of the cylinder through an exhaust port, p_0 is the cylinder pressure and p_T is the exhaust system pressure.

The value of C_D and the choice of reference area are linked together: their product $C_D A_R$ is the effective flow area of the valve and port assembly A_E. The discharge coefficient of a port is, in effect, the ratio between the actual mass flow rate through the port and the theoretical mass flow rate assuming an isentropic expansion. Figure 5-20 shows how, for a simple radial port, the discharge coefficient depends on the pressure ratio across the port and the port open fraction. The discharge coefficient varies significantly with pressure ratio. The effect of the port open fraction depends on the port geometric details. The effects of scavenge port open fraction and port geometry on coefficient of discharge for square and circular ports, with sharp and rounded entries, are shown in Fig. 5-21. Geometry effects are most significant at the larger open fractions.

Tangentially inclined inlet ports are used when a swirling flow is desired within the cylinder to improve scavenging, or when jet focusing or impingement within the cylinder off the cylinder axis is required. The discharge coefficient decreases as the jet tangential inclination increases (see Fig. 5-22). The jet angle and the port angle can deviate significantly from each other depending on the details of the port design and the open fraction.[16]

Discharge coefficients for an exhaust port have been measured[33] as a function of the open fraction of the port, the pressure ratio across the port, the temperature of the incoming charge (an important parameter when a burned mixture is considered), and the port geometry. Although the unsteady flow discharge coefficient is usually smaller than the steady flow coefficient, and the difference between these increases for small pressure differences across the port and high engine speeds,[34] steady flow model studies have been found to yield results close to the real situation.[35] Benson[33] measured the discharge coefficient of a piston-controlled single rectangular port under steady flow conditions. Figures 5-23 and 5-24 show some of his results.

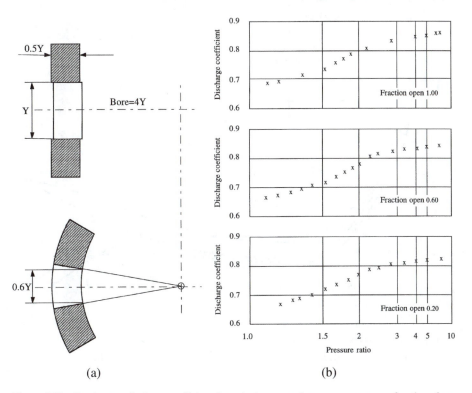

(a)   (b)

Figure 5-20   Steady-state discharge coefficient for a single rectangular scavenge port as a function of pressure ratio and open fraction.[14]

These data show that in piston-controlled exhaust ports, the pressure ratio has a significant effect on the exhaust port discharge coefficient. This is an important conclusion because the pressure ratio across the exhaust ports varies substantially during the exhaust blowdown process. However, the effect of gas temperature is modest, except when the gas velocity is low.

The angle of the issuing exhaust jet is another parameter which may affect the exhaust duct design near the exhaust port opening. While in thick exhaust ports the walls are usually tapered to allow the outward flow to diffuse, in thin-walled ports Jante found that the angle is practically invariant to the open fraction except for the short period when the open fraction is lower than 15%.[16] Typical values for a flat piston crown and sharp-edged thin cylinder wall, are shown in Fig. 5-25.[14] The changes in exit jet angle largely explain the effects of increasing open fraction and pressure ratio. The

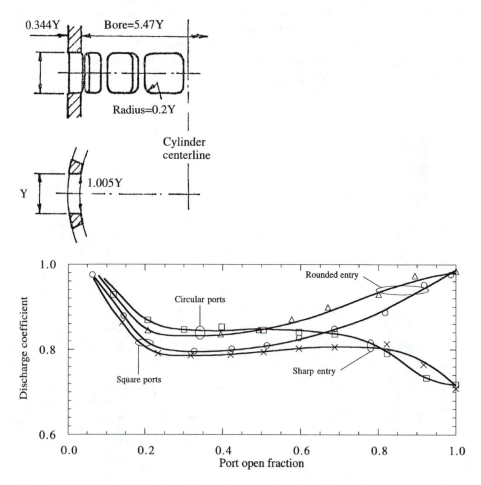

Figure 5-21 Steady-state discharge coefficient of square and circular radial scavenge ports as a function of port open fraction.[14,31,32] Reproduced with permission of The McGraw-Hill Companies.

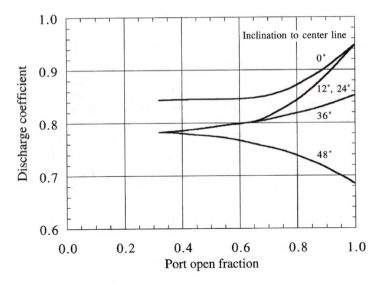

Figure 5-22 Steady-state discharge coefficient of tangentially inclined scavenge ports as a function of port open fraction[14]: pressure ratio = 1.306, air/fuel ratio = 13.

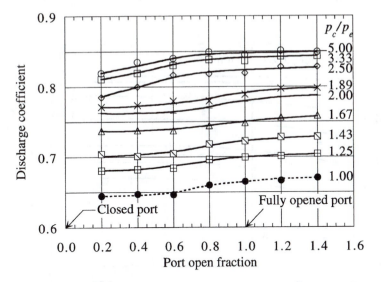

Figure 5-23 Steady-flow discharge coefficient of a single rectangular exhaust port (7.6 mm wide × 12.7 mm high × 12.7 mm thick) in the wall of a 51-mm-bore cylinder as a function of open fraction and pressure ratio.[33]

suggested geometry for a thick-walled port, based on this information, is also shown. Many two-stroke port scavenging systems now use a valve in the exhaust port to control the effective exhaust flow area as a function of engine speed and load. This additional exhaust flow control is discussed in Section 5-5-4.

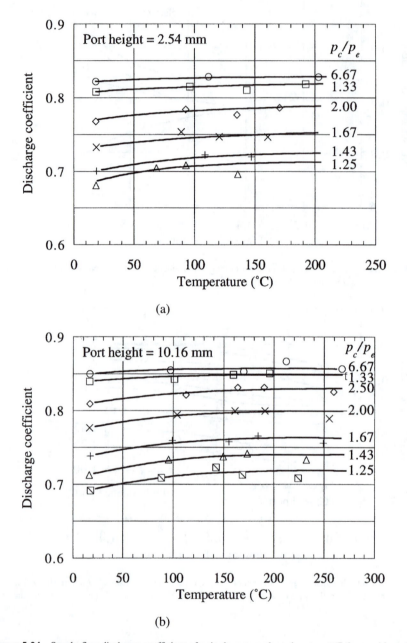

Figure 5-24 Steady-flow discharge coefficient of a single rectangular exhaust port (7.6 mm wide × 12.7 mm high × 12.7 mm thick) in the wall of a 51-mm-bore cylinder as a function of gas temperature and pressure ratio, at two different open heights: (a) 2.54 mm; (b) 10.16 mm.[33]

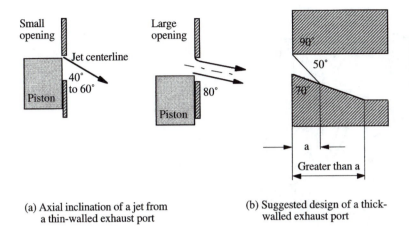

(a) Axial inclination of a jet from
a thin-walled exhaust port

(b) Suggested design of a thick-
walled exhaust port

Figure 5-25 Inclination of jet exiting from an exhaust port: (a) elevation angle of the jet from a thin-walled exhaust port; (b) suggested design for a thick-walled exhaust port.[14]

5-3-3 Piston Deflectors

With simple cross-scavenged two-stroke engines, a deflector on the piston crown is often used to force the scavenging flow toward the cylinder head and away from the exhaust port. Figure 5-26 shows the typical shape of such a deflector.† This type of piston, with cross-scavenging, has good scavenging characteristics when the engine is throttled at low power, and at low engine speed. At higher throttle openings the scavenging efficiency is not especially good. Also, the impact of the deflector on combustion chamber shape, and the engine's knock and surface ignition limits is detrimental. Usually these engines use a lower compression ratio than loop-scavenged engines. However, the cross-scavenged port design simplifies the block casting and machining significantly, providing manufacturing and packaging advantages. Also, narrow ports can be used, which permits the piston rings to operate without pinning, which evens out and reduces ring wear.[37]

Figure 5-27 shows the major geometric features of a piston deflector. Blair[37] recommends the following geometric proportions. The deflector open or flow area A_d, the area between the liner and vertical deflector surface, should be slightly larger than the total scavenging port open area [given by $n_P A_P$, where A_P is given by Eq. (5-2)]. Thus the entering flow will be turned smoothly through 90° and flow axially up the cylinder. A deflection ratio can be defined:

$$R_d = A_d/(n_P A_P) \tag{5-5}$$

which should be between 1.1 and 1.2. The deflector height is the other critical parameter. It should be higher than the port by about 50%, but no higher than necessary. The deflector wall (or ears) is placed about midway between the exhaust and scavenge ports,

---

†Note that piston crown shaping is sometimes done for other reasons: for example, to control the development of the fuel–air spray in direct-injection engines.

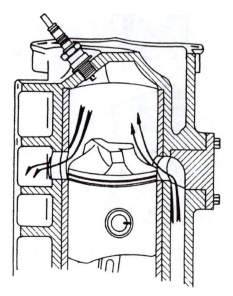

Figure 5-26 Typical shape and operation of a piston flow deflector in a cross-scavenged two-stroke engine.[36]

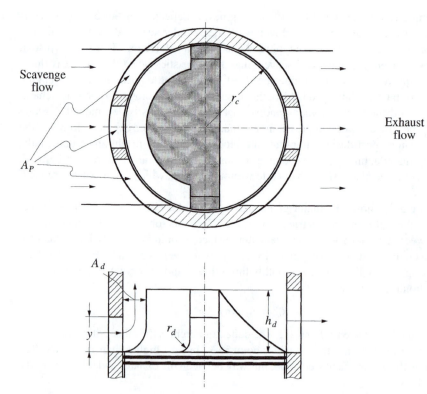

Figure 5-27 Geometric features of port and piston deflector in cross-scavenged engine. Adapted from Blair.[37]

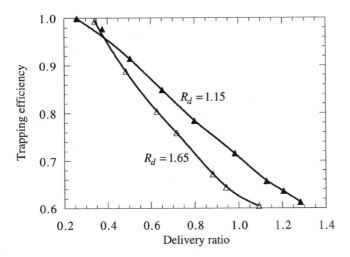

Figure 5-28 Effect of piston deflector deflection ratio (R_d) on the trapping efficiency versus delivery ratio relationship. Adapted from Blair.[37]

or slightly closer to the scavenge ports. Its sides are usually sloped at about 45°, or curved, so it will fit into the combustion chamber at TC with adequate clearance. The radius of the deflector as it joins the piston is usually about one-third the height.[37]

Figure 5-28 shows the effect of deflection ratio [R_d in Eq. (5-5)] on the relationship between trapping efficiency and delivery ratio. Smaller deflector flow areas ($R_d = 1.1$–1.2) deflect the flow better, reduce short-circuiting, and hence improve charge trapping.

5-4 SCAVENGE AND EXHAUST VALVES

5-4-1 Poppet Valves

Poppet valves are widely used in large-bore diesel engines and also in some recent designs of medium-sized gasoline two-stroke engines. In large-bore engines, a uniflow scavenging system is commonly employed. Scavenging is usually performed from the bottom of the cylinder to the top, where the scavenging ports are controlled by the piston and the exhaust flow by one or more poppet valves (e.g., Fig. 1-7). When poppet valves have been used in medium-sized two-stroke engines instead of exhaust and scavenging ports, it is to retain the standard wet crankcase and oil sump systems and the valve train of conventional four-stroke cycle engines (see Fig. 1-26). These valved two-stroke engines then use an external compressor (often a roots blower) to compress the scavenging air.

Poppet valves have also been used in small- and medium-sized two-stroke gasoline engines to implement a double charging system as an alternative to direct in-cylinder fuel injection. The IAPAC concept shown in Fig. 2-25 is an example of this approach.[38] The main scavenging process is performed with pure air (or sometimes with lean fuel–air mixture) through the cylinder scavenging and exhaust ports, whereas the final stage in the cylinder charging process is performed with a rich mixture via a poppet valve situated in the cylinder head (see Section 2-3-6 for a discussion of charge stratification methods). In the IAPAC engine, reed valves are installed at both the inlet and secondary (for the air–fuel mixture) scavenge ports, and an exhaust port flow control device is also employed.

In large two-stroke diesels employing uniflow scavenging, a single poppet valve in the cylinder head is used to control the burnt gas flow into the exhaust. Because this valve is subjected to very high thermal and mechanical loads, its design is especially challenging.[39] The low-speed high-load operating conditions of these turbocharged long-stroke engines creates this demanding environment. Figure 5-29 shows the combustion chamber and exhaust valve assembly for the larger-bore RTA Sulzer two-stroke diesel engines. The valve is mounted in a cage with a cooled valve seat. An impeller on the valve stem ensures continuous rotation of the valve. Each valve is actuated by hydraulic pressure provided by hydraulic pumps driven by the engine camshaft.

A combination of efficient valve cooling and appropriate choice of valve alloy provide the required high-temperature valve strength and corrosion resistance. Nimonic 80A alloy is used; this has sufficient hardness so that coating its seating face is not required. The valve seating surfaces are maintained well below 450°C through water cooling of the valve seat region. The valve, stem guide, and seat have full rotational symmetry to ensure even thermal and mechanical loading and to avoid distortion. Valve return is effected by an air spring, rather than by the helical steel spring used in smaller valved engines. This air spring contributes to smooth dynamic behavior of the whole valve system, and permits sufficient valve lift for minimum exhaust flow restriction. The air spring comprises the piston attached to the valve stem near its upper end, sliding in the fixed cylinder incorporated into the valve cage structure (see Fig. 5-29). The air supplied to the air spring cylinder incorporates a small oil feed. This lubricates the air spring piston, and bleeds air down the valve stem guide to prevent exhaust gases blowing back into the stem guide. These design features result in excellent performance, reliability, and life.[39]

As mentioned above, four-stroke engine poppet valve technology has been developed for use with automotive two-stroke cycle engines. The advantages of valved two-stroke designs are (a) use of standard four-stroke lubrication systems that avoid the problems of substantial oil consumption and white smoke, and piston ring and crankshaft bearing durability, (b) reduced bore distortion and piston temperature through use of a water jacket around the full cylinder and oil cooling of the underside of the piston, (c) asymmetric intake and exhaust valve timing, and (d) potential for higher boosting of the scavenging air with an external blower. The disadvantages are (a) the valve motion limits the maximum engine speed more than does port behavior, (b) the valve train adds extra friction, (c) the engine weight and bulk size for a given maximum power are greater, and (d) the cost is higher, due to the valve train and external blower.[40]

The analysis of flow through poppet valves (and their ports) is well developed[14,32] as is their dynamic behavior. Most of this work relates to four-stroke engine applications, and there are some important two-stroke engine differences. The frequency of opening and closing is twice as fast in the two-stroke application for a given engine speed. Also, the open duration (~120° in the two stroke) is about half that in the four

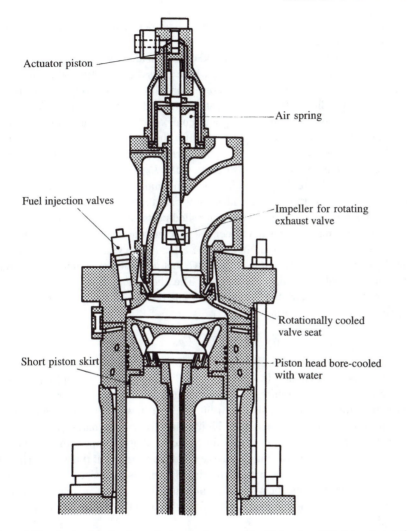

Actuator piston

Air spring

Fuel injection valves

Impeller for rotating exhaust valve

Rotationally cooled valve seat

Short piston skirt

Piston head bore-cooled with water

Figure 5-29 Combustion chamber and exhaust valve assembly of larger-bore RTA Sulzer two-stroke diesel engines.[39] (Courtesy Wärtsilä NSD Switzerland Ltd.)

stroke ($\sim 240°$). With poppet valves, it is difficult to realize the same open area readily achieved with piston-controlled ports because the valve and port diameters are more constrained by the cylinder bore than are liner port areas by the liner circumference. Note that in four-stroke applications, the exhaust and intake valves do not open at the same time (the valve lift during the valve overlap period is small). In two-stroke applications, however, both exhaust and scavenge or cylinder intake valves open almost simultaneously. With inclined valve stems, which are used to increase the valve head diameters that fit into a given cylinder bore, physical interference between the valves usually limits the maximum valve lifts that can be used. For these reasons, valved two-stroke engines usually employ four or even five valves per cylinder to achieve the maximum feasible geometric flow area.

The geometric details of poppet valve and port open areas are given in Heywood.[32] For low valve lifts (typically $L_v/D_v \leq 0.12$ where L_v is the valve lift and D_v the valve head diameter), the minimum flow area corresponds to a frustum of a right circular cone, which is perpendicular to the valve seat. For larger lifts, the minimum flow area is still the slant surface of a frustum of a right circular cone, but this surface is no longer perpendicular to the valve seat. At high enough lifts, the minimum flow area is no longer between the valve head and the seat, it is the port area (minus the valve stem cross-sectional area, a small correction). This occurs when $L_v/D_v \geq [1 - (D_P/D_s)^2]/4 \approx 0.25$, where D_P and D_s are the port and valve stem diameters, respectively.

Equations (5-3) and (5-4) for the mass flow rate through a flow restriction, used in Section 5-3-2 for flow through piston controlled ports, also apply to poppet valves. Several different geometric areas have been used as the reference area that, when multiplied by the discharge coefficient, defines the effective flow area. These include the valve head area $\pi D_v^2/4$, the port area at the valve seat $\pi D_p^2/4$, the geometric minimum flow area, and the valve curtain area $\pi D_v L_v$. The choice is arbitrary, although some of these choices allow easier interpretation than others. The most convenient reference area in practice is the valve curtain area

$$A_C = \pi D_v L_v \tag{5-6}$$

because it varies linearly with valve lift and is simple to determine.

For flows into the cylinder through poppet valves, the discharge coefficient based on valve curtain area is a discontinuous function of the valve lift/diameter ratio, because different flow regimes occur. As shown in Fig. 5-30, at low lifts the flow remains attached to the valve head and seat, giving high values for the discharge coefficient. At intermediate lifts the flow separates from the valve head at the inner edge of the seat as shown, and an abrupt decrease in discharge coefficient occurs at this point. The discharge coefficient then increases with increasing lift because the size of the separated region remains roughly constant while the minimum flow area is increasing. At high lifts, the flow separates from the inner edge of the valve seat as well, and the discharge coefficient decreases.[14,32]

Discharge coefficients for port and valve assemblies are normally measured with steady air flows, whereas the engine flow process is periodic and time varying. The change points between flow regimes in Fig. 5-30 shift slightly under dynamic operation, and the pressure upstream of the valve varies during the scavenging process. However, over the normal engine speed range, steady flow discharge-coefficient results can be used to predict dynamic performance with reasonable precision.[14]

In addition to valve lift, the performance of a poppet valve assembly is influenced by the following: valve seat width, valve seat angle, rounding of the seat corners, port design, cylinder head shape. Also, the port and valve assembly may be used to generate a rotational motion (*swirl*—rotation about the cylinder axis—or *tumble*—rotation about an axis perpendicular to the cylinder axis) inside the engine cylinder during the process; or the cylinder head can be shaped to restrict the flow through one side of the valve open area to generate such rotations; or the valve can be shrouded to achieve similar results.

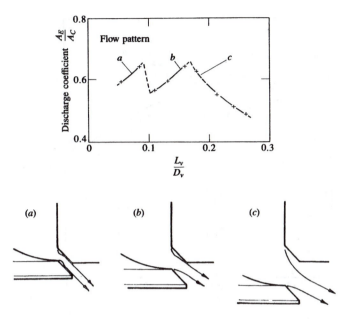

Figure 5-30 Discharge coefficient [(effective flow area)/(valve curtain area)] of typical poppet valve as a function of valve lift. Different segments correspond to flow regimes indicated.[14,32] Reproduced with permission of The McGraw-Hill Companies.

Generating swirl or tumble significantly reduces the valve (and port) flow coefficient because the open area all around the valve circumference is not then fully used.

For well-designed poppet valve ports, the discharge coefficient of the port and valve assembly need be no lower than that of the isolated valve (except when the port is used to generate swirl or tumble). However, if the cross-sectional area of the port is not sufficient, or the radius of the port surface at the inside of the bend is too small, a significant reduction in C_D for the assembly can result.[14] At high engine speeds, unless the valve is of sufficient size, the flow during part of the scavenging process can become choked (i.e., the gas velocity reaches sonic velocity at the minimum valve flow area). Choking substantially reduces the delivery ratio or normalized air flow.

In flows from the cylinder through an exhaust poppet valve, different flow regimes at low and high lift occur, also. At low lifts the flow remains attached to the valve and seat, and separates from the downstream edge of these surfaces. At high lifts the flow separates from the upstream edges of the valve and seat surfaces. Values of C_D based on the valve curtain area, for several different exhaust valve and port combinations, are shown in Fig. 5-31. A sharp-cornered isolated poppet valve, with straight pipe downstream, gives the best performance. At higher lifts, $L_v/D_v \geq 0.2$, the breakaway of the flow reduces the discharge coefficient. The port design significantly affects C_D at higher valve lifts. Good designs can approach the performance of isolated valves, however. Exhaust valves operate over a wide range of pressure ratios (1–5). For pressure ratios greater than about 2 the flow will be choked, but the effect of pressure ratio on discharge coefficient is small and confined to higher lifts (e.g., $\pm5\%$ at $L_v/D_v = 0.3$).[14]

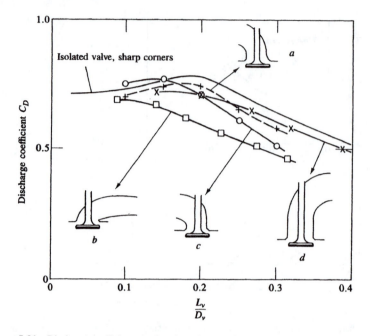

Figure 5-31 Discharge coefficient as a function of valve lift for several exhaust valve and port designs.[14,32] Reproduced with permission of The McGraw-Hill Companies.

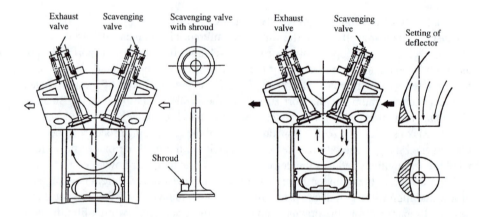

Figure 5-32 Examples of valve shroud and port defector used to generate V-shaped scavenging flow.[41,42]

We now describe some examples of valved two-stroke gasoline engines to illustrate the issues involved in achieving good scavenging. Most important is directing the scavenging flow as it enters the cylinder away from the exhaust valve or valves, to hold any short-circuiting to a minimum. A U-shaped scavenging flow is normally the desired objective, although this may make it difficult to scavenge the residual gas from the center of the cylinder effectively.

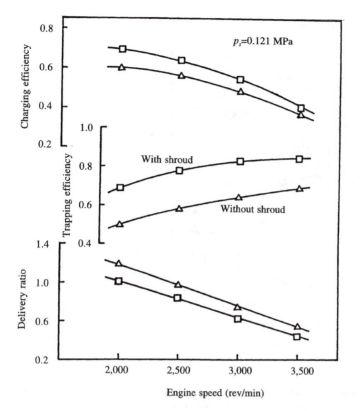

Figure 5-33 Effect of intake valve shroud on delivery ratio trapping and scavenging efficiencies of a single-cylinder two-stroke gasoline engine, as a function of engine speed. Scavenging pressure is 1.21 bar.[42]

Figure 5-32 shows two approaches: shrouding the scavenging valves and shaping the scavenging port with a deflector. These two arrangements were tested on remodeled commercially available four-stroke four-valve single-cylinder engines.[41,42] The camshaft rotation speed of the four-stroke engine was doubled for two-stroke operation, and the exhaust valve lift was reduced to prevent physical interference between the exhaust and scavenging valves. The shroud on the two scavenging valves significantly reduces the flow through that part of the open valve flow area, forcing the scavenging flow down into the cylinder as indicated on left in Fig. 5-32. The rotation of the valve must be prevented to maintain the desired U-shaped flow. A shroud angle of about 90° (relative to the full valve circumference of 360°) gives the best scavenging performance.[41] However, the limited height of the shroud makes this approach less effective at high valve lifts. Figure 5-33 shows the effect of the shroud on the engine's scavenging process.[42] At a given engine speed (and wide-open throttle) the shroud, as expected, improves the charging and trapping efficiencies. However, due to the flow resistance it introduces, shrouding the valves reduces the delivery ratio by about 20%.

Use of shrouded valves is not feasible in a practical engine due to durability and manufacturing issues. However, the scavenging port walls can be appropriately shaped

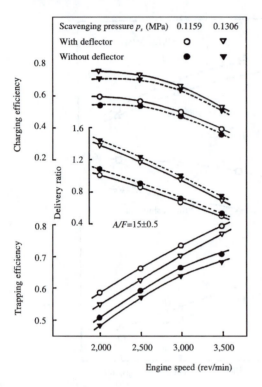

Figure 5-34 Effect of flow deflector in the scavenging valve intake port on delivery ratio, trapping efficiency, and scavenging efficiency of a single-cylinder two-stroke gasoline engine, as a function of engine speed for two scavenging pressures, 1.16 and 1.31 bar.[41]

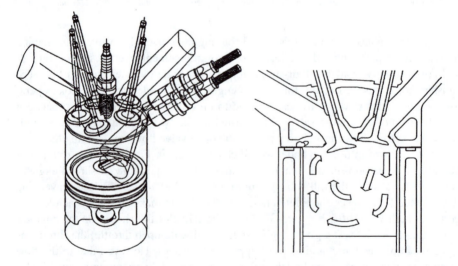

Figure 5-35 Cylinder head, valve, and port layout of the Toyota S-2 prototype valved two-stroke gasoline engine, and the cylinder head valve-shrouding approach used to produce the U-shaped in-cylinder scavenging flow shown.[40]

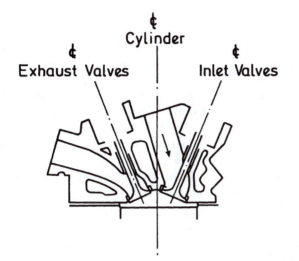

Figure 5-36 Cylinder head, valve, and port design of Ricardo prototype two-stroke gasoline engine.[43,44] The transfer port (with flow direction arrow) directs the scavenging flow down the cylinder and away from the exhaust valve.

to direct the flow as it enters the cylinder. Figure 5-32 (right) shows how a deflector installed in the port close to the valve can achieve the U-shaped in-cylinder flow. Figure 5-34 shows the impact on the scavenging performance.[41] Again, the charging and trapping efficiencies are increased, but the delivery ratio is reduced at a given speed by adding the deflector.

Two more practical approaches to achieving the desired U-shaped in-cylinder flow, for valved two-stroke gasoline engines, are shown in Figs. 5-35 and 5-36. The Toyota S-2 concept[40] uses five valves per cylinder—three exhaust and two scavenging valves—to obtain the minimum practical valve flow area for a given cylinder bore. The scavenging valves are effectively shrouded by recessing the valve seat substantially on the cylinder axis side of the valves. The shroud is designed into the cylinder head rather than attached to the valve. This achieves the U-shaped scavenging flow as shown. This was a direct-injection two-stroke engine with two fuel injectors: one for stratified operation at light load, the other for more homogeneous operation at high load. Good performance was achieved: the scavenging efficiency was about 80% at a delivery ratio of unity, and the value of bmep at the maximum torque point (3000 rev/min) was 922 kPa. In the prototype cylinder head geometry developed at Ricardo,[43,44] shown in Fig. 5-36, the desired scavenging cylinder-entry flow direction is achieved by directing the transfer port down the cylinder axis and away from the exhaust valve location. This produces the desired U-shaped scavenging flow. This prototype valved two-stroke engine produced about 90 kW/liter at 5000 rev/min and a maximum bmep at 2500 rev/min of about 1000 kPa.†

†With 100 Research Octane Number fuel: that is, with a gasoline of high enough quality to avoid any knock problem.

5-4-2 Sleeve Valves

During the first decades of this century, the conventional four-stroke automobile engines were very noisy in operation, largely due to their poppet valves. To reduce this noise, Knight (one of the major engine manufacturers in those days) produced engines where the valve functions were performed by means of two concentric ported sleeves (Fig. 5-37). These sleeves were interposed between the cylinder and the piston and reciprocated by means of small cranks on a half-speed shaft. Inside the cylinder liner there was a thin cast-iron sleeve in which ports for the inlet and exhaust were cut on opposite sides. Within this sleeve there was another sleeve with ports near its upper end. The piston was located within this inner sleeve. Ports in the cylinder wall corresponding to those in the sleeve valves were provided. To the authors' knowledge, no sleeve valves have been installed in commercial two-stroke engines, although this concept could pro-

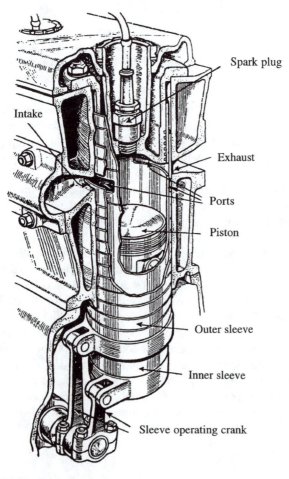

Figure 5-37 Knight double sleeve valve engine. Two concentric ported sleeves are reciprocated by small cranks on a half-speed shaft. Adapted from Day.[45]

vide the desirable asymmetrical opening and closing of the relevant ports for optimum scavenging.

5-5 INTAKE AND EXHAUST SYSTEM DESIGN

The flow of the fresh charge (air, with direct fuel injection, or air plus fuel in premixed-charge engines) in the intake system of a two-stroke engine involves a pulsating flow and a path of complex geometry. The intake system flow path normally includes these elements: air filter, carburetor (if used), throttle, intake manifold, often an element to restrict backflow (e.g., reed valve, disk valve, in some concepts a piston-controlled intake port), a crankcase volume, and scavenging ducts. For engines with port fuel-injection and carburetors, liquid fuel atomization, deposition on the walls, fuel vaporization, and air–fuel mixing phenomena also occur in the intake system. Important intake system design criteria are low flow resistance, even distribution of air and fuel between cylinders, runner and branch lengths that take advantage of ram and tuning effects (see below), minimum backflows, and an appropriate delivery ratio.

The exhaust system typically consists of an exhaust manifold, exhaust pipe, a catalytic converter, and a muffler or silencer. Some systems include an exhaust flow control valve in the exhaust port. This section deals with the important intake and exhaust system design issues that are specific to the two-stroke cycle engine. For a comprehensive discussion of the fundamentals of fuel metering, and intake and exhaust system flow phenomena (carburetors, fuel injection, feedback systems, flow past a throttle plate, and unsteady flow in intake and exhaust systems) we refer the reader to Heywood[32] and Annand and Roe.[14]

5-5-1 Ram and Pipe Tuning Effects

The pressure in the intake manifold varies during the intake process due to the varying piston velocity and intake port open area, and the unsteady gas-flow effects that result from these geometric variations. In crankcase-scavenged engines, the mass of charge inducted into the crankcase determines the pressure level in the crankcase at the start of the scavenging process, and hence the delivery ratio. At higher engine speeds, the momentum of the gas in the intake system toward the intake port, especially as the port is closing, increases the pressure in the port and hence the mass in the crankcase. This effect, which is termed the ram effect, becomes progressively greater as engine speed is increased (as $\bar{S}_p^2$).

The time-varying intake flow into the crankcase causes weak expansion (rarefaction) waves to be propagated back into the intake manifold. These expansion waves travel down the intake manifold at the local sound speed to the open end (or to any plenum) and are reflected as compression waves. If the wave travel time and engine speed are such that a compression wave arrives at the intake port as the port is closing, the pressure in the port is raised above the nominal intake pressure and additional air will flow into the crankcase. An intake system that takes advantage of this effect is termed a tuned intake system.

Similar phenomena occur in the exhaust system. Each exhaust blowdown process generates a pressure wave that propagates (at the local sound speed) along the exhaust system. This compression wave interacts with the pipe junctions, bends, exhaust pipe area changes, muffler components, and other elements, and is reflected back toward the engine cylinder partly as a pressure wave (e.g., when reflection occurs from change in direction of propagation or a reduction in pipe cross section), and partly as an expansion wave (when reflection occurs from an open duct end). In two-stroke engines in which the exhaust port closes after the scavenge ports close, the purity of the exhaust gases (i.e., their fresh charge fraction) at the end of the gas exchange process is high. If an expansion wave arrives just before EC, then substantial fresh charge, as well as burnt gas, will be sucked out of the cylinder. However, earlier in the scavenging process, it is advantageous to decrease the pressure in the exhaust port with an expansion wave because this pulls low-purity gas from the cylinder. In these two-stroke engines, the exhaust is said to be tuned when the exhaust system is designed so that an expansion wave reaches the exhaust port during scavenging, and a compression wave reaches the exhaust port as the exhaust is closing. In four-stroke engines, an expansion wave arriving at the port as the exhaust closes is an advantage because reducing the pressure at the port pulls additional residual gas out of the cylinder before the induction process commences. Tuning the exhaust system is especially important in high-performance two-stroke engines. It can have a major impact (up to a factor of 2) on the amount of fresh charge trapped inside the cylinder, and hence on the engine's power output.

Although ram and tuning effects are the most important phenomena in determining the performance of intake and exhaust systems, two additional phenomena are significant: flow friction and heat transfer. Both intake and exhaust systems are made up of many components, each of which presents some resistance to fluid flow. For each component, this frictional pressure drop Δp_i is given by

$$\Delta p_i = \xi_i \rho V_i^2 \qquad (5\text{-}7)$$

where ξ is the resistance coefficient and V_i is the characteristic flow velocity for that component. Assuming a quasi-steady flow,

$$V_i \doteq \overline{S}_p (A_p / A_i) \qquad (5\text{-}8)$$

Hence the total quasi-steady frictional pressure loss is

$$\Delta p_{\text{total}} = \sum_i \Delta p_i \approx \rho \overline{S}_p^2 \sum_i \xi_i \left(\frac{A_p}{A_i} \right)^2 \qquad (5\text{-}9)$$

It is proportional to piston speed squared.

Heat transfer affects the delivery ratio through changes in the charge density. Any heating of the air in the intake or in the crankcase increases its temperature, which reduces its density. At low speeds, when the flow is close to quasi-steady, the pressure is essentially unchanged so the air density and temperature are inversely proportional. At high engine speeds, compressibility effects can become important in the intake system; then the air density varies as $\rho_a \sim 1/\sqrt{T_a}$. Of these two phenomena, flow friction is the more important.

5-5-2 Analysis of Unsteady Intake and Exhaust Flows

Gas dynamic models have been in use for many years to study engine gas exchange processes. These models use the mass, momentum, and energy conservation equations for the unsteady compressible flow in the intake and exhaust. Normally, the one-dimensional unsteady flow equations are used. These models often use a thermodynamic analysis of the in-cylinder processes to link the intake and exhaust flows. In the past, the method of characteristics was used to solve these gas dynamic equations. Finite difference techniques are used in more recent intake and exhaust flow models. An extensive discussion of such models lies beyond the scope of this book. We refer the reader to several standard references.[14,32,37,46] We will, however, review the basic approach and assumptions of these models.

The basic equations used are the one-dimensional unsteady conservation equations for mass, momentum, and energy, for gas flow in a duct.[32] If flow friction and heat transfer effects can be neglected these equations can be considerably simplified. Under such conditions, the flow is isentropic (usually termed a homentropic flow). However, unsteady gas dynamic phenomena are most important at higher speeds where these effects are usually significant. Two types of methods have been used to solve these equations: (1) the method of characteristics and (2) finite difference procedures. The characteristic methods have a numerical accuracy that is first order in space and time, and they require a large number of computational points if resolution of short-wavelength variations is important. Finite difference techniques can be made higher order and prove to be more efficient: this approach is now usually preferred.

The method of characteristics is a well-established mathematical technique for solving hyperbolic partial differential equations. With this technique, the partial differential equations are transferred into ordinary differential equations that apply along so-called characteristic lines. Pressure waves are the physical phenomenon of practical interest in the unsteady intake flow, and these propagate relative to the flowing gas at the local sound speed. In this particular application, the one-dimensional unsteady flow equations are rearranged so that they contain only the local fluid velocity U and local sound speed c.

Because the absolute velocity of small-amplitude sound waves is $U + c$ in the direction of flow and $U - c$ in the reverse flow direction, the lines of slope $U \pm c$ are the *position* characteristics of the propagating pressure waves that define the position x of the pressure wave at time t. *Compatibility conditions* accompanying the position characteristics relate U to c. The compatibility relationships are expressed in terms of variables (called Riemann invariants) that are constant along the position characteristics for constant-area homentropic flow, although they vary if these restrictions do not apply. Thus, the solution of the mass and momentum conservation equations for this one-dimensional unsteady flow is reduced to the solution of a set of ordinary differential equations.

The equations are usually solved numerically using a rectangular grid in the x and t directions. The intake or exhaust system is divided into individual pipe sections that are connected at junctions. A mesh is assigned to each section of pipe between junctions. From the initial values of the variables at each mesh point at time $t = 0$, the values of the Riemann variables at each mesh point at subsequent time steps are then determined. Gas

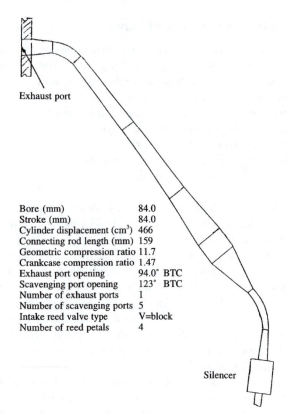

Exhaust port

Bore (mm)	84.0
Stroke (mm)	84.0
Cylinder displacement (cm³)	466
Connecting rod length (mm)	159
Geometric compression ratio	11.7
Crankcase compression ratio	1.47
Exhaust port opening	94.0° BTC
Scavenging port opening	123° BTC
Number of exhaust ports	1
Number of scavenging ports	5
Intake reed valve type	V=block
Number of reed petals	4

Silencer

Figure 5-38 Details of exhaust system configuration and two-stroke gasoline engine used for assessing unsteady exhaust flow calculations.[47]

pressure, density, and temperature can then be calculated from the energy conservation equation and the ideal gas law. See Benson[46] for additional details.

Finite difference methods for solving the one-dimensional unsteady flow equations in intake and exhaust manifolds are proving more efficient and flexible than the method of characteristics. The conservation equations are rearranged and written in matrix form: see Heywood.[32] Several finite difference methods can then be used to solve these equations (e.g., the Lax–Wendroff method). These finite difference solution methods usually require the introduction of some form of dissipation or damping to prevent instabilities and large nonphysical oscillations from occurring with such a nonlinear problem with large gradients (e.g., a shock wave in the exhaust system). Amplification of the physical viscosity and the addition of artificial viscosity and damping are frequently used techniques.

The boundary conditions at pipe ends and junctions are obtained from the appropriate conservation equations and pressure relations. Outflows and inflows obviously transfer mass and energy. For the flow through a restriction, the flow upstream of the restriction is isentropic. For pipe junctions, the conservation equations are applied to a

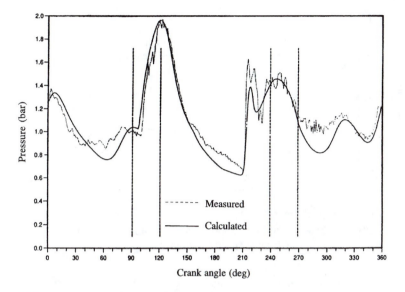

Figure 5-39 Comparison between calculated and measure pressure (in bars) in the exhaust pipe, for the system in Fig. 5-38, at 4500 rev/min.[47]

control volume around the junction. The pressure boundary conditions are most easily estimated by modifying the simple constant-pressure assumption with pressure losses at each pipe exit or entry, calculated from experimentally determined loss or resistance coefficients.

Calculations of intake and exhaust flows using these techniques predict the variations in intake and exhaust manifold pressure with crank angle in single-cylinder and multicylinder engines with acceptable accuracy. Measured scavenging flow variations with engine speed, intake and exhaust system design, and port dimensions and timing are adequately predicted also. Some example results illustrate the phenomena just discussed. Figure 5-38 shows an exhaust pipe for a single-cylinder two-stroke engine used to assess exhaust and scavenging flow calculations. The exhaust incorporates a tuned expansion chamber and a silencer–muffler. Figure 5-39 shows the calculated and measured pressure in the exhaust pipe (60 mm from the exhaust port) at 4500 rev/min. The calculation adequately replicates the blowdown generated pressure wave starting at EO, and the reflected pressure wave arriving at the exhaust port at about 210°. This exhaust port pressure variation, at this engine speed, is close to the desired variation: the pressure is low during the middle of the scavenging period but becomes high toward the end of the process. At lower engine speeds, the reflected pressure wave arrives at the exhaust port earlier in the scavenge period, which is not the desired situation because it reduces the delivery ratio, and the pressure as the scavenge process ends is also lower, which reduces the trapping efficiency: see Fig. 5-40.[47]

These results demonstrate the important role that intake and exhaust system gas dynamics play in the two-stroke engine gas exchange process. The analysis methods reviewed here are effective tools for developing and designing intake systems, and es-

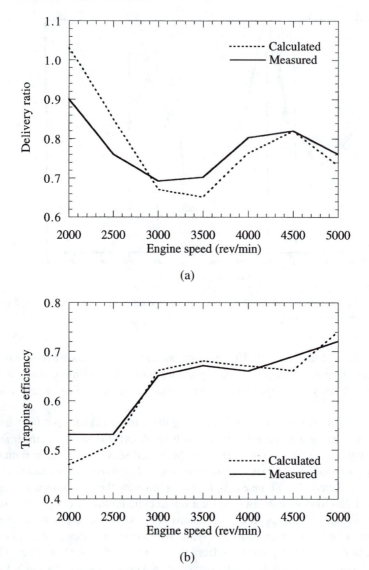

Figure 5-40 Comparison between calculated and measured delivery ratio and trapping efficiency, for the exhaust system and engine in Fig. 5.38, as a function of engine speed.[47]

pecially exhaust systems, that take full advantage of the benefits these unsteady flow phenomena offer.

5-5-3 Intake and Crankcase System Interactions

In crankcase-scavenged two-stroke engines, the selection of the crankcase volume and the inlet and exhaust port dimensions are important decisions that determine the power characteristics of the engine. Figure 5-41 shows how the delivery ratio (measured with short exhaust and inlet pipes of 17.5 and 13.2 cm, respectively) depends on the engine's speed and crankcase volume, for a crankcase-scavenged 199 cm^3 two-stroke engine (see Table 5-3 for engine details). The engine speed at which the maximum delivery ratio is obtained, $N_{\Lambda,max}$, changes approximately in inverse proportion to the square root of the crankcase clearance volume, $V_{c,c}$.[3] The effects of inlet and exhaust tuning on the maximum delivery ratio as a function of crankcase clearance volume, at an engine speed of 2400 rev/min, are shown in Fig. 5-42.

The benefit from exhaust tuning is larger than that of intake tuning for high crankcase clearance volumes, whereas the latter is larger for small crankcase volumes. When both intake and exhaust pipes are tuned, the improvement in the maximum delivery ratio is almost independent of the crankcase clearance volume. When the intake pipe is tuned, the delivery ratio decreases rapidly with increasing crankcase clearance volume, due to the reduction in the ram effect of the intake pipe flow because the intake length becomes shorter as the crankcase volume increases. The effect of a diffuser fitted to the end of the exhaust pipe (with a cone angle of 8° and area ratio of 9) is substantial when only the exhaust pipe is properly tuned, whereas it is small when a properly tuned intake pipe is used.

When the delivery ratio is plotted against $N_{\Lambda,max}/N$, the ratio between the engine speed at which the maximum delivery ratio is obtained and the actual engine speed, a single curve is obtained independent of the ratio between the crankcase clearance volume and the volume displaced by the piston (see Fig. 5-43). Also shown in this figure are the effects of tuning only the exhaust pipe, then only the inlet pipe, and finally both pipes. The extent of improvement in delivery ratio obtained by the tuned exhaust pipe (curve C) over that obtained by the shortest pipe (D) depends weakly on engine speed, whereas the effect of tuning the intake pipe (B) is much higher at lower engine speeds. Nagao and Shimamoto examined the effect of the intake and exhaust pipe tuning on the pressure in the crankcase volume at IO and IC, and concluded that their effect on the delivery ratio is closely related to their effect on the crankcase pressure.[3] Tuning the exhaust pipe increases the crankcase pressure about 15% at intake port opening; tuning the intake pipe gave increases between about 10% at low engine speed and 25% at high speed, at intake port closing.

The intake port open period also affects the delivery ratio under certain conditions; see Fig. 5-44. The most sensitive conditions are lower speeds with a tuned intake, and higher speeds with a short untuned intake.[3]

The effect of the ambient conditions on the optimum crankcase compression ratio (or alternatively the crankcase/engine swept volume ratio) is shown in Fig. 5-45. The data suggest that the ratio that gives the highest indicated power varies from 2.8 at high altitude and 6000 rev/min, to 4.6 at sea level and 4000 rev/min. However, the curvature of the constant-altitude lines is modest; the effect of the speed change is more important. The indicated specific fuel consumption at the higher engine speed depends on the crankcase-to-swept-volume ratio: a lower ratio results in a higher isfc. The indicated power does not deteriorate at low volume ratio, so the degradation in isfc may be attributed to the fuel losses due to a poorer scavenging process at high engine speed. Note

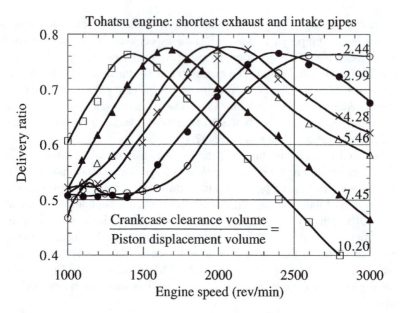

Figure 5-41 Delivery ratio as a function of engine speed and crankcase clearance to piston displacement volume ratio at wide open throttle. See Table 5-3 for Tohatsu engine specifications.[3]

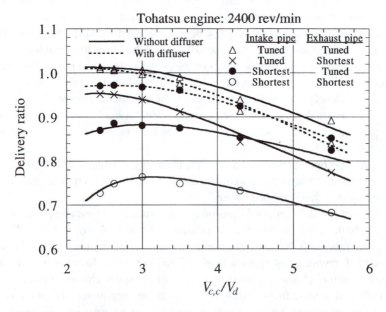

Figure 5-42 The effects of the intake and exhaust tuning on the maximum delivery ratio, as a function of the ratio of crankcase clearance volume to piston displacement volume, at an engine speed of 2400 rev/min. See Table 5-3 for Tohatsu engine specifications.[3]

Table 5-3 Engine specifications for Figs. 5-41–5-44; these are single-cylinder crankcase-scavenged Schnürle-type air-cooled two-stroke engines with symmetrical port opening and closing

Engine specification	Tohatsu TEA-65	Kawasaki KF-3-G
Displacement (cm^3)	199	148
Bore and stroke (cm)	6.5×6.0	5.8×5.6
Rated output (kW)	4.1	2.6
at engine speed (rev/min)	3600	3600
Crankcase compression ratio	1.41	1.51
Period of inlet port opening (degrees)	132	125
Period of scavenging port opening (degrees)	111	113
Period of exhaust port opening (degrees)	145	134
Inlet pipe diameter (cm)	2.9	2.6
Exhaust pipe diameter (cm)	4.0	2.8

that the scavenging efficiency depends strongly on the cylinder port geometry and the time variation of the mass flow through the scavenge port.

Manifold resonance is also an effective means of improving the delivery ratio under some operation conditions. Equsquiza and Virto[49] have studied the effect on the delivery ratio of including a side or branch pipe in the intake of a motored crankcase-scavenged engine (see Fig. 5-46). The engine (a 125-cm^3 single-cylinder Ossa engine) was tested without its cylinder head to avoid any exhaust system effects. They concluded that including a side resonator results in a decrease in the delivery ratio over the entire engine speed range. When a closed branch pipe of fixed length was attached instead of a resonator, the delivery ratio at medium speeds increased, depending on the length of the branch. At high speeds the branch pipe caused the delivery ratio to decrease. By varying the length of the branch pipe it was possible to increase the delivery ratio up to 40% over the speed range where the delivery ratio is normally a minimum. With a variable-length branch pipe it was possible to increase the delivery ratio over the entire engine speed range; a substantial power increase would be expected.

Hata et al.[50] have studied the effect of a branch pipe in the intake on the downstream side of the throttle valve of a Yamaha 123-cm^3 crankcase-scavenged engine installed with a reed valve. A chamber was connected to the end of the branch pipe as shown in Fig. 5-47. Hata et al.[50] concluded that such an arrangement could improve the delivery ratio by 20% under the partial-load operation conditions where engine performance is usually poor.

5-5-4 Exhaust System Control

We review here methods for controlling the exhaust process in two-stroke engines with piston-controlled ports. Because the opening and closing times of the ports in a simple

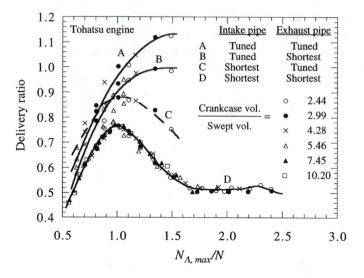

Figure 5-43 The effect of tuning the intake and exhaust on the delivery ratio, as a function of the ratio of engine speed at which the maximum delivery ratio is obtained to actual engine speed. See Fig. 5-41, and Table 5-3 for Tohatsu engine specifications.[3]

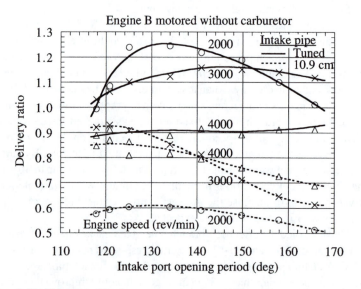

Figure 5-44 The effect of intake port open period on delivery ratio at various engine speeds. Kawasaki engine details in Table 5-3.[3]

two-stroke engine are fixed and symmetrical with respect to BC, the charging efficiency deteriorates significantly when short-circuiting of fresh charge through the exhaust port occurs. Poor fuel consumption and high HC emissions under such operation conditions are then inevitable. Two basic approaches to exhaust flow control have been used: one

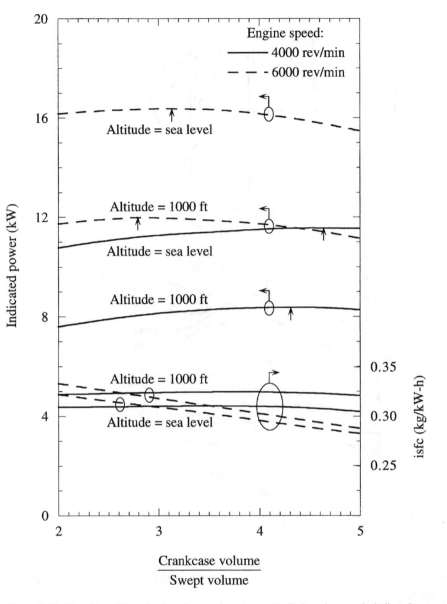

Figure 5-45   The effect of the ratio of crankcase volume to swept cylinder volume on the indicated power and specific fuel consumption of a crankcase-scavenged Schnürle-type two-cylinder 274-cm^3 two-stroke engine, at wide-open throttle.[48] Reproduced by permission of the Council of the Institution of Mechanical Engineers.

is to control the exhaust flow by raising the pressure in the exhaust port by throttling the exhaust; the other is to use an exhaust port timing control valve to vary the port opening angle (and also the port open area). These exhaust flow controls are normally used under light load conditions, which is when poor fuel consumption and high HC emissions are

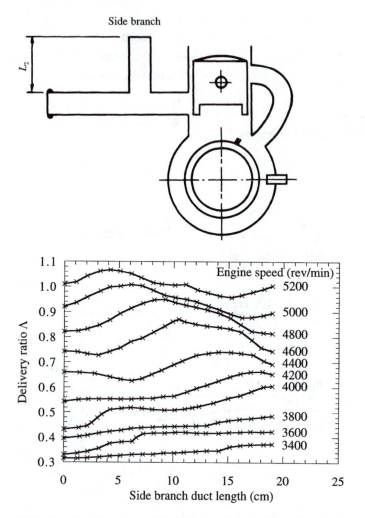

Figure 5-46 The effect of engine speed and side branch pipe length on the delivery ratio of a crankcase-scavenged engine.[49]

usually of most concern; or they are used in high-performance engines, at lower speeds, to improve the engine's output.

One simple method for improving the scavenging performance under these conditions by controlling the pressure in the exhaust port is with a butterfly exhaust valve. Sher et al.[51] showed that the hydrocarbon emissions of a 124.5-cm³ engine can be reduced by about 30% when an exhaust restriction is applied. The area contraction ratio for minimum HC emissions was found to be quite close to that which gave the minimum bsfc. Both effects were found to be a function of engine load, and to a lesser extent a function of the engine speed. The bsfc could be reduced by about 15%, which correlated well with the fuel lost through the short-circuiting mechanism. Cycle-to-cycle combus-

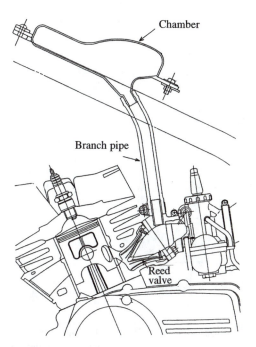

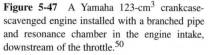

Figure 5-47 A Yamaha 123-cm^3 crankcase-scavenged engine installed with a branched pipe and resonance chamber in the engine intake, downstream of the throttle.[50]

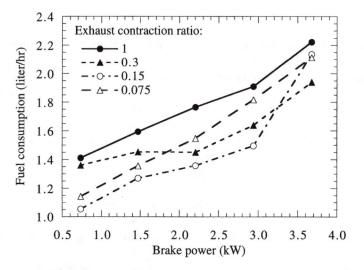

Figure 5-48 Effect of exhaust pipe contraction on the fuel consumption of a Yamaha two-cylinder 399-cm^3 crankcase-scavenged two-stroke engine installed with a reed valve and an exhaust pipe butterfly valve.[52] Engine operated at 4000 rev/min, $A/F = 14$.

tion variability was also substantially reduced, particularly under low engine loads. Figure 5-48 shows how an exhaust pipe contraction may improve the fuel consumption of a Yamaha two-cylinder 399-cm^3 two-stroke engine installed with a reed valve and an

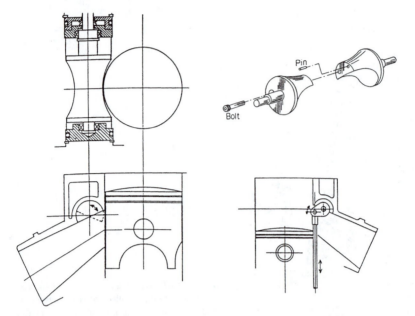

Figure 5-49 Yamaha power valve system to control exhaust port flow.[53,54] Top and left drawings show how valve functions; bottom right shows linkage which rotates valve in response to centrifugal governor on crankshaft.

exhaust butterfly.[52] Note, however, that when maximum engine power is required any exhaust contraction should be removed.

Controlling the area and the timing of exhaust port opening by means of a mechanical valve system has been used by several engine manufacturers. By lowering the effective exhaust port height, the closed-cycle portion of the expansion stroke is increased; this increases the trapped expansion ratio and the work done on the piston before blowdown, and hence the engine's efficiency. It also increases exhaust gas retention in the cylinder, which reduces NO_x emissions. Yamaha, for example, has developed a power valve system that operates in response to engine speed.[53,54] The valve system was connected to a centrifugal governor that converted engine speed to a linear displacement, which rotated the valve to partly close the exhaust port as shown in Fig. 5-49. A similar approach is a Yamaha guillotine exhaust valve that provides precise definition of the port opening time when fully down, but only slightly throttles the flow when partially shut.[55]

A rotating-lever type of exhaust port flow control valve has been used in several recent two-stroke engine designs: for example, see the Oribital engine in Fig. 1-24, and the IFP IAPAC engine in Fig. 2-25.

A rotating cylindrical valve arrangement was also employed by Honda in a single-cylinder 246-cm^3 displacement reed-valve engine.[56] The exhaust opening angle was determined by the engine speed (based on a nonlinear function). A significant improvement in fuel consumption, and HC emissions (60%), compared with the basic engine, was reported. Some exhaust port valve concepts vary the valve position, and hence the port open area, as a function of crank angle. A time-varying piston-type exhaust control valve, combined with in-cylinder air–fuel injection, was employed by Murayama et al.[57] Another example of a time-varying exhaust port valve is the lever-type valve and drive mechanisms proposed by Blundell and Sandford[9] (Fig. 5-9). They combined this variable-area exhaust port valve with a double-chamber rotary scavenging port valve in a 400-cm^3 single-cylinder engine. The exhaust valve is driven at engine speed via a simple eccentric. The valve is fully open (retracted) during the exhaust blowdown process, moves downward to start restricting the flow as the piston approaches BC, and continues downward to the end of its travel (maximum restriction) at scavenge port closure.

REFERENCES

1. Schweitzer, P.H., *Scavenging of Two-Stroke Cycle Diesel Engines*, Macmillan, New York, 1949.
2. Tsuchiya, K., Nagai, Y., and Gotoh, T., "A Study of Irregular Combustion in 2-Stroke Cycle Gasoline Engines," SAE Paper 830091, *SAE Trans.*, vol. 92, 1983.
3. Nagao, F., and Shimamoto, Y., "The Effect of Crankcase Volume and the Inlet System on the Delivery Ratio of Two-Stroke Cycle Engines," SAE Paper 670030, *SAE Trans.*, vol. 76, 1967.
4. Fleck, R., Blair, G.P., and Houston, A.R., "An Improved Model for Predicting Reed Valve Behavior in Two-Stroke Cycle Engines," SAE Paper 871654, *SAE Trans.*, vol. 96, 1987.
5. Flaig, J.D., and Broughton, G.L., "The Design and Development of OMC V-8 Outboard Motor," SAE Paper 851517, 1985.
6. Bostock, P.B., and Simpson, D., "The Two-Stroke Engine: Recent Development and Future Potential," British Petroleum Report, MKG-MISC330, 1974.
7. Okanishi, N., Fukutani, I., and Watanabe, E., "Volumetric Characteristics of a Crankcase-Supercharged 4-Stroke Cycle Engine with Rotary Disc Valves," SAE Paper 900174, *SAE Trans.*, vol. 99, 1990.
8. Heldt, P.M., *High-Speed Combustion Engines: Design; Production; Tests*, P.M. Heldt, Nyack, NY, 1944.
9. Blundell, D.W., and Sandford, M.H., "Two-Stroke Engines—The Lotus Approach," SAE Paper 920779, *SAE Trans.*, vol. 101, 1992.
10. Adams, J.W., Leydorf, G.F., Schrader, M.J., and Stein, R.A., "Initial Evaluation of a Spill Valve Concept for Two-Stroke Cycle Engine Light Load Operation," SAE Paper 901663, 1990.
11. Sher, E., "Improving the Performance of a Crankcase-Scavenged Two-Stroke Engine with a Fluid Diode," *Proc. Inst. Mech. Eng.*, vol. 196, pp. 23–34, 1982.
12. Nagao, F., Nishiwaki, K., Kato, K., and Iijima, T., "Prevention of Backflow in a Crankcase-Scavenged Two-Stroke Engine by Fluid Diodes," *Bull. JSME*, vol. 14, pp. 483–492, 1971.
13. Sher, E., and Yafe, H., "Flattening the Torque Curve of a Two-Stroke Engine with a Fluid Diode Installed at the Scavenging Port," SAE Paper 830092, 1983.
14. Annand, W.J.D., and Roe, G.E., *Gas Flow in the Internal Combustion Engine*, G.T. Foulis, Sparkford, Somerset, UK, 1974.
15. Bell, A.G., *Two-Stroke Performance Tuning*, Haynes (G.T. Foulis), Sparkford, Yoevil Somerset, UK, 1983.
16. Jante, A., "Scavenging and Other Problems of Two-Stroke Cycle Spark-Ignition Engines," SAE Paper 680468, *SAE Trans.*, vol. 77, 1968.
17. Ishihara, S., Murakami, Y., and Ishikawa, K., "Improvement of Pilot Tube Set for Obtaining Scavenging Pictures of Two-Stroke Cycle Engines," SAE Paper 880171, 1988.
18. Sung, N.W., and Patterson, D.J., "Air Motion in a Two-Stroke Engine Cylinder—The Effects of Exhaust Geometry," SAE Paper 820751, *SAE Trans.*, vol. 91, 1982.
19. Sher, E., Hossain, I., Zhang, Q., and Winterbone, D.E., "Calculations and Measurements in the Cylinder of a Two-Stroke Uniflow-Scavenged Engine under Steady Flow Conditions," *Exp. Ther. Fluid Sci.*, vol. 4, pp. 418–431, 1991.
20. Durst, F., Melling, A., and Whitelaw, J.H., *Principles and Practice of Laser Doppler Anemometry*, Academic Press, New York, 1981.

21. Ikeda, Y., Ito, T., and Nakajima, T., "Cyclic Variation of CO and CO_2 Emissions and Scavenging Flow in a Two-Stroke Engine," SAE Paper 940392, *SAE Trans.*, vol. 103, 1994.

22. Bopp, S., et al., "In-Cylinder Velocity Measurements with Mobile Fiber Optic LDV System," SAE Paper 900055, *SAE Trans.*, vol. 99, 1990.

23. Obokata, T., Matumoyo, N., and Hirao, Y., "LDV Measurement of Pipe Flow in a Small-Two-Cycle Spark-Ignition Engine," SAE Paper 840425, *SAE Trans.*, vol. 93, 1984.

24. Ikeda, Y., Kakemizu, K., and Nakajima, T., "In-Cylinder Flow Measurement and Its Application for Cyclic Variation Analysis in Two-Stroke Engine," SAE Paper 950224, *SAE Trans.*, vol. 104, 1995.

25. Bardsley, M.E.A., Gajdeczko, B., Boulouchos, K., Cherhroudi, B., and Bracco, F.V., "Measurements of the Three Components of the Velocity in the Intake Ports of an I.C. Engine," SAE Paper 890792, 1989.

26. Lorenz, N., and Prescher, K., "Cycle Resolved LDV Measurements on a Fired S.I. Engine at High Data Rate Using a Conventional Modular LDV-System," SAE Paper 900054, *SAE Trans.*, vol. 99, 1990.

27. Weclas, M., Melling, A., and Durst, F., "Characteristics of Scavenging Flow in Transfer Ports of a Motored Two-Stroke Engine," *J. Automobile Eng. Proc. Inst. Mech. Eng.*, vol. 211, pp. 301–317, 1997.

28. Fujikawa, T., et al., "Visualization Study of Misfire Phenomena Under Light Load Condition in a Two-Stroke Engine," *Trans. JSAE*, vol. 61, p. 590, 1995.

29. McKinley, N.R., Fleck, R., and Kenny, R.G., "LDV Measurement of Transfer Port Efflux Velocities in a Motored Two-Stroke Cycle Engine," SAE Paper 921694, *SAE Trans.*, vol. 101, 1992.

30. Ekenberg, M., and Johansson, B., "The Effect of Transfer Port Geometry on Scavenging Flow Velocities at High Engine Speed," SAE Paper 960366, *SAE Trans.*, vol. 104, 1996.

31. Wallace, W.B., "High-Output Medium-Speed Diesel Engine Air and Exhaust System Flow Losses," *Proc. Inst. Mech. Eng.*, vol. 182, part. 3D, pp. 134–144, 1967–1968.

32. Heywood, J.B., *Internal Combustion Engine Fundamentals*, McGraw-Hill, New York, 1988.

33. Benson, R.S., "Experiments on a Piston Controlled Port," *The Engineer*, Nov. 25, pp. 875–880, 1960.

34. Komotori, K., and Watanabe, E., "A Study of the Delivery Ratio Characteristics of Crankcase-Scavenged Two-Stroke Cycle Engines," SAE Paper 690136, *SAE Trans.*, vol. 78, 1969.

35. Dekker, B.E.L., and Hung, S.C., "Transient Discharge from a Cylinder at High Rates of Volume Expansion," SAE Paper 850082, *SAE Trans.*, vol. 94, 1985.

36. Kee, R.J., Blair, G.P., and Douglas, R., "Comparison of Performance Characteristics of Loop and Cross-Scavenged Two-Stroke Engines," SAE Paper 901666, *SAE Trans.*, vol. 99, 1990.

37. Blair, G.P., *Design and Simulation of Two-Stroke Engines*, SAE, Warrendale, PA, 1996.

38. Duret, P., and Ecomard, A., "A New Two-Stroke Engine with Compressed-Air Assisted Fuel Injection for High Efficiency Low Emissions Applications," SAE Paper 880176, *SAE Trans.*, vol. 97, 1988.

39. Brown, D., "Modern Sulzer Marine Two-Stroke Crosshead Diesel Engines," Report Re/22.18.07.40, Sulzer Brothers Ltd., Switzerland, 1986.

40. Nomura, K., and Nakamura, N., "Development of a New Two-Stroke Engine with Poppet-Valves: Toyota S-2 Engine," *A New Generation of Two-Stroke Engines for the Future?* (ed. P. Duret), Proceedings of the International Seminar, Rueil-Malmaison, France, Editions Technip, Paris, 1993.

41. Nakano, M., Sato, K., and Ukawa, H., "A Two-Stroke Cycle Gasoline Engine with Poppet Valves on the Cylinder Head," SAE Paper 901664, *SAE Trans.*, vol. 99, 1990.

42. Sato, K., Ikawa, H., and Nakano, M., "A Two-Stroke Cycle Gasoline Engine with Poppet Valves in the Cylinder Head—Part II," SAE Paper 920780, 1992.

43. Hundleby, G.E., "Development of a Poppet-Valved Two-Stroke Engine—The Flagship Concept," SAE Paper 900802, *SAE Trans.*, vol. 99, 1990.

44. Stokes, J., Hundleby, G.E., Lake, T.H., and Christie, M.J., "Development Experience of a Poppet-Valved Two-Stroke Flagship Engine," SAE Paper 920778, *SAE Trans.*, vol. 101, 1992.

45. Day, J., *Engines—The Search for Power*, St. Martin's Press, New York, 1980.

46. Benson, R.S., *The Thermodynamics and Gas Dynamics of Internal Combustion Engines* (ed. J.H. Horlock and D.E. Winterbone), Clarendon Press, Oxford, 1982.

47. Dabadie, J.C., Perotti, M., and Keribin, Ph., "Prediction of Two-Stroke Engine Performance by Unsteady Gas Dynamic Calculations," *A New Generation of Two-Stroke Engines for the Future?* (ed. P. Duret), Proceedings of the International Seminar, Rueil-Malmaison, France, Editions Technip, Paris, 1993.

48. Sher, E., "The Effect of Atmospheric Conditions on the Performance of an Airborne Two-Stroke Spark-Ignition Engine," *Proc. Inst. Mech. Eng.*, part D, vol. 198, no. 15, pp. 239–251, 1984.

49. Equsquiza, E., and Virto, L., "Delivery Ratio Changes with Branched Inlet Systems Attached to a Two-Stroke Cycle Engine," SAE Paper 820158, 1982.
50. Hata, N., Fujita, T., and Matsuo, N., "Modification of Two-Stroke Engine Intake System for Improvements of Fuel Consumption and Performance through the Yamaha Energy Induction System (YEIS)", SAE Paper 810923, *SAE Trans.*, vol. 90, 1981.
51. Sher, E., Hacohen, Y., Refael, S., and Harari, R., "Minimizing Short-Circuiting Losses in Two-Stroke Engines by Throttling the Exhaust Pipe," SAE Paper 901665, 1990.
52. Tsuchiya, K., Hirano, S., Okamura, M., and Gotoh, T., "Emission Control of Two-Stroke Motorcycle Engines by the Butterfly Exhaust Valve," SAE Paper 800973, *SAE Trans.*, vol. 89, 1980.
53. Hata, N., and Lio, T., "Improvement of Two-Stroke Engine Performance with the Yamaha Power Valve System (YPVS)," SAE Paper 810922, *SAE Trans.*, vol. 90, 1981.
54. Nomura, K., Hirano, S., Gotoh, T., and Motoyama, Y., "Improvement of Fuel Consumption with Variable Exhaust Port Timing in a Two-Stroke Gasoline Engine," SAE Paper 850183, *SAE Trans.*, vol. 94, 1985.
55. Thornhill, D., and Fleck, R., "Design of a Blower-Scavenged, Piston-Ported, V6, Two-Stroke Automotive Engine," SAE Paper 930980, 1993.
56. Ishibashi, Y., and Asai, M., "Improving the Exhaust Emissions of Two-Stroke Engines by Applying the Activated Radical Combustion," SAE Paper 960742, 1996.
57. Murayama, T., Sekiya, Y., Sugiarto, B., and Chikahisa, T., "Study on Exhaust Control Valves and Direct Air–Fuel Injection for Improving Scavenging Process in Two-Stroke Gasoline Engines," SAE Paper 960367, 1996.

CHAPTER SIX

COMBUSTION

In this chapter, we develop the relevant fundamentals and terminology required to understand two-stroke cycle engine combustion. Many of the important features of both spark-ignition and compression-ignition (diesel) combustion processes in two-stroke engines are similar to those in their four-stroke counterparts. The reader is referred to Heywood[1] for an extensive discussion of the latter topic. The important combustion process differences between these two engine cycles are either a consequence of the two-stroke's scavenging and mixture preparation process or a result of its firing frequency being twice that of the four-stroke. The most important of these differences are the following: high dilution of the fresh charge with residuals, which results in lower flame and burned gas temperatures; higher mixture temperatures prior to combustion due to this high dilution; substantial mixture nonuniformity (in carburetted or premixed-charge SI engines); larger cycle-by-cycle variations in the combustion process; hotter combustion chamber walls—especially the piston crown; larger variation in exhaust gas temperature with load due to charge short-circuiting and its variation with load.

This chapter discusses first the two-stroke cycle spark-ignition engine combustion process, and then the diesel combustion process. This then leads on to the next chapter, which describes the emissions characteristics of two-stroke cycle engines. The emissions formation processes in spark-ignition and diesel engines are closely linked to the essential features of these engines' combustion processes.

6-1 COMBUSTION FUNDAMENTALS

The combustion process is an especially important part of the internal combustion engine's operating cycle. Combustion must release the chemical energy of the fuel—the primary source of energy for the engine—in a relatively short time period between the compression and expansion processes, thereby producing the high-pressure, high-temperature, burned gases that then expand within the cylinder, thereby transferring work to the piston. A robust combustion process is important for smooth and reliable engine operation. The combustion process must be fast, that is it must occupy a small fraction of the total cycle time so that the engine's energy conversion process is efficient, and highly repeatable so that variations from one cycle to the next are small enough to be barely noticeable.

Combustion is also important for its impact on several other engine characteristics or requirements. The processes by which the engine's emissions form within the cylinder are closely linked with the details of the combustion process. The antiknock and volatility requirements of fuels for spark-ignition engines and the ignition quality of diesel fuels are dictated by each engine's combustion process. Even the gas exchange process of the engine, and hence its power, are affected by what the designer must do to achieve a good combustion process.

Combustion in engines takes place in a *flame*. A flame is the (confined) region within which the fuel oxidizing reactions and chemical energy release occur. Engine flames are normally thin: the actual reaction region is a fraction of a millimeter thick. In the standard spark-ignition engine, the fuel-vapor–air charge is well mixed so the flame is a *premixed flame*. The speed at which such a premixed flame propagates into the unburned gas ahead of it is called the *flame speed* or burning velocity. In a laminar or well-ordered flow, this velocity S_L is a characteristic property of the unburned mixture ahead of the flame, depending on the composition of the mixture, the properties of the fuel, and the mixture temperature and pressure. The value of S_L of the order of, and usually less than, 1 m/s. Figures 6-1 and 6-2 show how the laminar burning velocity for various fuels, at standard atmospheric conditions, varies with mixture composition: the relevant mixture variables are relative fuel/air equivalence ratio and burned residual gas fraction in the unburned mixture. Note that the maximum laminar burning velocity or speed occurs with slightly richer than stoichiometric mixtures where the temperature of the burned gases produced by combustion is a maximum. Dilution of fuel–air mixture (of fixed relative fuel/air equivalence ratio) with burned gases steadily and substantially decreases the laminar flame speed. Both these trends result from the decrease in temperature and reaction species concentration gradients within the flame (which result from the reduction of burned gas temperature) as mixtures become leaner, or richer, or more dilute, and from flame thickening. Heat conduction and diffusion down these gradients are the fundamental processes responsible for premixed flame propagation. The higher-than-ambient pressures and temperatures of the mixture in the cylinder during combustion in engines modifies these laminar burning velocities. The following power law fits the data well:

$$S_L = S_{L,0} \left(\frac{T_u}{T_{u,0}} \right)^\alpha \left(\frac{p}{p_0} \right)^\beta \tag{6-1}$$

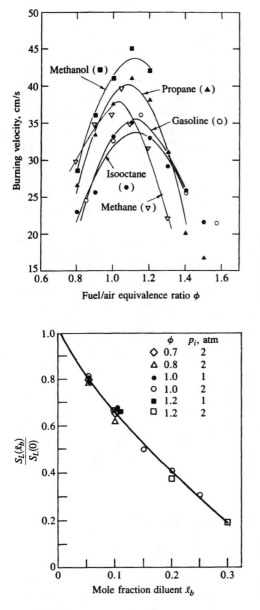

Figure 6-1 Laminar burning velocities for several fuels as a function of fuel–air equivalence ratio, at 1 atm and 300 K.[2] Reproduced from Heywood[1] with permission of The McGraw-Hill Companies.

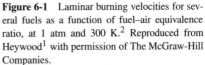

Figure 6-2 Laminar burning velocity as a function of the burned gas mole fraction $\tilde{x}_b$ in the unburned mixture (fuel is gasoline).[3] Reproduced from Heywood[1] with permission of The McGraw-Hill Companies.

where $S_{L,\,0}$ is obtained from Figs. 6-1 and 6-2, and α and β vary with fuel/air equivalence ratio and are about 2 and -0.2, respectively.[2,3] The data in Fig. 6-2 are correlated by the relation

$$S_L(\tilde{x}_b) = S_L(\tilde{x} = 0)\left(1 - 2.06\tilde{x}_b^{0.77}\right) \qquad (6\text{-}2)$$

where $\tilde{x}_b$ is the mole fraction of burned gas in the unburned mixture.

Premixed engine flames become turbulent as they develop, and their propagation rate is then defined by the turbulent flame speed. The flow inside the cylinder during combustion both convects and distorts or wrinkles the thin reaction sheet flame through its combination of bulk or mean flow pattern and local turbulence or random velocity fluctuations. The higher the local turbulence, the faster the flame propagates and becomes wrinkled. If engine flames were not turbulent, with a turbulent flame speed several times the laminar flame speed, engines could not operate satisfactorily because combustion and the pressure rise rate it produces would be much too slow. Because engine turbulence scales with piston speed,[1] combustion also scales (almost) with piston speed and so is fast enough throughout the full crankshaft speed range desired for satisfactory engine operation. However, the turbulent flow primarily affects the gross or larger-scale characteristics of the flame (on the scale of several mm), such as its average burning velocity. Locally, the thin reaction zone propagates forward into the adjacent unburned mixture at the laminar flame speed.

Combustion in the compression-ignition or diesel engine is more complicated and occurs through a different sequence of processes. Fuel is injected directly into the engine cylinder toward the end of the compression process, just before the desired start of combustion. The injected liquid fuel jets atomize, the fuel droplets move through the high-temperature high-pressure air in the combustion chamber, vaporize, and the fuel vapor mixes with air. The high temperature fuel-vapor–air mixture thus produced then ignites spontaneously and burns rapidly. Combustion then continues as a combustible mixture continues to be prepared by the above mixing process. Thus, for much of the diesel's combustion process, the fuel burning rate is controlled by the rate at which fuel mixes with air. Such "flames" are usually termed a *diffusion flame*, because the (gaseous) fuel and the air must diffuse into each other and then react in a thin reaction sheet close to where they reach the stoichiometric mixture proportions. Because the in-cylinder flow is turbulent, the diesel combustion process occurs primarily as a turbulent (unsteady) diffusion flame. Spontaneous ignition due to the high air temperature in the cylinder when injection commences, and that fuel–air mixing controls the burning rate, are the distinguishing features of the diesel combustion process.

6-2 COMBUSTION IN SPARK-IGNITION ENGINES

In a traditional crankcase-scavenged two-stroke engine, the fuel and air are mixed together in the intake system, inducted through the inlet port into the crankcase volume, compressed, and then introduced into the cylinder through the scavenging ports. The fresh charge entering the cylinder pushes out some of the residuals from the previous cycle that are still in the cylinder after blowdown, through the exhaust ports. However, a considerable part of these burnt gases mixes with the fresh gas and is still inside the cylinder when compression begins. Toward the end of the compression stroke, combustion of the trapped charge is initiated by an electric discharge. This charge, which is composed of air, fuel, and largely burned residuals, usually still has a large-scale mean bulk motion and substantial turbulence. A flame forms around the small ($\sim$1-mm diameter) hot gas kernel formed between the spark plug electrodes by the electrical discharge. The flame, initially laminar-like, then grows and increasingly interacts with the turbulent flow field as it propagates outward from the spark discharge center. This flame transitions to a turbulent flame as it propagates across the combustion chamber. The turbulence increases the flame speed to values (~5 m/s) well above laminar values (~0.7 m/s), and wrinkles or distorts the flame significantly. The increased flame speed results from the increased convection rates the turbulence produces. The flame wrinkling produces a relatively thick turbulent flame "brush" between the leading plane of this wrinkled thin reaction sheet and its trailing plane, thereby substantially increasing the amount of reaction sheet area that fits into the engine combustion chamber. Then, as the flame reaches the combustion chamber walls, it extinguishes as each part of the wrinkled flame sheet comes in contact with the chamber walls.

6-2-1 Cylinder Pressure, Spark Timing, and Mass Fraction Burned

During flame propagation, the pressure inside the cylinder steadily rises above the value it would have in the absence of combustion. Figure 6-3 shows a motored (nonfiring) cycle pressure versus crank angle curve, and a firing cycle pressure curve. The spark discharge commences at 40° BTC, but the two traces diverge some 10° later when the mass of mixture burned becomes significant (a few percent). The pressure reaches a maximum after TC (typically around 15° ATC), and then decreases as the cylinder volume continues to increase during the remainder of the expansion stroke. Figure 6-3 also shows how the burned mass fraction of the in-cylinder charge grows during combustion. Half of the charge is burnt at about 10° ATC, and combustion essentially dies out some 20° or more later. When combustion ends is not well defined.

Because the flame growth rate depends on local mixture motion and composition, the flame development and subsequent propagation vary from one cycle to another,

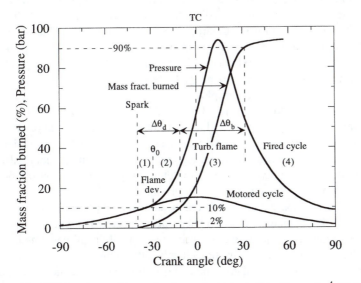

Figure 6-3 Different phases of the combustion process in a spark-ignition engine[4]: (1) flame kernel formation; (2) kernel becomes turbulent; (3) main turbulent propagation phase; (4) flame termination.

and from cylinder to cylinder in multicylinder engines. Especially significant factors in this variability are mixture composition and motion in the vicinity of the spark plug at time of ignition. Cycle-by-cycle variations in the combustion process result in cycle-by-cycle variations in cylinder pressure and, therefore, in torque. These variations have long been recognized as limiting the operating regime of spark-ignition engines (this subject is discussed in more detail in Section 6-2-5).

Figure 6-3 indicates that the combustion process in a spark-ignition engine can be divided into four distinct phases: (1) spark onset and the development of a (small) flame kernel; (2) the development of this flame kernel into a turbulent flame (a few centimeters in diameter); (3) the main, turbulent, flame propagation stage; and (4) the flame inter-action with the combustion chamber walls as combustion ends. The combined duration of the flame development and propagation process is typically between 40 and 90 crank angle degrees and depends on engine design details, mixture composition, and engine load and speed. We will return to these phases in more detail in Section 6-2-4.

The timing or phasing of this combustion process relative to the motion of the piston determines the indicated mean effective pressure and hence the brake torque of the engine. For a given throttle position and engine speed, the optimum spark timing is that which gives the maximum brake torque: it is accordingly termed MBT spark timing. The value of MBT spark timing depends on the duration of the flame development process, the rate of flame propagation during the main burning phase, the dimensions of the combustion chamber, and the spark plug location. Thus the unburned mixture composition and thermodynamic state, and the fluid bulk and turbulent motion inside the cylinder, are important factors in determining MBT timing.

Note that MBT timing is usually the appropriate reference spark timing for a given engine and operating condition. More advanced spark timing, when combustion starts earlier than with MBT timing, produces lower imep and torque though higher (and earlier) peak gas pressures and temperatures. All of these changes are detrimental and "over advanced" spark timing is never used. Retarded spark timing, when combustion starts later than at MBT timing, also produces lower imep and torque but with lower peak gas pressures and temperatures. Because lower peak pressures, and especially temperatures (these scale together), have a significant beneficial effect on knock onset and NO_x emissions, some spark retard is frequently employed in practice. A retard of 5° from MBT timing results in about a 1% loss in torque; 10° of retard results in a 4% loss in torque. Thus modest spark retard often offers useful improvements in engine operation with little loss in torque. The relative loss in torque with spark retard is almost independent of other engine design and operating variables. In summary, spark timing should always be referenced to the optimum (MBT) timing to define fully the actual phasing of the combustion process relative to the motion of the piston.

We have been discussing the normal SI engine combustion process. An important abnormal spark-ignition engine combustion phenomenon is called *knock*. Knock is the result of rapid autoignition of a portion of the mixture ahead of the propagating flame.[1] As the flame advances across the combustion chamber, the cylinder pressure rises, and the unburned mixture ahead of the flame (the end gas) is compressed and its temperature rises. At high enough end-gas temperatures and pressures, some parts of the end gas may undergo spontaneous ignition before the arrival of the flame front. Knock occurs when the induction time of the end-gas mixture (the time it takes for spontaneous ignition) is shorter than the time it takes for the flame front to arrive. The induction time depends on mixture composition and state and on the molecular structure of the fuel molecules. The antiknock resistance of a fuel is quantified by its *octane number*. The octane number of a fuel is determined by standard test procedures, in an engine designed for this purpose, that match the knock resistance of the fuel tested to that of a set of standard reference fuels. Research and motor octane numbers are obtained from different engine test conditions, the motor octane test being the more severe. Gasolines typically have research octane numbers in the range 90–100, and motor octane numbers 8–10 numbers lower; the average of the two is usually quoted as the fuel's octane number because it scales best with in-use antiknock behavior.

A different abnormal combustion phenomenon, which often leads to engine destruction, is *surface ignition*. Surface ignition is ignition of the mixture by a hot spot on the surface of the combustion chamber. Overheated valves or spark plug, or hot carbonaceous deposits are possible sources of surface ignition. Surface ignition may occur before spark ignition (preignition) or after spark ignition (postignition). Surface ignition causes a flame to originate at a location, and most important at a time, different from the normal spark. With preignition, combustion starts earlier than intended resulting in higher cylinder pressures and gas temperatures. This may cause knock to occur, which produces still higher temperatures. Higher gas temperatures lead to higher heat transfer rates to the combustion chamber walls, so that the hot spot becomes hotter leading to earlier preignition, which further exacerbates the problem. Thus surface ignition through this escalating and uncontrolled combustion development can rapidly lead to engine destruction (usually by melting the center of the piston crown). For a detailed review of abnormal combustion phenomena in spark-ignition engines, the reader is referred to Heywood.[1]

6-2-2 Analysis of Cylinder Pressure Data

Estimates of the mass fraction burned, the rate of chemical energy or heat release, and the gas temperature distribution are often required. These are needed to evaluate variables such as start of combustion, end of combustion, cylinder head temperature for thermal stress analysis and cooling system design, wall temperature for tribology purposes, bulk motion effects during combustion (for better design of port geometries), and predictions of engine performance under different operating conditions. Cylinder pressure is usually measured with a piezoelectric pressure transducer. This type of transducer contains a quartz crystal that generates an electric charge proportional to the applied pressure. A charge amplifier then converts this charge into a voltage output. With careful attention to details, accurate cylinder-pressure data can be obtained with these transducers.[1,5]

The cylinder pressure is affected by cylinder volume change, fuel chemical energy release during combustion, heat transfer to the combustion chamber walls, gas flow into narrow crevice regions connected to the combustion chamber (e.g., gaps between the piston, rings, and liner), and leakage (primarily to the crankcase). Not all these phenomena are of equal importance: the first two have the greatest impact; the last two the least impact. Obtaining accurate estimates of the fuel's chemical energy release rate (or the mass burning rate, a closely comparable but not precisely the same quantity) requires, first, accurate cylinder-pressure data, and then appropriate models for the more important of the remaining phenomena.

The simplest commonly used approach, that developed in the 1930s by Rassweiler and Withrow,[6] incorporates only the effect of volume change and yields a relative estimate of the mass of mixture burned. It is widely used (e.g., Bopp et al.[7]), although it does not provide any information about the *amount* of fuel trapped within the cylinder. The method works as follows.

When cylinder pressure versus cylinder volume data are plotted on a log–log plot, both the compression and expansion processes prior to and after combustion, respectively, are closely approximated by a polytropic, pV^n = constant, where n is about 1.3.† If, during combustion, when the pressure is p, the cylinder volume V has unburned and burned volumes V_u and V_b, respectively ($V = V_b + V_u$), then scaling the unburned gas volume back to spark onset (state 0, when all the mass was unburned) via pV^n = constant yields

$$V_{u,0} = V_u(p/p_0)^{1/n} \qquad (6\text{-}3)$$

Scaling the burned gas volume forward to the end of combustion, state f, when all the mass is burned, yields

$$V_{b,f} = V_b(p/p_f)^{1/n} \qquad (6\text{-}4)$$

At spark, the ratio $V_{u,0}/V_0$ is the mass fraction unburned $(1-x_b)$. At end of combustion, the ratio $V_{b,f}/V_f$ is the mass fraction burned, x_b. Thus

$$x_b = 1 - \frac{V_{u,0}}{V_0} = \frac{V_{b,f}}{V_f} \qquad (6\text{-}5)$$

and Eqs. (6-3)–(6-5) then yield

$$x_b = \frac{p^{1/n}V - p_0^{1/n}V_0}{p_f^{1/n}V_f - p_0^{1/n}V_0} \qquad (6\text{-}6)$$

The difficulty in using this technique is choosing an appropriate value for n that varies somewhat with engine operating conditions. An average of the compression and expansion slopes from the log p–log V plots is a commonly used approach (and checks the quality of the pressure data). This technique is useful for defining the often used burn parameters: crank angle interval from 0 to 10% mass fraction burned, $\Delta\theta_{0-10\%}$, the *flame development angle*; crank angle interval from 10 to 90% mass fraction burned, $\Delta\theta_{10-90\%}$, the *main* or *rapid burn angle* (see Fig. 6-3).

Another well-established technique for estimating the rate of fuel chemical energy release (usually called heat release analysis) from pressure and volume data is based

---

†In the absence of any heat losses, these processes would be an adiabatic and reversible compression and expansion: then, for fixed mass of gas in the cylinder, pV^γ = constant. In the actual engine, the effect of heat losses is relatively modest, so the functional form still holds and $n \approx \gamma$.

on an overall energy balance for the cylinder contents. A major advantage of such an approach is that the rate of energy release by combustion can be explicitly expressed in terms of the pressure changes. Thus the total amount of energy released can be estimated.

When the cylinder ports are closed and leakage is negligible, the cylinder volume may be considered a closed system where no mass crosses the system boundary. An energy balance for this in-cylinder gas system that contains unburned gas and burned gas, separated by a flame, yields

$$\frac{\delta Q_n}{dt} = \frac{dE}{dt} + \frac{\delta W}{dt} \tag{6-7}$$

Here $\delta Q_n/dt$ is termed the net heat release rate, which is the difference between the rate of chemical energy release by combustion ($\delta Q_{ch}/dt$) and the rate of heat loss to the walls ($\delta Q_w/dt$); thus

$$\frac{\delta Q_n}{dt} = \frac{\delta Q_{ch}}{dt} - \frac{\Delta Q_w}{dt} \tag{6-8}$$

Here dE/dt is the rate of change of the internal energy of the cylinder contents, and $\delta W/dt$ is the rate at which work is done by the system. Although the system consists of two different gases—unburned and burned—with two very different thermodynamic states, it is assumed that the system can be analyzed as a single-component gas that follows the ideal gas law with heat capacity c_v (either assumed constant or a simple function of temperature). Then

$$\frac{dE}{dt} = mc_v \frac{dT}{dt} \tag{6-9}$$

where m is the mass of the gas system. Using the ideal gas law, we change variables as follows:

$$\frac{d(mT)}{dt} = \frac{1}{R}\left(p\frac{dV}{dt} + V\frac{dp}{dt}\right) \tag{6-10}$$

The rate at which work is done by the system due to expansion is given by

$$\frac{\delta W}{dt} = p\frac{dV}{dt} \tag{6-11}$$

where p is the instantaneous cylinder pressure and V its volume. Combining Eqs. (6-7)–(6-11) yields

$$\frac{\delta Q_{ch}}{dt} = \frac{1}{\gamma - 1}\left(\gamma p\frac{dV}{dt} + V\frac{dp}{dt}\right) + \frac{\delta Q_w}{dt} \tag{6-12}$$

where γ is the specific heat ratio c_p/c_v.

The heat transfer through the cylinder walls $\delta Q_w/dt$ can be evaluated from an engine heat transfer model. Analysis of heat transfer data from various types of engines[8–10] shows that most of the heat transfer to the cylinder walls is due to convection. The mechanisms of engine heat transfer have been found to be very similar to that occurring when hot gas flows through a pipe. One well-established model[9] suggests that the heat transfer through the walls may be estimated with sufficient accuracy by linearly adding heat transfer by convection and by radiation to yield

$$\frac{\delta Qw}{dt} = A\left[a\frac{k}{B}R_e^b(T - T_w) + c\sigma\left(T^4 - T_w^4\right)\right] \tag{6-13}$$

where the instantaneous Reynolds number is defined by

$$R_e = \frac{S_p B}{v} \tag{6-14}$$

Here, a, b, and c are calibration constants; A is the surface area of the combustion chamber through which heat is transferred; T_w is the wall temperature; k is the thermal conductivity of the gas and v its kinematic viscosity; S_p is the instantaneous piston velocity and B its diameter (the bore); and σ is the Stefan–Boltzmann constant $(5.67 \times 10^{-8}$ W/m^2-K$^4)$. Since we assume a single-zone model for the system, T may be interpreted as the average temperature of the cylinder contents. Because, in two-stroke cycle engines, combustion occurs every crank rotation (compared with four-stroke engines, where it occurs every other rotation), it is expected that for the same engine speed the heat transfer to the chamber walls will be higher in the two-stroke engine. Based on a large amount of published data from various types of engines, Annand[9] has suggested the following values for a, b, and c: for four-stroke cycle engines,

$$a = 0.26 \qquad b = 0.75 \pm 0.15 \quad \text{and} \quad c = 0.69$$

for two-stroke cycle engines,

$$a = 0.76 \qquad b = 0.64 \pm 0.1 \quad \text{and} \quad c = 0.54$$

Note that, when using a single-zone combustion model, the average temperature inside the cylinder is obviously much lower than that of the burned gas that radiates. So the radiation heat transfer scaling suggested by Eq. (6-13) may underestimate any trend. However, in SI engines radiation heat transfer is relatively small compared with convection, and is usually neglected.

Equation (6-12) may also be used to evaluate the mass burning rate. In a premixed charge engine, the trapped mass of the fuel is

$$m_f = m(1 - x_r)/\left[(A/F) + 1\right] \tag{6-15}$$

where x_r is the trapped residual mass fraction, and A/F is the air/fuel ratio (this assumes there is negligible unburned fuel in the residual, and that the trapped and delivered A/F are the same). Because

$$\frac{\delta Q_{ch}}{dt} \doteq Q_{HV}\frac{dm_{f,b}}{dt} = Q_{HV}\frac{m_f}{m}\frac{dm_b}{dt} \tag{6-16}$$

then the mass fraction burning rate is given by

$$\frac{dx_b}{dt} = \frac{1}{m}\frac{dm_b}{dt} \doteq \frac{[(A/F) + 1]}{m(1 - x_r)Q_{HV}}\frac{\delta Q_{ch}}{dt} \tag{6-17}$$

The approximation in this equality enters when dissociation in the burned gases is significant; then the chemical energy released is less than the product of mass of fuel burned and fuel heating value Q_{HV}.

Figure 6-4 shows heat release rate and mass fraction burned profiles, obtained with the above methodology from cylinder pressure data from a loop-scavenged engine.[11] As the piston travels upward toward top center, the pressure inside the cylinder increases due to compression until the energy or heat release by combustion becomes significant, just before TC. The piston then travels downward while the heat release rate peaks at around 10° ATC. The pressure inside the cylinder continues to increase until the chemical energy release rate falls below the energy transfer rate out of the cylinder—the work done by the piston due to the expansion of the cylinder contents and heat transfer to the cylinder walls. Later, after about 400° (40° ATC), combustion is practically complete (mass fraction burned approaches 1) and the pressure decrease is due to the continuing work transfer to the piston and heat transfer to the walls.†

Note that more complete one-zone models have been developed for estimating the amount of chemical energy release and mass of mixture burned from cylinder pressure.[12] The major benefit of energy balance methods of obtaining combustion information is that the actual energy release or mass of fuel burned is calculated. This can usefully be compared with the chemical energy or mass of fuel within the cylinder obtained from fuel flow rate measurements and the trapping efficiency.

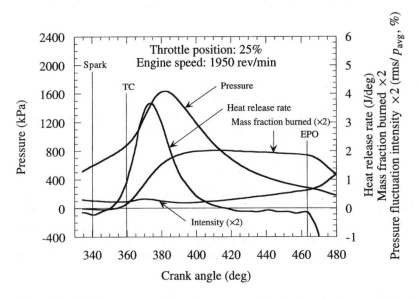

Figure 6-4 Cylinder pressure, net heat release rate, mass fraction burned and pressure fluctuation intensity in a crankcase-scavenged two-stroke cycle SI engine.[11]

†The negative heat release rates and decreasing mass fraction burned after about 420° are likely due to overestimating the heat losses, and perhaps some leakage.

6-2-3 Burn Rate Models

When numerical simulation of the two-stroke spark-ignition engine cycle is undertaken, a model for the mass burning profile is required. A Weibe function of the following form is often used:

$$x_b = 1 - \exp\left\{ - \left[\frac{c(\theta - \theta_{ign})}{\Delta\theta_b} \right]^{-b} \right\} \tag{6-18}$$

where θ_{ign} is the crank angle at start of combustion, $\Delta\theta_b$ is the crank angle interval from start to completion of combustion, c is an efficiency parameter, and b is a form factor. The values of b and c can be determined by matching Eq. (6-18) to mass fraction burned profiles obtained from experimental cylinder pressure measurements using the methods described in Section 6-2-2. Sher[13] has examined the combustion rate in three different types of small two-stroke cycle SI engines, and recommended a value of 5.0 for form factor b (for all of the three engines), and 3.0 for the efficiency parameter c (3.2 for the older engine studied).

The flame development time, or crank angle interval from spark onset to detectable combustion generated pressure rise ($\theta_{spark} - \theta_{ign}$), and the time or interval from start of combustion-generated pressure rise to combustion completion ($\Delta\theta_b$) may be estimated using such data at any reference operating condition, from a correlation developed by Hires et al.[14]:

$$\frac{\theta_{spark} - \theta_{ign}}{(\theta_{spark} - \theta_{ign})_{ref}} = \left(\frac{N}{N_{ref}} \right)^{1/3} \left(\frac{S_{L,ref}}{S_L} \right)^{2/3} \tag{6-19}$$

where N is engine speed and S_L is the laminar burning velocity of the fresh charge at the conditions prevailing inside the cylinder when half of the mixture is burned ($x_b = 0.5$). The crank angle interval from start of combustion-generated pressure rise to combustion completion ($\Delta\theta_b$) may be estimated in a similar way[14]:

$$\frac{\Delta\theta_b}{\Delta\theta_{b,ref}} = \left(\frac{N}{N_{ref}} \right)^{1/3} \left(\frac{S_{L,ref}}{S_L} \right)^{2/3} \tag{6-20}$$

The laminar burning velocity S_L can be estimated from Eqs. (6-1) and (6-2), which include the effects of relative air/fuel ratio, residual burned fraction, and mixture temperature and pressure. For his three types of two-stroke cycle engines, Sher recommended a combustion duration $\Delta\theta_{b,\,ref}$ of $55° \pm 5°$, for an engine reference speed of 4000 rev/min, and a crank angle interval from spark onset to detectable combustion ($\theta_{spark} - \theta_{ign}$) of $10 \pm 2°$.[13]

6-2-4 Physics of the SI Engine Combustion Process

We now discuss the essential features of the spark-ignition engine combustion process in more detail, identifying the critical underlying physical processes.[15] This will prove especially valuable when in Section 6-2-5 we review one of the major two-stroke SI engine combustion issues: cycle-by-cycle combustion variability.

Combustion is initiated by the spark discharge, which generates, on a sub-microsecond time scale, a high-temperature kernel of order 1-mm diameter between the spark plug electrodes. The size of the spark-generated kernel depends on the breakdown energy that the ignition system delivers to the gas. The extremely high discharge-generated temperatures of order 60,000 K result in rapid heat conduction to the surrounding gas and the electrodes.[16] The chemical energy of the fuel–air mixture heated by the discharge is released. An outward propagating flame kernel results, as shown in the Schlieren photographs of an engine ignition process in Fig. 6-5a, during the first 300 μs after spark onset.[17]

This flame kernel growth is initially laminar-like, at least at low to mid-speed engine operating conditions. However, the electrical energy released in the kernel and heat losses from the kernel to the spark plug make this early flame development process nonadiabatic. Also, when the flame radius is less than or of order 1 mm, flame curvature effects on the growth rate are significant.

A gradual transition to a turbulent flame then occurs, as the local flow increasingly distorts the thin laminar-like reaction sheet flame. The local bulk flow can also convect the flame kernel center away from the spark plug electrodes. Both these phenomena are apparent in the Schlieren photograph in Fig. 6-5b, which shows two orthogonal views of the flame in three different cycles, 0.84 ms after spark onset. Both the bulk flow (through flame kernel convection), and the turbulence in the vicinity of the spark discharge, affect this stage of the flame development process. The flame reaction sheet becomes increasingly wrinkled and distorted as the enflamed region grows and the local turbulence at various length scales has time to influence the flame. The convection of the kernel center becomes less significant as the flame grows beyond about 10 mm radius because the enflamed mass is then too large to be transported by the local fluid motion.

The flame propagates outward in an approximately spherical manner from its center under normal in-cylinder flow conditions. As the flame grows, the flame front contacts the combustion chamber walls and then, locally, quenches. The walls obviously prevent continuing flame growth where the enflamed region is in contact with the combustion chamber walls. Figure 6-5c shows the flame front location as a function of time during this flame growth and wall interaction period. The engine design and its operating conditions, medium load and speed, are typical.

These flame front contours define the fraction of the combustion chamber volume enflamed. However, the mass fraction of the mixture that has burned is much less than the volume fraction enflamed because, for typical engine operating conditions (stoichiometric fuel–air mixture and part-throttle operation), the ratio of the unburned gas density to the burned gas density is about 4 and varies little during the combustion process. Thus, for example, with the spark plug located on the cylinder axis in a disc-shaped combustion chamber, only some 10% of the charge has burned when the flame radius is half the cylinder radius.

As the flame kernel grows, the thin reaction-sheet flame is wrinkled by the turbulent motion at scales smaller than the flame radius. Scales of turbulence comparable to and larger than the flame radius only distort the overall flame shape, or convect the flame, respectively. Experimental studies of flame growth[18] shows that, once the flame has grown to more than about 5 mm in radius, it can appropriately be described as a turbulent flame. At engine speeds of 1500–3000 rev/min, it takes about 5–10 crank angle degrees after spark onset to make this transition. Now turbulent convection is the dominant

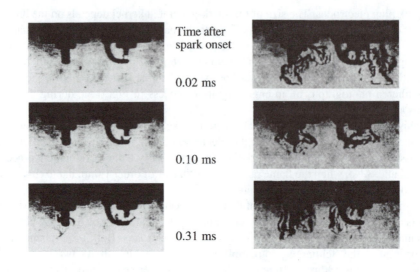

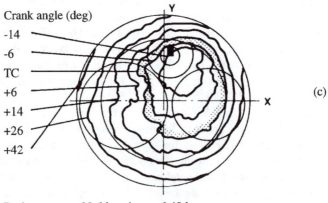

Peak pressure=28.6 bar, imep=6.40 bar

Engine operating conditions: Engine speed=3,000 rev/min, bmep=5.0 bar,

Air/fuel ratio=1.3, Ignition time=45° BTC

Figure 6-5 Images from photographic and optical diagnostic studies of spark-ignition engine flames: (a) Schlieren photos of ignition, and (b) early flame development[17]; (c) flame contours illustrating flame propagation[18]; (d) Schlieren photo of developed turbulent wrinkled reaction-sheet flame[1]; (e) simultaneous flame radiation (top) and planar laser-induced-fluorescence image of unburned HC, 4° before TC.[19]

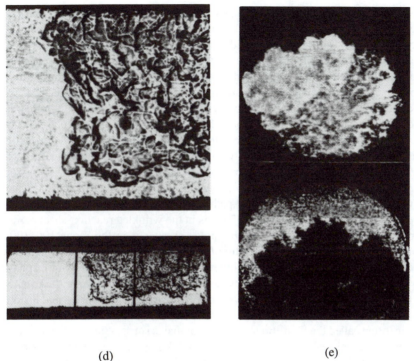

(d)   (e)

Figure 6-5   (Continued).

transport process, and the flame speed approaches that of a freely propagating developed turbulent flame.

Figure 6-5d and e shows results from three different experimental techniques often used to visualize engine flames in their turbulent stage. Figure 6-5d is a Schlieren image of a turbulent engine flame in a square-cross-section visualization engine.[1] The flame shown started at the spark plug electrode gap on the right-hand side of the lower photo and is propagating downward toward the piston crown and to the left toward the far cylinder wall. A portion of this photo is shown enlarged at the top. Schlieren photographs are integrated images along the optical path of the parallel light beam. Given that this flame is propagating outward in a roughly spherical manner, the Schlieren image of the flame moving to the left can be interpreted as a tangential view of the turbulent wrinkled reaction-sheet flame, where those sheet regions that have moved ahead of the mean sheet location appear as "hills," stacked beside and behind each other on the "horizon." That portion of the Schlieren image that is viewed perpendicular to the plane of the mean flame front appears as an "aerial view" of "hilly terrain," where the "bottoms of the valleys" appear as lines. A feature of the wrinkled flame-sheet model, where the turbulence distorts and convolutes the reaction sheet, is that the local laminar forward diffusion of the sheet rounds the leading flame sheet regions and forms cusps at the trailing flame sheet regions.

Figure 6-5e shows simultaneous images obtained from flame radiation and planar laser-induced fluorescence from fuel vapor in the unburned mixture, in an optically accessible spark-ignition engine with a disc-shaped combustion chamber.[19] The flame radiation technique produces an outer flame image similar to that produced by the Schlieren technique. Two-dimensional planar imaging of flames has become a well-developed and valuable technique for examining the details of the flame structure. The example shown, which identifies the unburned mixture region in a plane through the clearance height, shows finger-like regions of unburned gas that have penetrated into the turbulent flame and similar regions of burned mixture that have moved ahead of the mean reaction-sheet location. These two-dimensional planar images usually show a more irregular boundary between the burned and unburned regions than is apparent with the Schlieren visualization technique, which integrates density-gradient-produced refractive index changes along the optical path.

The turbulent spark-ignition engine flame is in the wrinkled reaction-sheet regime of turbulent combustion. The turbulent Reynolds number is in the range 100–1000. The Damköhler number, which is the ratio of the turbulent eddy turnover time to the residence time in the laminar flame, is of order 100 (the fast chemistry regime), and the smallest turbulent eddy size is larger than the laminar flame thickness so the reaction sheet structure is little affected by the turbulence. The turbulence wrinkles and distorts the flame, increasing the amount of thin reaction sheet contained within the turbulent flame zone (often called the flame brush) by a factor that is substantial and that increases with increasing turbulence. Most important, this wrinkling increases the mass burning rate. An opposing effect that wrinkling also produces is the stretching of the flame sheet, which slows down the molecular diffusive processes within the flame. Whether this flame-stretching effect is significant depends on strain rate and mixture composition.

The turbulence intensity u'—the root mean square value of the fluctuating in-cylinder flow velocity—at time of combustion, has a value about equal to the mean piston speed.† Hence, the ratio of reaction sheet area (often called the laminar flame area) to the turbulent flame frontal area (for which flame envelope area is a useful term) increases with increasing engine speed. It is primarily for this reason that the mass burning rate, which is proportional to the reaction sheet area, increases almost linearly with engine speed, thereby maintaining the angle required to burn from 10 to 90 percent of the in-cylinder charge almost constant over the full speed range of the engine. This ratio of laminar to turbulent flame area typically has a value of about 10 at 2000 rev/min.

The final stage of combustion, when the flame approaches the wall and extinguishes, has been much less extensively examined. There is a substantial literature on laminar flame quenching at a cool wall leading to the formation of quench layers—a thin layer of unburned fuel–air mixture adjacent to the wall (for a summary, see Heywood[1]). It is often assumed that the local (laminar) regions of the wrinkled turbulent flame that come in contact with the wall quench as just described, while the remainder of the turbulent flame brush burns out as if the wall is not there. This inherently assumes that the

---

†In four-stroke cycle engines, u' is equal to about half $\overline{S}_p$. The limited data on two-stroke cycle engines shows comparable values of turbulence intensity to four-stroke cycle engines. However, the scavenging process is less obviously driven by piston speed than is the intake process in four-stroke cycle engines.

unburned gas thermal boundary layer thickness is small relative to the turbulent flame thickness. However, as the flame approaches the wall, only successively smaller scales of turbulence are available to affect the flame so the final burn-up process will be slower than would occur in the developed turbulent flame.

In terms of the engine *combustion system*—the chamber geometry and plug location, and the geometry of the scavenging ports, all of which determine the in-cylinder flow during combustion—the primary quantities of interest are as follows:

1. the mass burning rate of the unburned mixture because this, through energy conservation, is directly linked to cylinder pressure and hence engine performance;
2. the cycle-by-cycle variation in the mass burning rate because this, via torque fluctuations, determines the stability of engine operation;
3. the burned gas heat transfer areas, which have a significant, though secondary, effect on the energy balance.

At a more fundamental level, the factors that affect the mass burning rate are as follows:

1. the mean flame frontal (or envelope) area contained within the boundaries of the combustion chamber;
2. the relevant bulk and turbulence flow parameters during combustion—the local mean flow, the local turbulence intensity, and the wrinkling length scales;
3. the local unburned mixture composition and state, which determine the local laminar burning velocity.

6-2-5 Cyclic Variations in Combustion

It has long been known that spark-ignition engines, apparently operating under steady-state conditions, do not maintain perfectly stable operation. A comparison between one cycle and another reveals substantial variations in the in-cylinder peak pressure, and indicated mean effective pressure. These variations are random in some instances and periodic in others. These variations are caused by variations in the engine's combustion process, cycle-by-cycle. They are particularly noticeable when an engine is operated with a lean and/or highly diluted in-cylinder mixture, and at low engine speeds and loads where these irregularities are usually discernible.

Periodic variations, often called surging, produce low-frequency engine speed fluctuations. The major reason for surging is a poor cylinder charging process that results in irregular combustion involving firing and misfiring cycles. Surging is a two-stroke engine phenomenon most common under throttled conditions when the delivery ratio and hence the scavenging efficiency are low. Under these conditions, parts of the mixture inside the cylinder are highly diluted by the residual and may not be ignitable, and a misfiring cycle can then occur. In the following cycle, the in-cylinder residual then contains substantial unburned fuel–air mixture, so the scavenging efficiency is therefore significantly higher, and the probability of successful ignition is much increased. Under extreme versions of these conditions, successful ignition occurs only after two or more unsuccessful ignitions, and combustion becomes periodic.

Tsuchiya et al.[20] have shown that the extent of surging can usefully be quantified by the engine speed fluctuation rate. This property, designated by C_f, is measured in revolutions per second-squared (rev/s^2). Figure 6-6 shows the performance map of a 200-cm^3 displacement single-cylinder loop-scavenged engine. Surging is most significant in the low to medium engine speed range and under light load conditions. Its severity is well characterized by the engine speed fluctuation rate, which these authors showed correlated well with the standard deviation in imep under these light load operating conditions. Note that σ_{imep} can be substantial under these conditions: of order 100 kPa at a C_f of 15 rev/s^2. It was found that surges whose C_f was 5 rev/s^2 or less would not affect riding comfort.

Note that an autoignition region exists on the performance map shown, where some of the unburned fuel, air, and residual gas mixture spontaneously ignites toward the end of the combustion process, as it is compressed to temperatures greater than about 1200 K as the cylinder pressure approaches its maximum. At higher speeds and mid-loads, the substantial mass of trapped hot residual gas and the high piston crown temperatures at these conditions, which increase the fresh charge temperature during scavenging due to convective heat transfer to the fresh charge, both increase the maximum (unburned) end-gas temperature.

Observation of individual-cycle cylinder pressure versus crank angle measurements, shows that the extent of variations, cycle-by-cycle, varies significantly with the operating conditions of the engine. Figure 6-7 shows a typical example of how the in-cylinder pressure varies from one cycle to another, at light load, for a 100-cm^3 displace-

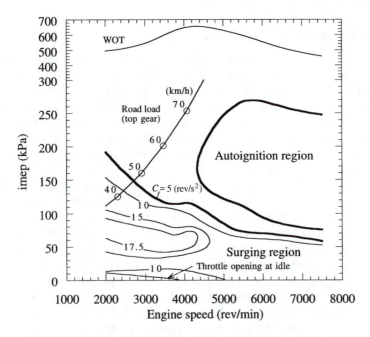

Figure 6-6 Surging and autoignition regions on a two-stroke cycle SI engine performance map.[20]

ment crankcase-blown loop-scavenged two-stroke engine, and Fig. 6-8 shows the measured distributions (the probability density functions) of a peak pressure and its crank
angle location, for three different delivery ratios in a 666-cm^3 displacement crankcase-
scavenged two-stroke engine. Figure 6-7 shows normal combustion cycles, slow burning cycles, partial burning cycles, and misfiring cycles. Note how some cycles have a
different compression stroke pressure; these are presumably cycles that follow misfires
or largely incomplete combustion cycles, where the in-cylinder mixture composition is
substantially different from normal. These figures show that both the peak pressure and
its location vary significantly; for a delivery ratio of 0.6, for example, the peak pressure
varies between 1.7 and 2.2 MPa in the 50% of the cycles around the average cycle (peak
pressure = 1.97 MPa), and the angle of peak pressure varies between 16 and 22° ATC
for the 50% of the cycles around the average cycle (peak pressure angle = 18.1° ATC).
For a delivery ratio of 0.8, the shape of the histograms are similar. For a delivery ratio of
0.4, the peak pressure distribution is highly skewed, and the crank angle of peak pressure is bimodal. This behavior is attributed to two very different combustion modes.[22]
For the mode characterized by later crank angles of peak pressure, combustion is faster
and the heat release rate after TC is significantly greater than the rate at which work
is done during expansion. For the other mode, combustion is slower so the heat release
rate is much lower, and the pressure history is only slightly increased above the motored
cycle pressures.

For two-stroke engines, cyclic variability is a serious problem in terms of smooth
operation, fuel consumption, and pollutant emissions, under low-load operation conditions. In multicylinder engines, in addition to these variations in each individual cylinder, there can be significant differences in the pressure–time diagrams between the
cylinders. Because the pressure development depends strongly on the combustion process, substantial variations cycle-by-cycle on the pressure–time diagrams largely result

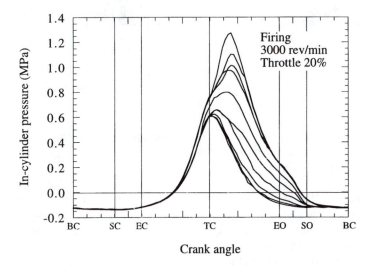

Figure 6-7 Cylinder pressure variations, cycle-by-cycle, under partial-load conditions.[21]

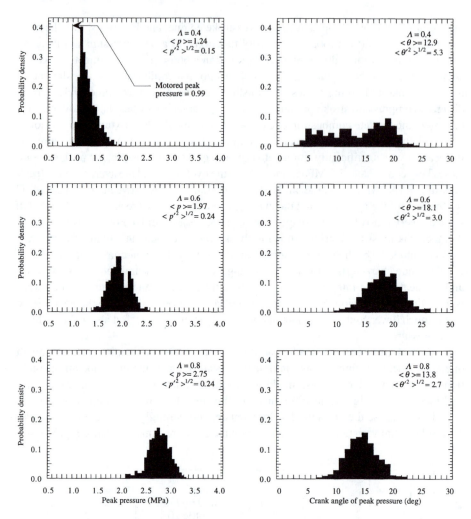

Figure 6-8 Histograms of peak pressure and its crank angle location, for three different delivery ratios in a 666-cm^3 crankcase-scavenged two-stroke engine[22]: delivery ratio Λ, mean value $\langle\ \rangle$, and standard deviations are given.

from substantial cycle-by-cycle variations in the combustion process. The major consequences of cycle-by-cycle variations in the combustion process and hence cylinder pressure development are the following[23]:

1. Relative to the mean cycle, faster burning cycles are ignited too early and slower burning cycles are ignited too late: thus only a fraction of the cycles will be burning with the optimum ignition timing. As a consequence, both the torque or power output and the efficiency are reduced relative to an engine with negligible variability. If cyclic combustion variability could be eliminated, there would be about a 10–15% increase in the power output for the same fuel flow rate.

2. Different amounts of work are produced in each cycle. Fluctuations in the torque output or engine speed result. Significant torque or speed variations cause poor driveability.
3. At high loads, the faster burning cycles are more likely to knock because they produce higher end-gas compression. These cycles will determine the octane requirement of the engine (the minimum octane rating fuel required to just avoid knock); or for a given fuel, determine the maximum permissible compression ratio.
4. When the engine is operating at conditions where combustion is inherently slow, the slowest burning cycles may not have completely burned by the time the exhaust valve opens (this is known as a partial burn). Partial burning leads to high levels of unburned hydrocarbons and carbon monoxide.

The causes of cycle-by-cycle combustion variations have been most extensively studied in four-stroke SI engines.[24] Table 6-1 lists the phenomena that are believed to contribute significantly. Each phenomenon affects one or more of the different phases of the combustion process: flame development, flame propagation, late combustion stage, as indicated. These same phenomena occur in two-stroke cycle engines. However, the two-stroke scavenging process introduces additional variability. Note that the processes listed in Table 6-1 fall into two main categories: phenomena that occur in the vicinity of the spark plug, which primarily affect the early stages of combustion (the flame development), and phenomena whose impact is on a larger physical scale that affect the main flame propagation process. However, processes that change the rate of flame development do have an impact later in the combustion process. The time it takes the flame to develop to a certain size affects the piston position at that flame size. Hence, for different flame development rates, the combustion chamber geometry at a given flame radius and location will be different, which changes the flame envelope area contained within the combustion chamber. The phenomena listed in Table 6-1 originate in the flow and composition variations. Variations in the flow, both locally and on a larger or bulk scale, cause variations in the combustion process. Figure 6-9 shows how the flow near the spark plug can convect the developing flame kernel. It shows flame kernel convection histograms in a four-stroke cycle engine at two different spark timings.[25]

Table 6-1 Causes of SI engine cycle-by-cycle variations†

Spark energy deposition in gas (1)
Flame kernel motion (1, 3)
Heat losses from kernel to spark plug (1)
Local turbulence characteristics near plug (1)
Local mixture composition near plug (1)
Overall charge composition—air, fuel, residual (2, 3)
Average turbulence in the combustion chamber (2, 3)
Large-scale features of the in-cylinder flow (3)
Flame geometry interaction with the combustion chamber (3)

†Each of these, primarily, affects (1) the early stages of flame development, (2) the main flame propagation process, or (3) the later stages of combustion.

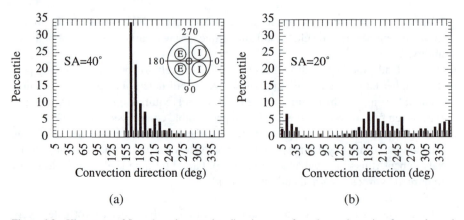

Figure 6-9 Histograms of flame kernel convection direction away from the spark gap in a four-stroke cycle SI engine with a tumbling in-cylinder flow[25]: (a) spark onset at 40° BTC; (b) spark onset at 20° BTC.

Each histogram represents the frequency of occurrence of kernel convection toward the indicated direction in a two-dimensional plane roughly parallel to the piston face. Figure 6-9a shows that with ignition at 40° BTC the flame kernel convection is consistently directed toward the 180° axis, most likely due to the large-scale intake-generated tumbling flow driving fluid past the spark gap at this point in the cycle. With ignition at 20° BTC, the flame kernel convection direction is much more random, presumably because the large-scale tumbling flow has disintegrated into smaller scale motions and turbulence.

These cyclic flow fluctuations, coupled with composition fluctuations due to imperfect mixing of the cylinder contents, can lead to substantial variations in the flame geometry and rate of flame kernel development. Figure 6-10 illustrates this: it shows engine cycles obtained from two orthogonal Schlieren photographs of the flame kernel as it developed from the spark plug electrodes in a special visualization (four-stroke) engine operating at typical light-load conditions.[26] These sequences of the flame images from several successive engine cycles illustrate important features of the spark-ignited flame kernel development process, and indicate that there are substantial differences between flame kernel shape, growth rate, and motion of the kernel center relative to the electrode gap, in each cycle, as the flame kernel is convected and distorted by the local flow field, which varies from cycle to cycle.

The variations in flame shape, center location, and growth rate shown in Fig. 6-10 both cause and are affected by variations of these additional factors: the electrical energy the spark discharge deposits in the developing flame kernel; heat losses from the kernel to the plug electrodes and insulator; the geometry of the flame front and its interaction with the combustion chamber wall, which defines the flame envelope area across which unburned mixture enters the flame, both as the flame develops and then propagates across the chamber.

The spark discharge typically lasts about 1–2 ms. If the flame kernel center moves away from the gap, the discharge is stretched which (provided it is not extinguished) increases the electrical energy deposited into the kernel. Also, the kernel loses heat to the

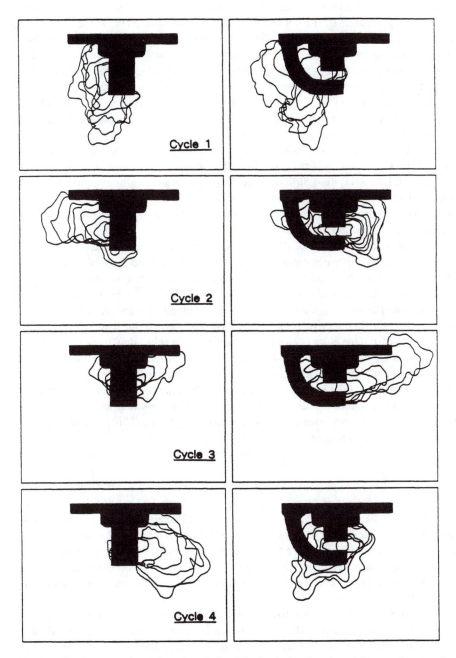

Figure 6-10   Sequence of outer flame boundaries of the developing flame kernel, for several consecutive cycles, in a spark-ignition engine at light load. Two orthogonal views are shown. Time between flame boundaries is 0.25 ms or 2 crank angle degrees.[26]

plug electrodes and insulator surfaces. The amount of heat lost depends on the surface area of the plug "wetted" by the hot kernel gases. This depends on the kernel shape and location. The rate of electrical energy dissipation in the kernel and the rate of heat loss to the electrodes are about comparable in magnitude (and opposite in sign). Only some 250 μs after spark onset does the rate of combustion energy release become greater than either of these.[27] Thus, the geometry of the spark plug, and the orientation of its ground electrode with respect to the average direction of gas motion (swirl, tumble, etc.) have a significant effect on combustion cyclic variability.[28] Spark energy and spark duration were found to be less important.[29] Studies with four-stroke cycle engines[30–33] have shown that the use of electrodes of different size and shape can significantly reduce the minimum spark energy required for ignition, the extent of cycle variability, and when misfire occurs. Ignition is most difficult under lean (excess air) or dilute (with substantial residual) operating conditions. Under these conditions, details that result in faster initial flame growth (e.g., smaller electrodes, longer discharge) reduce combustion variability and extend the engine's operating limits.

Misfire in a small fraction of the engine's combustion events defines the engine's operating limits. By misfire is meant an operating cycle with no discernible chemical energy release. More than about 1% misfires is unacceptable for smooth engine operation and low HC emissions. Two different misfire mechanisms have been observed.[27,34]

In one, the flame did not move away from the electrodes and therefore lost excessive heat to the electrodes. In the other, large portions of the flame were rapidly convected away from the electrodes and the flame was quenched due to excessive shear. Pischinger and Heywood[26] explained about two-thirds of the cyclic combustion variability measured in their four-stroke SI engine at part-load conditions by assuming this results from excessive heat loss to the spark plug. Extending this concept, Sher et al.[35] have proposed that ignition fails to occur when the heat loss rate to the spark plug ($\dot{Q}_{loss}$) exceeds the energy release rate by combustion ($\dot{Q}_{comb}$). This postulate leads to the following misfire criterion:

$$f_c/(1 - f_f) \geq 2 \qquad (6\text{-}21)$$

where f_c ($f_c \geq 0$) is a contact area ratio defined as the ratio between the surface area of the electrodes, insulator, and plug "wetted" by the hot kernel gases, and the flame surface area, and f_f ($0 \leq f_f < 1$) is the fraction of the flame area lost due to interruption by the electrodes. Figure 6-11 shows the successful ignition and misfire domains on a plot of contact area ratio versus lost flame area fraction. For the very early stages of ignition, when the characteristic length of the flame kernel is of the order of the electrode gap, the lost flame area fraction is comparable to the contact area ratio, and successful ignition will occur as long as $f_c < 2/3$.

Sher and Keck[36] have shown that, in general, the probability of initiating a self-sustained propagating reaction sheet in nonhomogeneous mixtures is higher when a larger initiating hot region is produced in a shorter time. The size of the hot region volume at a given time depends mainly on the burning velocity of the mixture, which depends on the pressure, temperature, local equivalence ratio, and residual (burnt) gas fraction. Sher and Keck have shown that variations of ±5% in the pressure affect the burning velocity by less than 1%. Similar magnitude variations in temperature, equiv-

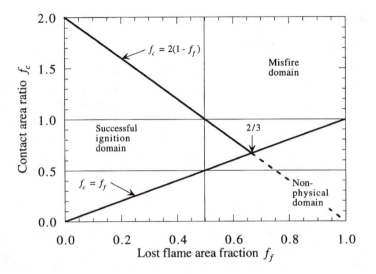

Figure 6-11 Successful ignition and misfire domains on a plot of flame contact area ratio [(plug contact area)/(flame envelope area)] versus lost flame area fraction.[35]

alence ratio, and residual fraction affect the burning velocity by 10%, 5%, and 13%, respectively. Under extreme conditions the burning velocity may vary by some $\pm 30\%$, which means that the initial volume of the hot gas may vary by a factor of about 6: i.e., $(1.3/0.7)^3$.

The gas exchange or scavenging process in the two-stroke SI engine is a significant contributor to mixture nonuniformity and flow effects on combustion variability. The two-stroke scavenging process results in cyclic variations in residual gas content that are typically larger than $\pm 5\%$.[37] Because, as explained above, the initial kernel growth rate is highly sensitive to this parameter, higher cycle-by-cycle combustion variations than occur in the four-stroke SI engine are expected. Thus variations in the residual gas fraction are largely responsible for the increased magnitude of cycle-to-cycle variations in two-stroke cycle SI engines.

Measurements of the instantaneous composition of the exhaust gases (CO, CO_2) at the exhaust port of a lightly loaded two-stroke cycle engine, have shown that the concentrations of CO and CO_2 peak before a misfiring cycle occur.[38] Also, a significant reverse flow through the exhaust port (an inflow into the cylinder) was observed during the gas exchange period that preceded the misfiring cycle. Under lightly loaded conditions, the cylinder is charged with a highly diluted mixture that is difficult to ignite. Partial burning or misfiring cycles are therefore likely to occur. The driving force for the observed reverse flow was not investigated. However, it is likely to be related to the effect that increased backflow of residuals into the scavenging ducts has on the trapped fresh charge. Suppressing the reverse flow through the exhaust port with a fluid diode installed at the exhaust port does reduce cyclic combustion variations in crankcase-scavenged two-stroke engines.[39]

6-3 AUTOIGNITION AND DIRECT-INJECTION COMBUSTION

6-3-1 Active Radical Combustion

Figure 6-6 shows that autoignition—spontaneous ignition due to sufficiently high un-burned mixture temperature—occurs over part of the two-stroke SI engine operating map. By controlling the speed and load at which autoignition occurs through controlling either the exhaust port or the transfer duct open area, this combustion mode can be used over part of the engine's operating map, a part important in normal engine use. Then, in the regions with autoignition as the combustion mode, very low cyclic variability can be achieved, and HC emissions and fuel consumption can be reduced. First proposed by Onishi et al.[40] and termed *active radical combustion*, it is effected by retaining a larger than normal amount of hot residual exhaust gas in the low to medium speed range, thereby raising the trapped in-cylinder unburned mixture temperature, so that toward the end of the compression stroke spontaneous ignition will occur. With appropriate levels of residuals, chemical energy release then takes place in a controlled highly re-peatable manner. More recent efforts by Honda (Activated Radical Combustion),[41,42] IFP (Active Thermo Atmosphere Combustion),[43] and others have developed practical two-stroke engine versions of this process.

Figure 6-12 shows how the in-cylinder mixture composition and state determine whether or not activated radical combustion occurs. The figure illustrates how the mean charge temperature at start of compression depends on the amount of residual gas and of fresh charge. It is difficult in region A to produce stable activated radical combustion because the gas temperatures are too low. Region B is most suitable due to the high charge temperature and the substantial dilution of the fresh charge with residual. Region D is where the normal spark-ignition flame propagation occurs; in C normal spark-

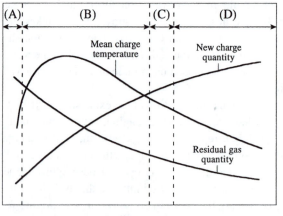

Delivery ratio

Figure 6-12 Different engine combustion regimes on a plot of new charge quantity, residual gas quantity, and mean (unburned) charge temperature, as a function of delivery ratio. Active radical combustion is feasible in region B, difficult in region A. Normal spark-ignition combustion occurs in D; both active radical and normal combustion can occur in C.[40]

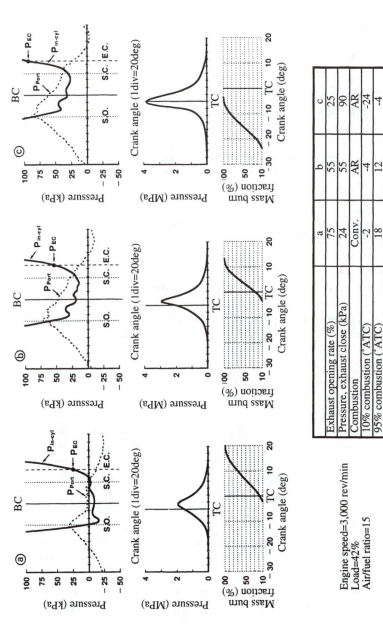

	a	b	c
Exhaust opening rate (%)	75	55	25
Pressure, exhaust close (kPa)	24	55	90
Combustion	Conv.	AR	AR
10% combustion (°ATC)	-2	-4	-24
95% combustion (°ATC)	18	12	-4
Specific heat ratio (-)	1.32	1.31	1.31
Crank angle max. pressure (°ATC)	14	10	-2
imep (MPa)	0.35	0.41	0.31

Engine speed=3,000 rev/min
Load=42%
Air/fuel ratio=15

Figure 6-13 The effect of throttling the exhaust, which increases the pressure at exhaust port closing, on combustion regime. As the effective exhaust open area is reduced (EOR, the exhaust opening rate is the throttled effective flow area divided by the unthrottled effective flow area), the transition from normal combustion (a) to active radical combustion [(b) and (c)] occurs.[41]

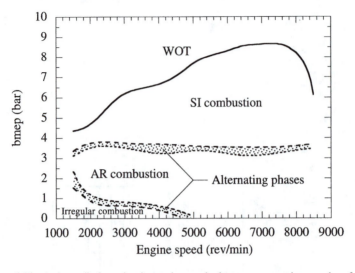

Figure 6-14 Active radical combustion region on the bmep versus engine speed performance map of a Schnürle-type loop-scavenged single-cylinder 246-cm³ displacement production engine, modified by adding an exhaust control valve with both open area and timing control.[42]

ignition and activated radical combustion are apt to occur interchangably.[40] Control of the mixture condition in region B (when the engine is operating at part throttle—a low delivery ratio) can be achieved by throttling the transfer duct(s) to reduce the amount of fresh charge entering the cylinder via the scavenging ports, relative to the amount of in-cylinder residual. Onishi et al. used this approach, claiming it also reduces the in-cylinder turbulence. More recent efforts to implement this concept have used exhaust port throttling[41] or an exhaust valve system that varies the port opening angle and rate[42] (see Section 5-5-4), to control the amount of residual and fresh charge.

In active radical combustion, the combustion reactions occur spontaneously at many points throughout the in-cylinder mixture and occur sufficiently slowly that the chemical energy release occurs in a time comparable to that required for normal combustion. Under conditions where this type of combustion process occurs (region B in Fig. 6-12), it has been shown that OH, CH, and C_2 radicals start to form relatively early in the compression stroke, as a result of the higher than normal charge temperature, and increase in concentration as compression proceeds.[40] Chemical energy release starts relatively slowly throughout the in-cylinder charge, occurring at many finely distributed locations, with small time differences in start of combustion between these many locations. The chemical energy release rate then increases smoothly to a maximum and then decreases, typically lasting overall about 30° crank angle. Because active radical combustion is a distributed phenomenon, it is important that the quantity and composition of the fresh mixture supplied to the cylinder each cycle be highly repeatable as must also be the scavenging flow, to ensure the necessary consistence of the in-cylinder mixture.[40]

The transition to active radical combustion from spark-ignited flame propagation as the exhaust port opening rate is reduced is illustrated in Fig. 6-13. The diagrams on the left show normal combustion with optimum spark timing. As the exhaust opening rate is reduced, and the pressure at EC and hence the amount of trapped residual increases, active radical combustion now occurs (middle diagram). Relative to normal combustion it is both much more repeatable (e.g., the standard deviation in imep is about 1–2% of the mean imep, rather than close to the 10% value typical of normal spark-ignition engine combustion at light load) and ends sooner. Both these factors increase the mean imep. At a still lower exhaust port opening rate (right-hand diagrams), the higher residual fraction and mixture temperature lead to earlier start and end of active radical combustion—too early in fact, so imep decreases. The delivery ratio at these light load conditions was 42%.

Figure 6-14 shows that region of the full operating map where a 246-cm^3 Schnürle loop-scavenged single-cylinder two-stroke motorcycle engine can operate in this active radical combustion mode. The advantages of this different mode are very repeatable combustion (thus higher imep for the same fuel input or, equivalently, better fuel consumption by as much as 50%) and lower hydrocarbon emissions (down to about one-half to one-third). The timing of active radical combustion onset is controlled by controlling the exhaust port opening rate to maintain the chemical energy release at the optimum time in the cycle.[41]

6-3-2 Direct-Injection Spark-Ignited Combustion

Direct-injection two-stroke spark-ignition engine concepts were introduced in Section 1-6 (see Figs. 1-24–1-26). An alternative approach to direct injection that produces a stratified fuel–air mixture in the cylinder is the IAPAC concept, which injects fuel at low pressure into a surge tank whose contents then blow into the cylinder after BC when the primary scavenging air flow has occurred. This concept is described in Section 2-3-6 (see Fig. 2-25). One major objective of these concepts is to prevent any significant short-circuiting of fuel. At high load, where full utilization of the air trapped in the cylinder is important, injection is done in such a way as to produce a relatively homogeneous mixture by the time combustion occurs. At light loads, where air utilization is not a constraint, creating a stratified fuel–air mixture within the cylinder may be advantageous. If the mixture is homogeneous, then combustion occurs in a manner similar to its occurrence in carburetted premixed-charge engines. A description of the fuel distribution in direct-injection two-stroke engines, illustrated by data from various optical diagnostic techniques, can be found in Section 3-3-2. As illustrated there, the spray penetrates into the cylinder, the fuel droplets vaporize (Fig. 3-25), and they produce at spark discharge a nonuniform fuel–air mixture distribution, with an easily ignitable region that envelops the spark-plug electrodes (Figs. 3-26 and 3-27). It is a challenging design task to achieve this situation over the full load and speed range of the engine. Any nonuniformities in the fuel–air mixture composition at the spark plug location, cycle-by-cycle, can have a significant effect on ignition and flame development. Additional examples of the fuel distribution in a two-stroke engine with an air-assist injector can be found in Ghandhi et al.[44] Figure 6-15 illustrates how the spray and the piston crown and cylinder head geometry in the Orbital combustion chamber design interact during the latter part of the

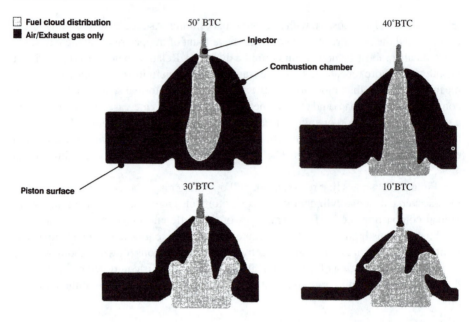

Figure 6-15 Computational fluid dynamic prediction of direct-injection fuel spray development with Orbital Engine Co. air-assist fuel injector and combustion chamber and piston crown geometry.[45]

compression stroke to produce a fuel-vapor-containing cloud surrounded by regions of air and residual but no fresh injected fuel.[45]

Insights into the features of the direct-injection engine's combustion process can be obtained from the data in Fig. 6-16 from a loop-scavenged two-stroke engine operating at 1500 rev/min and a light-load imep of 177 kPa. The delivery ratio is 0.23, the trapped fuel/air equivalence ratio is 0.69, and the residual gas fraction is 0.65. The mean and the maximum and minimum of 100 cycles of cylinder pressure data are shown. There is significant cyclic variability [the coefficient of variation (COV) in imep is 12%; in a fully warmed version of the special optical engine, the COV would be about 3%]. The lowest mass fraction burned curve is the average of partial burning cycles where less than 80% of the trapped fuel mass burned. The burning profile (lower left) shows an initial rapid burning phase where about 50% of the fuel burns in a partially premixed mode, followed by a slower mixing-controlled burnout phase. There is a substantial variability in the total mass of fuel burned. The midburn times (10–50% burned angles) vary more widely than the other burn angles shown, and correlate inversely with the final mass fraction burned. The poorest-burning cycles have unusually long 0–10% and 10–50% burn durations. Combustion video imaging showed that after the initial flame development, luminosity from around the injector region indicated that fuel-rich combustion was occurring. Combustion luminosity ended about 40° ATC.[46]

Because of its obvious benefit of eliminating short-circuiting of fuel, development of in-cylinder direct fuel injection technology for use in two-stroke engines is being actively pursued. Many options are available: dual fluid or air-assist injection systems; high-pressure injection systems; different geometry sprays (different spray cone or di-

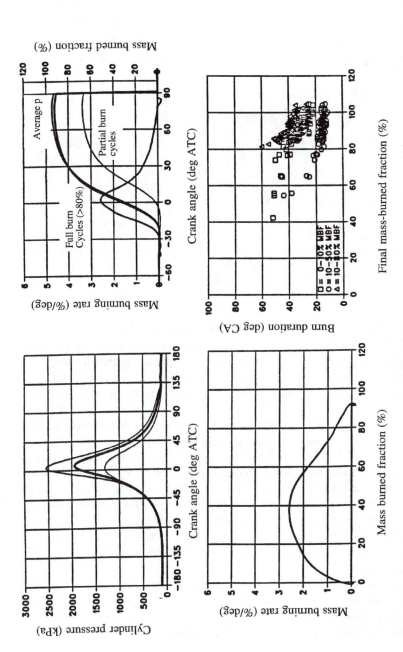

Figure 6-16 Experimental data for cylinder pressure (heavy curve is 100-cycle average, light curves are max. and min.) versus crank angle, with individual-cycle mass burned and burn rate obtained from pressure data (heavy curve is ensemble-averaged for all cycles burning more than 80% of the fuel in the cylinder, medium curve is calculated from mean pressure trace, light curve is an ensemble average for all partial burning cycles), burning rate versus mass-burned fraction, and individual cycle 0–10%, 10–50%, and 10–80% burn durations versus final mass-burned fraction for an air-assist injector direct-injection two-stroke engine, operating at 1500 rev/min, imep of 177 kPa, delivery ratio of 0.23, trapped fuel/air equivalence ratio of 0.69, and residual fraction of 0.65.[46]

241

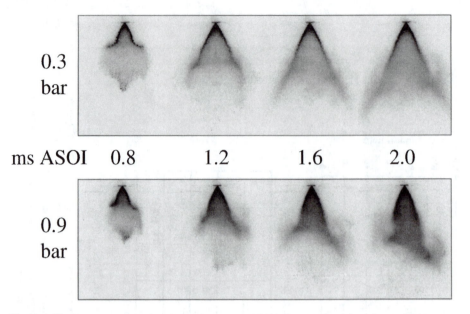

Figure 6-17 Spray patterns produced by high pressure (5 MPa) liquid-fuel swirl injector. Images are laser induced fluorescence from a laser sheet through the nozzle and spray axis. Fluorescence is dominated by emission from components in the indolene droplets in the spray, with darker regions corresponding to higher intensity emission. Top images: injection into air at 0.3 bar pressure. Bottom images: injection into air at 0.9 bar, with same injector and liquid fuel pressure. Times after start of injection (ASOI) in ms. (Courtesy Brad VanDerWege, MIT).

vergence angles, hollow cone sprays, multiple spray nozzles). The injector location, cylinder head shape, and piston crown shape if spray interaction with the piston is important, are additional geometric factors affecting spray behavior. Injection timing is an important variable: it controls the time available for spray penetration and liquid evaporation, and after exhaust port closing it affects the charge density into which the spray is injected.

High-pressure (about 5 MPa) liquid fuel injection technology is the simpler choice. Such fuel injection systems are being developed for four-stroke cycle direct-injection gasoline engines, also. With the injector nozzle passageway designed to generate swirl—rotation of the fluid about the injector axis—these injectors can produce a spray whose penetration and spreading depend on the in-cylinder gas density and injection pressure. Drop sizes that spread about an average diameter of 20 μm can be produced; these will mostly evaporate prior to combustion under normal engine operating conditions. Figure 6-17 shows the spray pattern from a typical liquid fuel injection system. Its hollow cone structure, which produces a more uniform fuel-vapor–air mixture within the spray region is apparent.

Air-assist (or air-blast) injection technology generally provides better spray characteristics, but at the cost of increased complexity, and a modest power penalty to compress the air. In a modern air-assist fuel injector, liquid fuel is fed to the injector at about 6 bar pressure and air at about 5 bar.[47] The pressure difference between air and fuel is controlled so precise metering of liquid fuel can be achieved. The appropriate amount of fuel is injected into the injector sac volume by activating the fuel solenoid that controls the fuel valve needle. A standard post fuel injector can be used for this purpose. The injector poppet valve is then opened by activating its solenoid, and the flow of high-pressure air that now occurs blows this metered ammount of fuel through the main injector poppet valve into the cylinder as a finely atomized spray.[47] Examples of the spray pattern from an optimized air-assist fuel injection system are shown in Fig. 6-18. With their poppet main-injection valve geometry, these injectors produce a hollow cone spray, typically with drop sizes spread about a mean diameter of about 8 μm (smaller than those obtained with high-pressure liquid injection), with about 85% of the mass of injected fuel below 10-μm drop size.

Figures 3-25–3-28 show how the fuel distribution develops between start of injection and start of combustion. With later injection, at lighter loads, the spray region spreads more slowly (Fig. 3-25) and has less time to penetrate. The fuel vapor cloud remains relatively confined; what is important is that the mixture composition around the spark plug be stoichiometric or slightly rich at time of ignition (Fig. 3-26). At higher load, a much more uniform fuel distribution throughout the cylinder is needed to ensure good air utilization. Earlier injection accomplishes this. Figure 6-19 shows Exciplex liquid and vapor images at (a) a low-speed light-load condition (800 rev/min, $\phi_{overall} = 0.15$, start of injection 130° ABC) and (b) a medium-speed medium-load condition (1600 rev/min, $\phi_{overall} = 0.5$, start of injection 70° ABC).[44] This diagnostic technique uses a laser sheet through the spray axis to cause two dopants added to the fuel to fluoresce. One of these controls the intensity of fluorescent emission from vapor-phase fuel; the liquid-phase fluorescence depends on the concentration of the excited complex formed from both of the dopants. The images are shown on an eight-level gray scale with white representing low intensities and black high intensities. The higher load (b) with start of injection at 70° ABC spreads out to essentially fill the combustion chamber with fuel vapor, relatively uniformly, as the piston approaches top center. For both higher load–earlier injection and lower load–later injection, the liquid fuel has almost completely vaporized by the time combustion starts. An advantage of air-assist injection is that it eliminates any dribbling of fuel or low-velocity drops leaving the injector as injection ends. However, the need to provide compressed air at 6 bar pressure increases the complexity and cost of this technology and requires additional compression work.

Nonuniformities in mixture composition, which are inevitable in direct-injection engines and are larger than in port-injected or carburetted premixed engines, contribute significantly to cyclic combustion variability. Observations in direct-injection two-stroke cycle engines indicate that cyclic combustion variability, eventually leading to misfires and partial burns, is associated with cyclic variations in the fuel concentration near the spark gap. Drake et al.[48] have applied laser-induced-fluorescence imaging of gasoline, high-speed video imaging of spectrally resolved combustion luminosity, and simultaneous exhaust HC sampling using a fast-response flame-ionization detector. Figure 6-20 shows measurements of cylinder pressure, mass burning rate, HC concentration, and HC exhaust mass flow rate. Results such as those shown suggest that most of the cylinder contents and a major part of the total HC mass are exhausted during

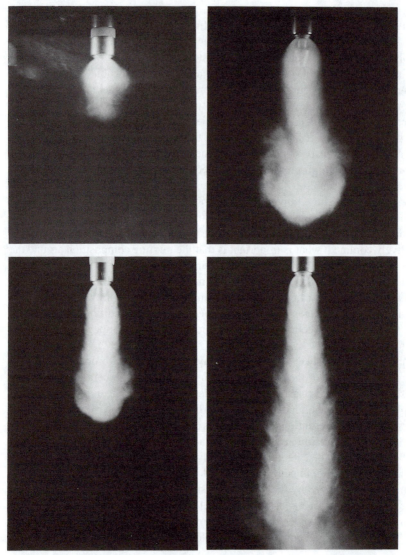

Figure 6-18 Spray patterns produced by air-assist fuel injection technology. Top photos: 300 kPa air pressure, 1 and 3 ms after start of air injection, with 3 ms delay between fuel metering and air injection. Bottom photos: 750 kPa air pressure, 1 and 3 ms after start of air injection, with 3 ms delay after fuel metering. 5 mg fuel injected. Air flow is typically a few percent of engine air flow. (Courtesy Orbital Engine Company).

the main scavenging phase when the exhaust mass flow and HC concentration are both high, and that the cyclic variability results primarily from incomplete combustion in randomly located pockets of unburned mixture. These pockets contain fuel-rich or fuel-lean mixture that mix and react too slowly to burn completely. The pockets are formed when injected fuel over mixes (forming regions of mixture that are too lean to burn) and are due to fuel that outgases from injector crevices.[48]

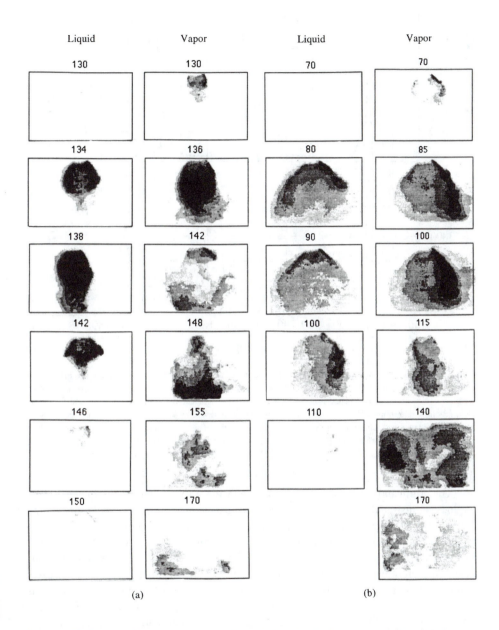

Figure 6-19 Exciplex laser-sheet images of liquid fuel-produced fluorescence (left) and fuel-vapor-produced fluorescence (right) for (a) low speed, light load, no swirl condition (800 rev/min, overall $\phi = 0.15$, injection timing 130° ABC) and (b) medium speed, medium load, no swirl condition (1600 rev/min, overall $\phi = 0.5$, injection timing 70° ABC) in an optical-access two-stroke engine, with a forced-air injection system.[44]

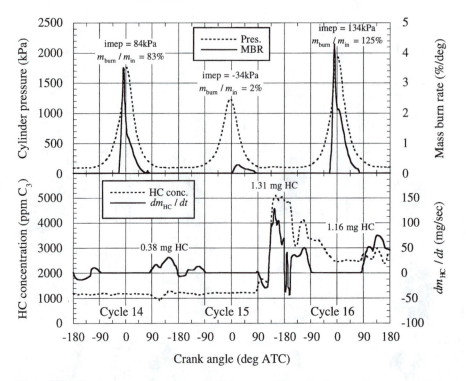

Figure 6-20 Simultaneous measurements of cylinder pressure, mass-burning rate, fast-response FID hydrocarbon exhaust concentrations, and exhaust HC mass flow rate in three successive engine cycles. Also shown are each cycle's imep, mass burned relative to average, and mass HC exhausted. Middle cycle misfires. The HC mass flow rate in normal cycles (e.g., 14) shows (i) a cylinder blowdown peak, (ii) a higher second peak as the transfer or scavenging ports are uncovered, (iii) a backflow from the exhaust to the cylinder around BC, and (iv) a final outflow of HC as the rising piston closes first the transfer, then the exhaust ports. Misfiring cycles (15) show an initial backflow, rather than a blowdown, and highly elevated HC concentrations.[48]

The origin and extent of cyclic combustion variability in direct-injection two-stroke engines is illustrated in Fig. 6-21, by the individual cycle flame envelope contours of the developing flame kernel a few millimeters in diameter and the larger flame a centimeter or more in diameter that rapidly (in 4° crank angle) develops from it.[49] Contours from six different cycles are shown. The cyclic variability results from both variations in mixture motion and mixture composition. This study of direct-injection two-stroke combustion showed that combustion of premixed close-to-stoichiometric regions, as well as locally much richer regions of a millimeter length scale, occurs. The fuel transport from the injector, and the mixing of fuel vapor with air and residual gas, depends strongly on the duration and timing of the air pulse in the blast atomizer system. The spray will contain both richer and leaner than average regions. This random mixture nonuniformity is a major cause of the different flame kernel growth rates indicated by the flame contours at 6° after spark onset. Variations in mixture motion are also important, and their impact through the different flame envelope shapes and flame center locations shown is evident. This affects heat loss to the plug electrodes, electrical energy

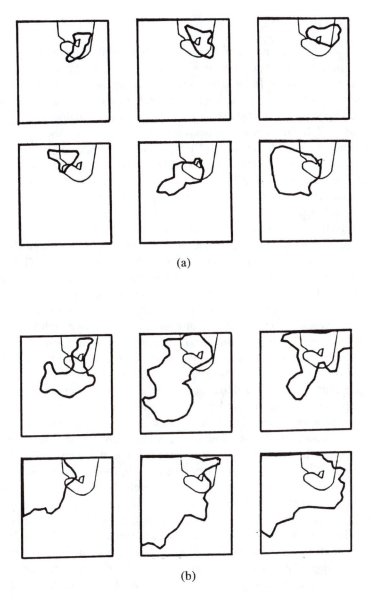

(a)

(b)

Figure 6-21 Tracings of flame kernel from six cycles, (a) 6° after spark and (b) 10° after spark. Thin line shows view of spark plug through optical window.[49]

deposited in the kernel, the rate at which the local turbulence impacts the flame growth, and the subsequent geometry of the flame interaction with the combustion chamber walls, as explained in Section 6-2-5. Substantial variations in direct-injection combustion, cycle-by-cycle, present a major design challenge at lighter operating loads where the spreading of the fuel spray must be carefully controlled.

6-4 DIESEL COMBUSTION

Larger two-stroke cycle diesel or compression-ignition (CI) engines are used as propulsion units in marine and large-scale power generation. Medium-sized two-stroke diesel engines have been less popular. In the past, only limited use has been made of small two-stroke diesels. There is, however, a growing interest in small two-stroke diesels due to their combination of high indicated efficiency, compact size and lighter weight, and lower engine friction.[50] In these diesel engines, fresh air is introduced to the cylinder by a supercharger, is compressed by the moving piston, and fuel is injected at high pressure directly into the main or an auxiliary combustion chamber just before combustion commences. Delivery ratios of 1.5–2.2 are commonly used to scavenge the burned gas of the previous cycle effectively from the cylinder. Higher delivery ratios in two-stroke diesels than are used in premixed spark-ignition engines are acceptable because only air, instead of fuel–air mixture, short-circuits the cylinder.

In diesel or compression-ignition engines, the fuel is introduced to the cylinder after the fresh air has been compressed to a sufficiently high pressure (about 5 MPa), and especially temperature (above about 850 K), for autoignition of the fuel–air mixture to occur. To achieve rapid mixing, the fuel is injected at a high pressure. Depending on the engine size and type of combustion chamber employed, high-pressure differences across the fuel injector nozzle are required to provide droplet velocities high enough to traverse the combustion chamber in the short time available, and droplet sizes small enough for effective liquid fuel distribution and rapid evaporation. Typical liquid fuel injection pressures range from 20 to 150 MPa. The role of the injection system is, therefore, to provide the appropriate quantity of fuel to each cylinder as (several) well-atomized fuel jets, at the appropriate time at the desired rate. The broader objective is to produce the required power output with the lowest fuel consumption, along with smooth and reliable operation, within the constraints of the allowable emission levels. The specific quantitative requirements depend on the particular engine size and type of combustion chamber, and the operating conditions of the engine (speed and load). In fact, there is not much difference between the requirements of a two-stroke cycle diesel engine combustion process compared with that of a four-stroke engine, and the fuel injection systems are closely comparable. For further reading on the basic principles of fuel injection systems for compression-ignition engines, the reader is referred to Heywood[1] and Bosch.[51,52]

The start of injection is set just before the desired start of combustion, and ends some 20–40° crank angle later. The fuel is injected through one or, more commonly, several small orifices in the injector tip, and disintegration of the liquid jet into small droplets occurs immediately due to shear stresses across the interface between the liquid and the surrounding air. The droplets penetrate into the combustion chamber, vaporize, and the fuel vapor then mixes with air to form a combustible mixture within each developing fuel spray. Because the mixture temperature and pressure are well above the fuel–air mixture ignition conditions, spontaneous ignition occurs after a short delay time in those regions of the spray where enough air has mixed with the fuel to be within the combustible limits. Following ignition, the cylinder pressure increases, the vaporization rate of the remaining liquid fuel increases, and the ignition delay of the mixing-produced fuel-vapor–air mixture is shortened. Fuel vaporization, fuel–air mixing, ignition, and combustion occur simultaneously in different portions of the fuel, while charge compression, continuing fuel injection, and piston movement all take place.

The diesel engine is inherently much more efficient than the spark-ignition engine, due primarily to its different method of ignition and combustion process. The diesel always operates lean (to prevent excessive exhaust black smoke) and runs extremely lean at light load (ϕ about 0.2) because the airflow is unthrottled. The compression ratio is higher, because the diesel is not limited by the knock constraint of the premixed spark-ignition engine. These differences combine to improve engine brake efficiency significantly. As mentioned previously, greater emphasis on engine efficiency in the future may increase the role of the two-stroke diesel.

6-4-1 Diesel Combustion Systems

For satisfactory combustion in diesel engines, both the fuel and air distribution and motion must be arranged so that they have ample opportunity to mix together and thus burn completely and cleanly. Any air or fuel that does not participate in a complete combustion process represents a waste of potential output. Unburned fuel or rich-mixture burning results also in hydrocarbon and smoke emissions. Fast enough mixing and complete burning of the fuel over the full range of operating conditions are the more important features of a well-designed combustion system. Other important qualities that depend on the combustion system design are the maximum output (bmep), minimum fuel consumption, maximum useful engine speed, and low noise. There is no single combustion system—geometry of piston and cylinder head, nozzle spray pattern, in-cylinder air motion—that combines all these qualities, for all engine sizes.

Compared with four-stroke cycle diesels, there is less freedom of choice with two-stroke engines because, in addition to providing the air and fuel motions that produce fast enough mixing and combustion, there is a need for a good scavenging flow pattern. Combustion chambers that produce a vigorous air motion may disrupt the scavenge pattern and so would be inferior to a more quiescent chamber.[53]

Diesel engines are divided into two basic categories according to their combustion system design: (1) *direct-injection* (DI) engines, which have a single open combustion chamber into which the fuel is injected directly; and (2) *indirect-injection* (IDI) engines where the combustion chamber is divided into two regions and the fuel is injected into a "prechamber" that is connected to the main chamber above the piston crown via a passageway or nozzle, or one or more orifices. Within each category, there are various different cylinder head and piston geometries, fuel spray arrangements, and airflow patterns.

The diesel combustion process is largely mixing controlled. The maximum crankshaft speed of engines, the speed at which the engine's maximum power is developed, varies inversely with engine stroke. This result follows from the fact that, at maximum power, engines are breathing-limited because the flow velocities in the most constrained regions of the intake system approach the sonic velocity. Thus smaller engines with their correspondingly higher maximum speed require faster fuel–air mixing rates to enable the duration of combustion in crank angle degrees to remain about constant. In direct-injection diesel combustion systems, faster mixing is obtained by higher air swirl, achieved by both higher intake-generated swirl and by deeper piston bowls with smaller diameters relative to the cylinder bore.

In the smallest engine sizes (with bores below about 80 mm), which operate at the highest engine speeds, historically it has proved impractical to achieve fast enough mixing in direct-injection combustion systems, even with very high air swirl. Indirect-injection combustion systems, which generate the extremely vigorous air motion required for fast enough mixing and combustion during the compression process, are used instead.

Direct-injection (DI) combustion systems. In this type of combustion system (Fig. 6-22), the combustion chamber shape is usually a bowl in the piston crown, a central multihole fuel injector is used, and, where necessary, the mixing between the air and the fuel is enhanced by generating a swirling air motion around the cylinder axis within the combustion chamber by means of a suitable geometry inlet port. The air swirl rate generated during intake is increased during compression as the piston approaches TC by forcing the air into the bowl-in-piston combustion chamber, toward the cylinder axis. In the larger diesels, which operate at the lowest engine speeds ($\sim$100 rev/min), the momentum of the fuel jets is usually sufficient to produce fast enough mixing between the fuel vapor and air. Then, simple open combustion chamber geometries suffice and have the advantage of lower relative heat losses due to their lower surface-to-volume ratio and lower gas velocities. As engine size decreases, deeper and smaller diameter piston bowls are used to increase the amount of air swirl generated to achieve faster fuel–air mixing rates. Multiple fuel-jet injector nozzles are usually used to distribute the fuel throughout the air charge. Some four-stroke cycle diesels use the single-spray hemispherical bowl-in-piston with air swirl design in Fig. 6-22c, where the fuel is vaporized off the hot piston bowl wall, instead of within each of the multiple fuel sprays as in Fig. 6-22b. Use of direct-injection combustion systems in four-stroke cycle engines has recently been extended to smaller diesels: engines with bores of about 80 mm and maximum speeds of 4500 rev/min. This has been done through use of more swirl-generating intake ports, generating more vigorous air motion during compression with reentrant shaped piston bowls, and with higher pressure fuel injection and smaller nozzle holes. This (small) high-speed four-stroke DI diesel is rapidly entering the passenger-car diesel engine market.

Direct-injection two-stroke diesel engines are characterized by the following:

1. low relative fuel consumption, because (a) heat losses are relatively low (low surface-to-volume ratio and low levels of turbulence) and (b) in larger engines, with their lower engine speeds, the combustion duration in crank angle degrees is short and hence the operating cycle approaches the ideal constant-volume-combustion thermodynamic cycle more closely;
2. good scavenging characteristics in the larger engine sizes, which allow the use of low delivery ratios for the same scavenging efficiency; lower speeds allow more time for effective scavenging; also, low levels of turbulence result in less mixing between the fresh charge and the burned gas;

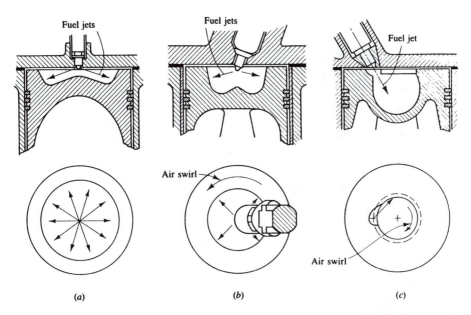

Figure 6-22 Different designs of direct injection (DI) chambers: (a) quiescent, shallow bowl chamber; (b) deep bowl in piston chamber with multihole nozzle and air swirl; (c) hemispherical bowl with single spray onto piston bowl surface and air swirl.[1] Reproduced with permission of The McGraw-Hill Companies.

3. easier cold starting because (a) the air into which the fuel is injected is at a higher temperature due to the lower heat losses during the compression process, (b) ignition occurs at the center of the combustion chamber (so heat losses during combustion are reduced), and (c) lower levels of turbulence prevail so there is less fluid shear during ignition.

In many of the large two-stroke marine diesel engines a different version of the quiescent chamber is used (see Fig. 6-23). Here the combustion chamber is essentially disc-shaped, the cylinder head is occupied by the exhaust valve, the swirling motion of the uniflow scavenging flow persists in the combustion chamber, and two, three, or four individual injectors (one is shown) each spray a single fuel jet toward the center of the chamber to be carried circumferentially by the swirling air.[54]

Indirect-injection (IDI) combustion systems. Inlet-generated air swirl, despite amplification in the piston bowl, has not in the past provided sufficiently high fuel–air mixing rates for the smallest diesel engines. Indirect-injection or divided-chamber engine combustion systems have been used instead, where the vigorous charge motion required during fuel injection is generated during the compression process. Two broad classes of IDI system are in current use: (1) swirl chamber systems and (2) prechamber systems, as illustrated in Fig. 6-24a and b, respectively. During compression, air is forced from the main chamber above the piston into the auxiliary chamber, through the nozzle or orifice (or set of orifices). Thus, toward the end of compression, a vigorous flow in the auxiliary chamber is set up. In swirl chamber systems the connecting passage

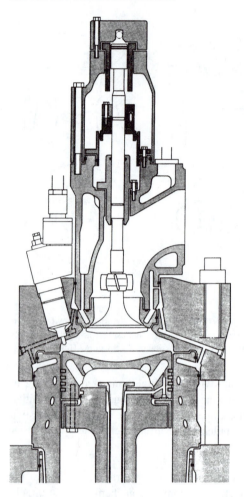

Figure 6-23 Direct-injection combustion chamber of large bore (from 380 to 840 mm), long stroke (from 1100 to 2400 mm) Sulzer engine, for marine and power generation applications (courtesy Wärtsilä NSD Switzerland Ltd.).[54]

and chamber are shaped so that this flow within the auxiliary chamber rotates extremely rapidly about the swirl chamber axis. This highly swirling prechamber flow produces fast enough mixing with the injected fuel spray to achieve fast enough combustion. The turbulent prechamber achieves fast enough fuel–air mixing by generating a highly turbulent flow within the prechamber as gas flows rapidly into the prechamber toward the end of compression through the one or more orifices connecting the prechamber to the space above the piston—the main chamber. In the example shown (Fig. 6-24b), a baffle across the middle of the prechamber generates additional turbulence.

Fuel is usually injected into the auxiliary chamber at lower injection-system pressure than is typical of DI systems through a pintle nozzle as a single spray. Owing to the high velocity of the air in the prechamber, the shear stresses between the air and the fuel are extremely high, and a finely atomized fuel spray is not necessary; a coarser spray is acceptable. Combustion starts in the auxiliary chamber; the pressure rise associated with combustion forces fluid (fuel vapor, air, burning and burnt gases) back into

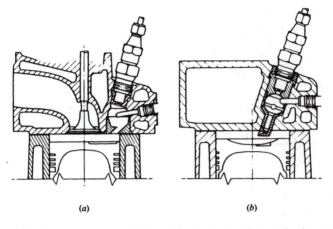

(a)   (b)

Figure 6-24 Two common types of indirect-injection (prechamber) combustion systems: (a) swirl prechamber; (b) turbulent prechamber.[1] Reproduced with permission of The McGraw-Hill Companies.

the main chamber, where the jet (or jets) issuing from the nozzle or orifices entrains and
mixes with the main chamber air. Swirl chambers are very effective at mixing the fuel
and consequently are capable of producing high bmep and clean combustion. However,
because of the high tangential velocities, heat transfer from the air to the chamber walls
is more significant than in more quiescent combustion chambers, and energy losses
due to cylinder head cooling are higher. For the same reason, poor cold starting is another characteristic of swirl chamber engines. Compared with four-stroke cycle diesels,
starting problems are generally more acute in two-stroke engines, and the high level of
turbulence created in the prechamber conflicts with the scavenging process objectives.
The glow plug shown on the right of the prechambers in Fig. 6-24 is a commonly used
cold-starting aid. The plug is heated prior to starting the engine to ensure vaporization
and ignition of fuel early in the engine cranking process.

Indirect-injection engines are characterized by the following:

1. good high-speed operation because (a) the mixing rate during combustion between
   the fuel and the air is extremely rapid and (b) the ignition delay (for fuel evaporation
   and mixing with air) is shorter due to the high turbulence intensity;
2. high maximum bmep (in naturally aspirated engines) because (a) the breathing
   penalty associated with creating swirl during intake is avoided and (b) rapid mixing in the prechamber enables higher air utilization;
3. low hydrocarbon and smoke emissions owing to the intense mixing process;
4. smooth operation at low engine loads because the fuel can be burned with the
   prechamber fraction of the inducted air at closer to stoichiometric conditions;
5. quiet operation due to lower rate of pressure rise and lower maximum pressure in the
   main chamber.

Figure 6-25 shows an example of a swirl prechamber two-stroke diesel. The details
of this supercharged Toyota S-2 prototype engine are given in Table 6-2.[55] The cylinder

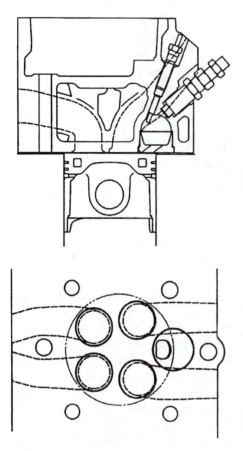

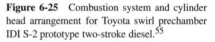

Figure 6-25 Combustion system and cylinder head arrangement for Toyota swirl prechamber IDI S-2 prototype two-stroke diesel.[55]

head contains two intake valves (on the right) and two exhaust valves for effective scavenging, and the swirl prechamber as shown. Compared with a four-stroke turbocharged diesel of equal displacement, the maximum power is about 25% higher, the maximum torque (bmep = 807 kPa) is about 40% higher, and the minimum bsfc is about equivalent. Relative to an equivalent four-stroke prechamber diesel, the compression ratio can be lowered (to 18) due to the increased amount of hot residual in the cylinder. This slightly improves efficiency and output (the optimum efficiency in practice occurs at a compression ratio of 15–16). Engine vibration is noticeably less than that of four-stroke diesels due to the doubling of the combustion frequency. Combustion noise is substantially reduced, presumably because the delay between start of fuel injection and onset of combustion is reduced by the higher in-cylinder charge temperature at start of injection due to the hotter two-stroke residual gas.[55]

Choice of combustion system. Clean burning, high bmep, low fuel consumption, ability to run at high speeds and low sound level, and a wide range of operation are the important features to be accomplished by a well designed diesel combustion system. Because there is no one combustion chamber approach that combines all these char-

**Table 6-2 Specifications and performance of
Toyota S-2 indirect-injection two-stroke diesel[55]**

Engine type	Water-cooled, in-line, 4-cylinder
Displacement	2489 cm^3
Bore × stroke	96 mm × 86 mm
Compression ratio	Geometric 18
Valve lift	Intake 7.2 mm
	Exhaust 8.6 mm
Valve timing	IVO 45° BBC
	IVC 50° ABC
	EVO 75° BBC
	EVC 65° ABC
Maximum engine speed	3600 rev/min
Maximum engine output	100 kW
Maximum engine torque	320 N-m
Minimum fuel consumption	274 g/kW-h

acteristics, the relative importance of these features determines the choice of the combustion chamber type. In two-stroke cycle engines, chamber designs that promote rapid mixing between the air and the fuel are usually less than ideal from the perspective of gas exchange because minimum mixing between fresh air and residual is required during the scavenging period. Combustion chambers that produce a vigorous air motion during intake may disrupt the desired scavenge flow pattern and may be inferior overall to a more quiescent chamber. For these reasons, provided the cylinder size is large enough and engine speeds are correspondingly low, simpler direct-injection combustion systems are the preferred choice. Weaknesses of the direct-injection chambers, compared with indirect-injection chambers, are their tendency to produce higher exhaust emissions of NO_x and smoke, and their demand for a high-pressure fuel injection system to produce more finely atomized sprays.

6-4-2 Essential Features of the Diesel Combustion Process

The combustion process in a compression-ignition two-stroke cycle engine is essentially the same as in a four-stroke diesel. In a typical direct-injection engine, four distinct phases in the combustion process can be identified. This description of diesel combustion comes from studies of high-speed photographic movies of various types (e.g., see Heywood[1] and Balles and Heywood[56]) data obtained from laser-sheet optical diagnostics (see Dec[57]), cylinder pressure diagrams (see Fig. 6-26), and heat release analysis. The concept of *heat release* is important to understanding our description. It is defined as the rate at which the chemical energy of the fuel is released by the combustion process. It can be calculated from cylinder pressure versus crank angle data as the energy

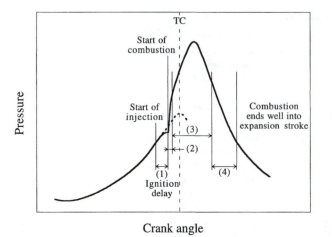

Figure 6-26 Direct-injection diesel cylinder pressure versus crank angle diagram showing typical location of the four phases of the diesel combustion process.

release required to produce the observed pressure change, using techniques discussed in Section 6-4-3.†

The four phases of the diesel combustion process are identified on the cylinder pressure versus crank angle diagram in Fig. 6-26:

1. *Ignition delay period*—This is the time between the start of fuel injection into the combustion chamber and the start of combustion energy release. The start of injection is usually determined from an injector needle-lift indicator. The onset of combustion can be determined from the p–θ diagram, as the crank angle where the cylinder pressure turns sharply upward in Fig. 6-26. During the ignition delay period, the injected liquid fuel jets disintegrate and atomize into small droplets. These move through, and entrain into each spray, the hot air in the combustion chamber. Droplet vaporization takes place, and mixing between fuel vapor and the air entrained into the spray occurs. The spray may impinge on the combustion chamber walls; normally, the liquid drops vaporize well prior to the interaction of each fuel-vapor–air spray with the walls. During this period, the cylinder pressure is slightly lowered below the value it would have in a motored engine due to the fuel vaporization and fuel–air mixing process (see Fig. 6-26). *Autoignition* then occurs in the fuel–air mixture that has mixed well enough, and for long enough, that energy-releasing chemical reactions commence. The air temperature into which the fuel is injected is a critical variable: it must result in high enough mixture temperatures for the spontaneous ignition chemistry to occur sufficiently rapidly to keep the ignition delay period short. Recent studies[57] have identified the onset of autoignition with the onset of chemilu-

---

†Depending on the analysis method used, the actual fuel chemical energy released, or that quantity less the heat loss to the walls, is obtained. The latter is usually called the net heat released.

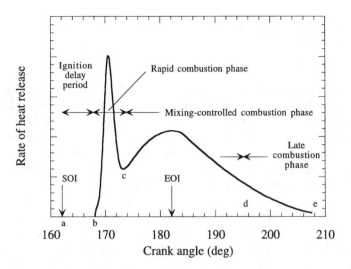

Figure 6-27   Rate of (net) energy or heat release in a direct-injection compression-ignition engine, showing the four phases of the diesel combustion process.

minescence radiation produced as the fuel molecules that have mixed with air in the leading portion of each spray break down and start to release their chemical energy.

2. *Rapid combustion phase*—This phase is characterized by a rapid rise in cylinder pressure (see Fig. 6-26) due to rapid chemical energy release from the fuel that has had time to vaporize and mix with air during the delay period. Pressure rise commences prior to onset of high luminosity (yellow–orange) emission due to energy-releasing chemical reactions in the relatively uniform rich fuel-vapor–air mixture ($\phi \sim 2$–4) that constitutes much of the leading portion of the spray. The chemiluminescence radiation from these reactions indicates where and when they occur.[57] Shortly thereafter, diffusion flames form in several locations close to each spray boundary and rapidly spread to surround each fuel spray, burning up the reaction products from the chemistry occurring in the fuel-rich leading portion of the spray as these products diffuse into the outer spray region where there is excess air.[56,57] If the cylinder pressure versus crank angle data are converted into a diagram of the rate of chemical energy (or heat) release (see Fig. 6-27), this combustion phase contributes the spike at the start of the energy or heat release process. In the past this has been called the "premixed burning phase." Although some of the energy release now appears to come from the burning of the "premixed" rich ($\phi \approx 2$–4) mixture in the leading portion of the spray, some of it comes from the burning of the products of this rich-mixture combustion in a diffusion flame positioned around the spray where, through diffusion, the overall mixture composition is stoichiometric. During this rapid burning (premixed) combustion phase, fuel injection and atomization continue, and droplet vaporization is enhanced by radiation heat transfer from the

diffusion flame, and the rising temperature of the air entrained into each spray as that air is compressed.

3. *Mixing-controlled combustion phase*—Once the fuel and air that mixed during the ignition delay have burned rapidly as described in phase 2, combustion continues with increasing and then gradually decreasing rate of energy release (Fig. 6-27). The combustion rate is controlled by the rate at which mixture becomes available for burning, which is primarily controlled by the mixing of fuel vapor with air. Recent studies[57] suggest that, during this phase, the liquid fuel atomizes, vaporizes, and mixes with enough air within the spray to produce fuel-rich mixture, which reacts within the spray to produce rich incomplete combustion products, which then burn to completion as mixing with oxygen occurs in the turbulent diffusion flame that now surrounds the spray. The rate of burning and energy or heat release during this phase depends on parameters that govern the fuel–air mixing rate (e.g., nozzle hole geometry, fuel injection pressure, air swirl rate) and not on parameters that primarily affect the chemical reaction rates (such as air temperature at the end of compression, fuel ignition quality as quantified by the fuel's cetane number).

4. *Late combustion phase*—Many experimental observations show that after injection ceases the heat release rate decreases monotonically well into the expansion stroke. Interpretation of photographs from high-speed movies of fuel sprays injected into swirling air flows in rapid compression machines[1,56] suggests that late in the combustion process, while only a small fraction of the fuel may not yet have burned, a larger fraction of the fuel energy is present unreleased in soot and fuel-rich combustion products. However, the expansion rate of the cylinder volume is now high, the temperature of the cylinder gases is falling rapidly, and the kinetics of the final burnout processes become much slower. This last stage of diesel combustion is not yet well understood. However, luminosity from the burnup of soot particles formed in the fuel-rich regions of each spray where combustion occurs, continues well into the expansion stroke.[1]

This description of the diesel combustion process is illustrated by the fuel spray schematics shown in Fig. 6-28. These are based on the conceptual model proposed by Dec[57] to explain the results of laser-sheet imaging studies of diesel combustion in a standard modern direct-injection four-stroke engine. This model significantly extends our earlier understanding of this process by explaining when and where soot forms in the diesel fuel spray, and where premixed and mixing-controlled combustion chemistry occurs. This model also helps explain the origin of the diesel engines particulate and nitrogen oxides emissions problems, the subject of Chapter 7.

The most important current two-stroke cycle diesels are large-bore direct-injection engines (e.g., Figs. 1-7 and 2-7). Although indirect-injection (prechamber) combustion systems are commonly used in small four-stroke engine sizes, they are increasingly being replaced by small high-speed direct-injection combustion systems with high air

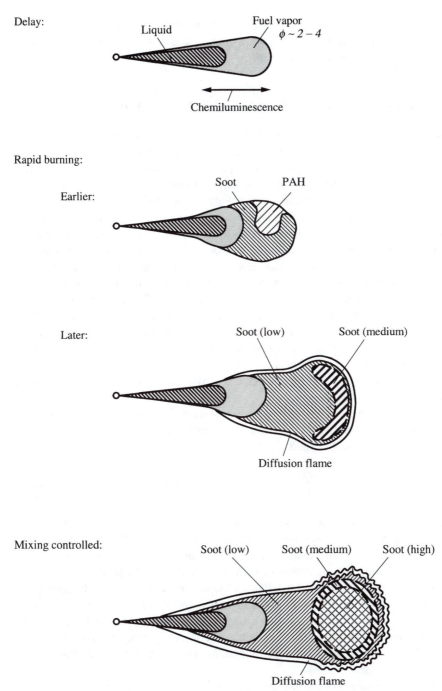

Delay:

Liquid

Fuel vapor
$\phi \sim 2 - 4$

Chemiluminescence

Rapid burning:

Earlier:

Soot PAH

Later:

Soot (low) Soot (medium)

Diffusion flame

Mixing controlled:

Soot (low) Soot (medium) Soot (high)

Diffusion flame

Figure 6-28 Schematics illustrating critical processes and regions of a diesel fuel spray during the delay period, rapid (premixed) burning, and mixing-controlled burning phases of phenomenological diesel combustion model (based on Dec[57]).

swirl. Any small two-stroke diesel developments are likely to use this more efficient high-speed DI approach. Accordingly we will not discuss the details of the combustion process in indirect-injection engines. We refer the reader to Heywood's text.[1]

6-4-3 Heat Release Analysis from Cylinder Pressure Data

Quantifying the rate at which combustion proceeds inside the engine cylinder helps us understand the critical processes that are occurring. Calculations relating the mass fraction burned, rate of heat release, and gas temperature distribution yield important information often needed to evaluate variables such as optimum start of injection, injection duration, cylinder liner temperature for tribology studies, bulk motion during combustion for better port geometry design, and predictions of engine performance under different operating conditions. Cylinder pressure is usually measured with a piezoelectric pressure transducer inserted through the cylinder head. Combustion rate information can be obtained from accurate cylinder pressure data provided appropriately accurate models for the other relevant phenomena that affect the pressure are available.

As is done in spark ignition engines (see Section 6-2-2), an overall energy balance for the cylinder contents is used to estimate the rate of chemical energy or heat release from cylinder pressure and volume data. Here we develop an approximate, yet useful and practical, method for direct-injection diesels. A major advantage of such an approach is that the rate of energy or heat release by combustion can be explicitly expressed in terms of the pressure changes. More accurate models may be found elsewhere.[1]

The combustion chamber can be considered an open system where mass is allowed to cross into or out of the system through ports, orifices, or nozzles. An energy balance for the combustion chamber yields

$$\frac{\delta Q_n}{dt} = \frac{dE}{dt} + \frac{\delta W}{dt} - \sum \dot{m}_i h_i \tag{6-22}$$

Here $\delta Q_n/dt$ is the net heat release rate, which is the difference between the chemical energy release by combustion and the heat losses to the wall, dE/dt is the rate of change of the internal energy of the system (the cylinder contents), $\delta W/dt$ is the rate at which work is done by the system, and $\dot{m}_i h_i$ is the rate at which enthalpy enters the system with any mass flow. Although the cylinder contents may comprise two different gases (air or fuel–air mixture, and burnt gas) with two very different thermodynamic states, it is assumed that the system contains a single gas that obeys the ideal gas law, that has a constant specific heat c_v, and negligible kinetic energy compared with its thermal energy. During compression and expansion, $\dot{m}_i h_i$ is the enthalpy entering the cylinder with the injected fuel.

The sensible energy of the gas in the cylinder can be expressed as

$$\frac{dE}{dt} = \frac{d}{dt}(mc_v T) \tag{6-23}$$

Because $pV = mRT$ and $c_v/R = (\gamma - 1)^{-1}$, where γ, the specific heat ratio (c_p/c_v), is assumed independent of temperature and mixture composition, Eq. (6-23) yields

$$\frac{dE}{dt} = \frac{1}{(\gamma - 1)} \frac{d}{dt}(pV) \tag{6-24}$$

The rate at which work is done by the system due to volume expansion is

$$\frac{\delta W}{dt} = p\frac{dV}{dt} \tag{6-25}$$

The net (or apparent) heat release rate is the difference between the rate of energy release by combustion $\delta Q_{ch}/dt$ and the rate of heat loss to the walls $\delta Q_w/dt$:

$$\frac{\delta Q_n}{dt} = \frac{\delta Q_{ch}}{dt} - \frac{\delta Q_w}{dt} \tag{6-26}$$

Substitution of Eqs. (6-24), (6-25), and (6-26) into Eq. (6-22) yields

$$\frac{\delta Q_{ch}}{dt} - \frac{\delta Q_w}{dt} = \frac{1}{(\gamma - 1)} \frac{d}{dt}(pV) + p\frac{dV}{dt} - \dot{m}_f h_f \tag{6-27}$$

where the last term represents the sensible or thermal enthalpy of the injected fuel.† This term is small compared with the other terms and may be neglected. Equation (6-27) may be simplified to yield the chemical energy or gross heat release rate as follows:

$$\frac{\delta Q_{ch}}{dt} = \frac{1}{(\gamma - 1)}\left(\gamma p \frac{dV}{dt} + V\frac{dp}{dt}\right) + \frac{\delta Q_w}{dt} \tag{6-28}$$

The heat loss rate to the wall $(\delta Q_w/dt)$ may be estimated through a suitable heat transfer model in a manner similar to that used for spark-ignition engines[9]:

$$\frac{\delta Q_w}{dt} = A\left[0.76\frac{k}{B}\left(\frac{S_p B}{\nu}\right)^{0.64}(T - T_w) + 0.54\sigma(T^4 - T_w^4)\right] \tag{6-29}$$

Here A is the surface area of the system (cylinder volume) through which heat is transferred, T_w is the wall temperature, k is the thermal conductivity of the gas, ν is its kinematic viscosity, S_p is the instantaneous piston speed, B is the cylinder bore, and σ is the Stefan–Boltzmann constant (5.67×10^{-8} W/m^2-K^4).

Thus, the rate of energy heat release due to combustion $(\delta Q_{ch}/dt)$ may therefore be evaluated from measurements of the pressure inside the cylinder and the cylinder volume versus time, and an estimate of the heat transfer. It should, however, be noted that the simplifying assumptions made, in particular the assumption that c_v is independent of temperature and gas composition, may introduce significant errors into the calculations. However, due to difficulties in dealing with the impact of the piston, ring, and liner crevices, and of accurately estimating the heat transfer, more sophisticated and rigorous models give only marginally better results for the interpretation of the combustion energy release profile.

---

†Note that the chemical energy of the fuel, which can be released by combustion, is the $\delta Q_{ch}/dt$ term.

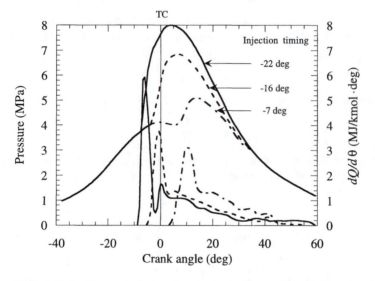

Figure 6-29   Indicator diagrams and rates of combustion in DI compression-ignition four-stroke cycle engine, at different injection timings.[58]

Equation (6-28) may also be used to estimate the fuel mass-burning rate (dm_f/dt) via the relation

$$\frac{dm_f}{dt} = \frac{1}{Q_{HV}} \frac{\delta Q_{ch}}{dt} \tag{6-30}$$

An indirect-injection diesel version of this DI engine analysis can be found in Heywood.[1]

Typical cylinder pressure diagrams and rates of combustion in a DI four-stroke cycle engine, at varying injection timings, are shown in Fig. 6-29. Depending on the injection timing, load, and engine speed, the shapes of the rate of combustion curves vary significantly. However they generally exhibit the basic two-stage combustion behavior identified as the rapid-burning or premixed phase and the mixing-controlled combustion phase (see Section 6-4-2), regardless of the operating conditions.

For practical purposes, when numerical simulation of a diesel cycle is required with the objective of predicting the cylinder pressure, the use of simple models for the rate of energy release by combustion is often appropriate. Watson et al.[59] have suggested that the rate of combustion in a DI diesel engine can be correlated with critical operating parameters using a double Weibe function. Miyamoto et al.[58] have extended this method to include IDI engines and have correlated the combustion rate with key operating and design variables. They carried out a systematic study of a wide range of diesel combustion conditions resulting from different types of combustion chambers, different fuels, and a range of engine speeds, loads, and injection timings. It is interesting to note that correlations for noise, smoke emissions, and thermal efficiency with the parameters of the double Weibe functions were also developed.[58] Although their recommendations are based on data taken with four-stroke cycle engines, their model should be suitable for two-stroke cycle engines as well because there are no fundamental differences between their combustion processes.

Miyamoto et al.[58] suggested the following correlation for the combustion energy release rate:

$$\frac{\delta Q_{ch}}{d\theta} = 6.9 \frac{Q_p}{\Delta\theta_p}(M_p+1)\left(\frac{\theta}{\Delta\theta_p}\right)^{M_p} \exp\left[-6.9\left(\frac{\theta}{\Delta\theta_p}\right)^{M_p+1}\right]$$
$$+ 6.9 \frac{Q_d}{\Delta\theta_d}(M_d+1)\left(\frac{\theta}{\Delta\theta_d}\right)^{M_d} \exp\left[-6.9\left(\frac{\theta}{\Delta\theta_d}\right)^{M_d+1}\right] \quad (6\text{-}31)$$

where the subscripts p and d refer to premixed and diffusive combustion parts, respectively, $\Delta\theta_p$ and $\Delta\theta_d$ are the duration of the premixed and diffusive combustion periods, and M_p and M_d are the relevant shape factors, respectively. Q_p may be interpreted as the chemical energy released during the rapid or premixed combustion phase, and Q_d as the difference between total energy release and Q_p.

Of the six parameters, the two shape factors and the premixed combustion duration were found to be nearly independent of operating conditions and fuel for both DI and IDI engines. Their recommended values are $M_p = 3.0$ for DI and IDI engines, $M_d = 0.5$ for DI engines, and 0.9 for IDI engines, and $\Delta\theta_p = 7°$ for DI and IDI engines. Because Q_d is the difference between the total energy released and Q_p, two additional parameters, $\Delta\theta_d$ (the duration of the diffusive combustion period) and Q_p (the heat released during the premixed combustion period) must be determined. Although these parameters were found to be specific to individual engines and, therefore, should be found by experiment, Miyamoto et al.[58] recommended the following correlations:

$$Q_p = \begin{cases} 0.5 Q_{id} & \text{for DI engines} \\ 0.8 Q_{id} & \text{for IDI engines} \end{cases} \quad (6\text{-}32)$$

where Q_{id} is the fuel injected during the ignition delay period, expressed in terms of energy of combustion, and

$$\Delta\theta_d = \begin{cases} 0.93 Q_d + 24.5° & \text{for DI engines} \\ 0.93 Q_d + 19.0° & \text{for IDI engines} \end{cases} \quad (6\text{-}33)$$

REFERENCES

1. Heywood, J.B., *Internal Combustion Engine Fundamentals*, McGraw-Hill, New York, 1988.
2. Metghalchi, M., and Keck, J.C., "Burning Velocities of Mixtures of Air with Methanol, Isooctane and Indolene at High Pressure and Temperature," *Combustion and Flames*, vol. 48, pp. 191–210, 1982.
3. Rhodes, D.B., and Keck, J.C., "Laminar Burning Speed Measurements of Indolene–Air–Diluent Mixtures at High Pressures and Temperature," SAE Paper 850047, *SAE Trans.*, vol. 94, 1985.
4. Harari, R., and Sher, E., "The Effect of Ambient Pressure on the Performance Map of a Two-Stroke SI Engine," SAE Paper 930503, 1993.
5. Lancaster, D.R., Kreiger, R.B., and Lienesch, J.H., "Measurement and Analysis of Engine Pressure Data," SAE Paper 750026, *SAE Trans.*, vol. 84, 1975.
6. Rassweiler, G.M., and Withrow, L., "Flame Temperatures Vary with Knock and Combustion-Chamber Position," *SAE Trans.*, vol. 36, pp. 125–133, 1935.

7. Bopp, S.C., Cousyn, B.J., Green, R.M., and Witze, P.O., "Experimental Study of Scavenging and Combustion Processes in a Two-Stroke Cycle Research Engine," SAE Paper 920183, *SAE Trans.*, vol. 101, 1992.

8. Taylor, C.F., and Toong, T.Y., "Heat Transfer in Internal Combustion Engines," ASME Paper 57-HT-17, 1957.

9. Annand, W.J.D., "Heat Transfer in the Cylinders of Reciprocating Combustion Engines," *Proc. Inst. Mech. Eng.*, vol. 177D, pp. 973–989, 1963.

10. Woschni, G., "A Universally Applicable Equation for Instantaneous Heat Transfer in the Internal Combustion Engine," SAE Paper 670931, *SAE Trans.*, vol. 76, 1967.

11. Rohrer, R., and Chehroudi, B., "Preliminary Heat Release Analysis in a Single-Cylinder Two-Stroke Production Engine", SAE Paper 930431, 1993.

12. Chun, K.M., and Heywood, J.B., "Estimating Heat-Release and Mass-of-Mixture Burned from Spark-Ignition Engine Pressure Data," *Combust. Sci. Technol.*, vol. 54, pp. 133–144, 1987.

13. Sher, E., "The Effect of Atmospheric Conditions on the Performance of an Airborne Two-Stroke Spark-Ignition Engine," *Proc. Inst. Mech. Eng.*, vol. 198D, no. 15, pp. 239–251, 1984.

14. Hires, S.D., Tabaczynski, R.J., and Novak, J.M., "The Prediction of Ignition Delay and Combustion Intervals for a Homogeneous Charge, Spark Ignition Engine," SAE Paper 780232, *SAE Trans.*, vol. 87, 1978.

15. Heywood, J.B., "Combustion and its Modeling in Spark-Ignition Engines," The Third International Symposium on Diagnostics and Modeling of Combustion in Internal Combustion Engines, July 11–14, Yokohama, Japan, COMODIA 94, pp. 1–15, 1994.

16. Sher, E., Ben-Ya'ish, J., and Kravchik, T., "On the Birth of Spark Channels," *Combustion and Flame*, vol. 89, pp. 186–194, 1992.

17. Pischinger, S., and Heywood, J.B., "A Study of Flame Development and Engine Performance with Breakdown Ignition Systems in a Visualization Engine," SAE Paper 880518, *SAE Trans.*, vol. 97, 1988.

18. Spicher, U., and Backer, H., "Correlation of Flame Propagation and In-Cylinder Pressure in a Spark Ignited Engine," SAE Paper 902126, 1990.

19. Winklhofer, E., Philipp, H., Fraidl, G., and Fuchs, H., "Fuel and Flame Imaging in SI Engines," SAE Paper 930871, 1993.

20. Tsuchiya, K, Nagai, Y., and Gotoh, T., "A Study of Irregular Combustion in 2-Stroke Cycle Gasoline Engines," SAE Paper 830091, *SAE Trans.*, vol. 92, 1983.

21. Ikeda, Y., Ohira, T., Takahashi, T., and Nakajima, T., "Misfiring Effects on Scavenging Port and Exhaust Pipe in a Small Two-Stroke Engine," SAE Paper 930498, 1993.

22. Miles, P.C., Green, R.M., and Witze, P.O., "In-Cylinder Gas Velocity Measurements Comparing Crankcase and Blower Scavenging in a Fired Two-Stroke Cycle Engine," SAE Paper 940401, *SAE Trans.*, vol. 103, 1994.

23. Stone, C.R., Brown, A.G., and Beckwith, P., "A Turbulent Combustion Model Used to Give Insights Into Cycle-by-Cycle Variations in Spark Ignition Engine Combustion," IMechE Conference on Computations in Engines, Dec. 1992.

24. Ozdor, N., Dulger, M., and Sher, E., "Cyclic Variability in Spark Ignition Engines: A Literature Survey," SAE Paper 940987, *SAE Trans.*, vol. 103, 1994.

25. Lord, D.L., Anderson, R.W., and Brehob, D.D., "The Effects of Charge Motion on Early Flame Kernel Development," SAE Paper 930463, 1993.

26. Pischinger, S., and Heywood, J.B., "How Heat Losses to the Spark Plug Electrodes Affect Flame Kernel Development in an SI-Engine," SAE Paper 900021, *SAE Trans.*, vol. 99, 1990.

27. Pischinger, S., and Heywood, J.B., "A Model for Flame Kernel Development in a Spark-Ignition Engine", 23rd International Symposium on Combustion, pp. 1033–1040, The Combustion Institute, 1990.

28. Ozdor, N., Dulger, M., and Sher, E., "An Experimental Study of the Cyclic Variability in Spark Ignition Engines," SAE Paper 960611, 1996.

29. Dulger, M., and Sher, E., "Experimental Study on Spark Ignition of Flowing Combustible Mixtures," SAE Paper 951004, 1995.

30. Ballal, D.R., and Lefebvre, A.H., "The Influence of Spark Discharge Characteristics on Minimum Ignition Energy in Flowing Gases," *Combustion and Flame*, vol. 24, pp.99–108, 1975.

31. Kono, M., Hatori, K., and Iinuma, K., "Investigation on Ignition Ability of Composite Sparks in Flowing Mixtures," 20th Symposium (International) on Combustion, The Combustion Institute, 1984.

32. De Soete, G.C., "Propagation Behavior of Spark Ignited Flames in Early Stages," IMechE Conference Publication C59/83, 1983.

33. Douaud, D., De Soete, G.C., and Henault, C., "Experimental Analysis of the Initiation and Development of Part-Load Combustion in Spark-Ignition Engines," SAE Paper 830338, *SAE Trans.*, vol. 92, 1983.

34. Matekunas, F.A., "Modes and Measures of Cyclic Combustion Variability," SAE Paper 830037, 1983.

35. Sher, E., Heywood, J.B., and Hacohen, J., "Heat Transfer to the Electrodes: A Possible Explanation of Misfire in SI Engines," *Combust. Sci. Tech.*, vol. 83, pp. 323–325, 1992.

36. Sher, E., and Keck, J.C., "Spark Ignition of Combustible Gas Mixtures," *Combustion and Flame*, vol. 66, pp.17–25, 1986.

37. Abraham, M., and Prakash, S., "A Theory of Cyclic Variations in Small Two-Stroke Cycle Spark Ignited Engines—An Analytical Validation of Experimentally Observed Behavior," SAE Paper 920426, 1992.

38. Ohira, T., Ikeda, Y., Ito, T., and Nakajima, T., "Cyclic Variation of CO and CO_2 Emissions and Scavenging Flow in a Two-Stroke Engine," SAE Paper 940392, *SAE Trans.*, vol. 103, 1994.

39. Sher, E., "Improving the Performance of a Crankcase-Scavenged 2-S Engine with a Fluid Diode," *J. Power Energy (Proc. Inst. Mech. Eng.)*, vol. 196, pp. 23–34, 1982.

40. Onishi, S., Jo, H.S., Shoda, K., Jo, P.D., and Kato, S., "Active Thermo-Atmosphere Combustion (ATAC)—A New Combustion Process for Internal Combustion Engines," SAE Paper 790501, *SAE Trans.*, vol. 88, 1979. (Also in *Two-Stroke Cycle Spark-Ignition Engines*, PT-26, pp. 221–230, SAE, Warrendale, PA, 1982.)

41. Ishibashi, Y., and Tsushima, Y., "A Trial for Stabilizing Combustion in Two-Stroke Engines at Part Throttle Operation," *A New Generation of Two-Stroke Engines for the Future?* (ed. P. Duret), Proceedings of the International Seminar, Nov. 29–30, Rueil-Malmaison, France, Editions Technip, Paris, 1993.

42. Ishibashi, Y., and Asai, M., "Improving the Exhaust Emissions of Two-Stroke Engines by Applying the Activated Radical Combustion," SAE Paper 960742, *SAE Trans.*, vol. 105, 1996.

43. Duret, P., and Venturi, S., "Automotive Calibration of the IAPAC Fluid Dynamically Controlled Two-Stroke Combustion Process," SAE Paper 960363, 1996.

44. Ghandhi, J.B., Felton, P.G., Gajdeczko, B.F., and Bracco, F.V., "Investigation of the Fuel Distribution in a Two-Stroke Engine with an Air-Assisted Injector," SAE Paper 940394, 1994.

45. Houston, R., Archer, M., Moore, M., and Newmann, R., "Development of a Durable Emissions Control System for an Automotive Two-Stroke Engine," SAE Paper 960361, *SAE Trans.*, vol. 105, 1996.

46. Drake, M.C., Fansler, T.D., and French, D.T., "Crevice Flow and Combustion Visualization in a Direct-Injection Spark-Ignition Engine Using Laser Imaging Techniques," SAE Paper 952454, *SAE Trans.*, vol. 104, 1995.

47. Blair, G.P., *Design and Simulation of Two-Stroke Engines*, SAE, Warrendale, PA, 1996.

48. Drake, M.C., French, D.T., and Fansler, T.D., "Advanced Diagnostics for Minimizing Hydrocarbon Emissions from a Direct-Injection Gasoline Engine," 26th International Symposium on Combustion, The Combustion Institute, 1996.

49. Abata, D., and Wellenkotter, K., "Characterization of Ignition and Parametric Study of a Two-Stroke-Cycle Direct-Injected Gasoline Engine," SAE Paper 920423, *SAE Trans.*, vol. 101, 1992.

50. Duvinage, F., Abthoff, J., Hardt, T., Kramer, M., and Paule, M., "The Two-Stroke DI-Diesel Engine with Common-Rail-Injection for Passenger Car Application," SAE Paper 981032, 1998.

51. *Bosch Automotive Handbook*, 3rd ed., Robert Bosch GmbH, 1993. (Distributed by SAE, Warrendale, PA.)

52. *Bosch Diesel Fuel Injection*, Robert Bosch GmbH, 1994. (Distributed by SAE, Warrendale, PA.)

53. Howarth, M.H., *The Design of High Speed Diesel Engines*, Constable, London, 1966.

54. Brown, D., "Modern Sulzer Marine Two-Stroke Crosshead Diesel Engines," Report Re/22.18.07.40, Sulzer Brothers Ltd., Winterthur, Switzerland, 1986.

55. Nomura, K., and Nakamura, N., "Development of a New Two-Stroke Engine with Poppet-Valves: Toyota S-2 Engine," *A New Generation of Two-Stroke Engines for the Future?* (ed. P. Duret), Proceedings of the International Seminar, Nov. 29–30, Rueil-Malmaison, France, Editions Technip, Paris, 1993.

56. Balles, E.N., and Heywood, J.B., "Spray and Flame Structure in Diesel Combustion," *J. Eng. Gas Turbines Power*, vol. 111, pp. 451–457, 1989.

57. Dec, J., "A Conceptual Model of DI Diesel Combustion Based on Laser-Sheet Imaging," SAE Paper 970873, 1997.
58. Miyamoto, N., Chikahisa, T., Murayama, T., and Sawyer, R., "Description and Analysis of Diesel Engine Rate of Combustion and Performance Using Wiebe's Functions," SAE Paper 850107, *SAE Trans.*, vol. 94, 1985.
59. Watson, N., Pilley, A.D., and Marzouk, M., "A Combustion Correlation for Diesel Engine Simulation", SAE Paper 800029, 1980.

CHAPTER SEVEN

EMISSIONS FORMATION AND CONTROL

7-1 POLLUTANT FORMATION MECHANISMS

7-1-1 Background

Internal combustion engines are an important source of urban air pollutants, and of noise. Although two-stroke cycle engines are nowhere near as numerous as four-stroke cycle engines, they are still a significant source of pollutants, and their emissions are subject to increasingly stringent regulations. Two-stroke gasoline engines are widely used in developing countries to power mopeds, scooters, motor bikes, and small three-wheeled vehicles, due to their cheapness, compact size, and light weight. This chapter discusses how air pollutants form in two-stroke spark-ignition and diesel engines, and what is being done to control these emissions both within the engine and in the exhaust system with catalysts. We also discuss engine noise. The air pollutants of concern are oxides of nitrogen (NO_x), carbon monoxide (CO), hydrocarbons (HC), aldehydes, particulates, and oxides of sulfur. Their formation inside the engine is strongly connected with the engine's combustion process (Chapter 6). Because the two-stroke engine combustion processes (spark ignition, and compression ignition or diesel) are similar to the respective processes in their four-stroke counterparts, the pollutant formation processes are similar, too. We will draw heavily on the available knowledge of four-stroke cycle pollutant formation processes (see Heywood[1] and Sher[2] for extensive reviews). There are several important differences between the two cycles, however. These result primarily from the two-stroke scavenging process, and the different exhaust gas stream it produces, from the higher levels of in-cylinder burned residuals, from the higher in-cylinder gas temperatures that result from a combustion process every crankshaft revolution, and from the method used to lubricate small two-stroke cycle spark-ignition engines—adding oil to the fuel or airstream.

The pollutants just listed are of concern for the following reasons. Oxides of nitrogen react with hydrocarbons in the atmosphere, in the warmer seasons and regions, energized by sunlight, to form photochemical smog that contains oxidizing species the most prevalent of which is ozone (O_3). Nitrogen dioxide (NO_2) is a toxic gas. It is emitted from combustion sources along with nitric oxide (NO), although in much smaller amounts, but also forms from NO in the atmosphere. NO_2 is soluble in water, is washed out of the atmosphere by rain, and increases the acidity of the soil. Hydrocarbon emissions come from engine exhaust as well as from the fuel system. The exhaust contains both unreacted fuel that did not burn or react at all, as well as organic compounds formed from partial reaction of the fuel molecules. Fuel vapor can be expelled from the fuel tank and the engine's fuel system, as the temperature of the fuel and air in the fuel distribution system changes significantly, or as fuel displaces air and fuel vapor as the fuel tank is filled, or as fuel is spilled in the filling process. These many hydrocarbon compounds react in the atmosphere with NO_x with widely varying reactivities: olefinic compounds are most reactive, aromatics next, paraffinic compounds least (and methane is unreactive). Ozone production reactivity ratings for the various hydrocarbon emissions are now well established. Aldehydes are toxic fuel-partial-reaction products, which are highly reactive in the atmosphere. Carbon monoxide is toxic: it combines with the hemoglobin in the human bloodstream to form carboxyhemoglobin, which prevents that hemoglobin from carrying oxygen and tissue anoxia results. The industrial exposure limit for CO is 35 molar parts per million (ppm) and an adverse level of CO for urban air has been set at 9 ppm for a period of 8 hours. Particulates originate as tiny soot particles formed during combustion on which the less volatile hydrocarbon compounds in the fuel, and in the oil, condense along with sulfates from the sulfur in the fuel. Internal combustion engine particulate emissions (especially from diesels) are of special health concern because they are submicron in size and therefore are drawn into the lungs. They also significantly impair visibility. Sulfur oxides contribute to the particulate emissions, as noted, and also to acid rain.

Internal combustion engines, and especially small two-stroke spark-ignition engines, are an important source of noise. Noise originates from the engine's exhaust system and intake system as pressure waves resulting from the exhaust and intake processes propagate into the atmosphere. The engine's component surfaces also radiate noise when they vibrate in response to the pressure fluctuations inside the cylinders, the exhaust, intake, and crankcase, and to the force fluctuations in the engine's connecting rod and crankshaft mechanisms. We review the topic of engine noise in the final section (7–6) of this chapter.

7-1-2 Air Pollutant Fundamentals

We will now review the basic mechanisms by which the major pollutants form within the engine. Additional details can be found in Heywood.[1]

Oxides of nitrogen. Nitric oxide (NO), the predominant component, forms in the high-temperature burned gases inside the engine cylinder during and shortly after combustion. The primary NO forming reactions are

$$N_2 + O = NO + N \tag{7-1}$$

$$N + O_2 = NO + O \tag{7-2}$$

$$N + OH = NO + H \tag{7-3}$$

although additional reactions (which also involve N_2O and NO_2) play a secondary role. Reaction (7-1) is the rate-controlling step, and only the more energetic collisions between N_2 molecules and O atoms break the nitrogen–nitrogen bond and form NO. The N atom produced by reaction (7-1) then rapidly via reaction (7-2) or (7-3) forms a second NO molecule. The characteristic reaction time for reaction (7-1) at typical engine burned gas conditions is significantly longer than the time scale of the hydrocarbon oxidation reactions that occurs in the engine flame, so NO formation primarily occurs in the postflame burned gases. The carbon, hydrogen, and oxygen-containing species in the engine burned gases, due to the much shorter chemical reaction times at engine conditions, are close to chemical equilibrium.

Applying the law of mass action to these reactions (product species formation rates are proportional to the product of reactant concentrations for these binary reactions), with the appropriate rate constants for each reaction in its forward (product-forming) and backward (product-removing) directions, and assuming that the species N_2, O_2, O, OH, and H are in chemical equilibrium, leads to an equation for the net rate of formation of NO. This net rate is the difference between the formation of NO [reactions (7-1)–(7-3) in the forward, left-to-right, direction] and NO decomposition (which occurs once significant NO concentrations have built up, via the backward, right-to-left, direction of these reactions). If the backward rate becomes equal to the forward rate, then NO is in equilibrium. Details of this deviation can be found in Heywood.[1]

The dependence of the NO formation rate on burned gas composition and state can be demonstrated by considering the initial value of $d[NO]/dt$, the formation rate when $[NO] \ll [NO]_{eq}$. Here, [ ] denote concentrations in moles per cubic centimeter, and subscript eq denotes equilibrium values. Because reaction (7-1) is rate-controlling, and either reaction (7-2) or (7-3) rapidly adds a second NO molecule, this initial formation rate is given by

$$\frac{d[NO]}{dt} = 2k_1^+ [O]_{eq}[N_2]_{eq} \tag{7-4}$$

where k_1^+ is the rate constant for reaction (7-1) in the forward direction. $[O]_{eq}$ can be related to $[O_2]_{eq}$ in the burned gases via the equilibrium relationship between atomic and molecular oxygen, to give

$$\frac{d[NO]}{dt} = \frac{6 \times 10^{16}}{T^{1/2}} \exp\left(\frac{-69,090}{T}\right)[O_2]_{eq}^{1/2}[N_2]_{eq} \quad \text{mol/cm}^3\text{-s} \tag{7-5}$$

where T is in degrees kelvin, and t is in seconds.[1] The strong dependence of NO formation rate on temperature (via the exponential term) and relative A/F or F/A ratio, which determines $[O_2]_{eq}$, is clear.

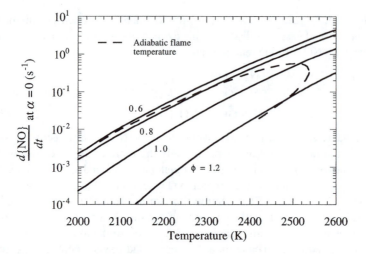

Figure 7-1 NO formation rate, in mass fraction per second, in equilibrium burned gases with α = {NO}/{NO}$_{eq}$ ≪ 1, as a function of temperature and fuel/air equivalence ratio ϕ. Dashed line shows adiabatic flame temperature for kerosene–air mixtures initially at 700 K and 15 bar.[3]

The impact of these two key variables is shown in Fig. 7-1, where the initial NO formation rate given by Eq. (7-5) is plotted (as mass fraction NO, {NO}) against burned gas temperature, for different fuel/air equivalence ratios (lean to rich).[3] Also shown are the adiabatic flame temperatures that result from burning hydrocarbon fuel–air mixtures, initially at 700 K and 15 bar, of various equivalence ratios, at constant pressure. This corresponds to mixture elements of various compositions, burning toward the end of the compression stroke. The NO formation rate peaks slightly lean of stoichiometric, and decreases rapidly as the mixture becomes leaner or richer. A leaner mixture results in a lower burned gas temperature, which explains the trend. A richer mixture does not significantly affect the burned gas temperature, which has a maximum about 10% rich of stoichiometric, until well rich of stoichiometric. It is the rapidly varying oxygen concentration, slightly lean to stoichiometric to rich, that explains this NO formation rate trend at fixed mixture temperature prior to combustion. Rich burned gas mixtures, in equilibrium, contain little O_2, and hence have much lower oxygen-atom concentrations than stoichiometric or slightly lean burned gas mixtures. Burned gas temperature and fuel/air ratio relative to stoichiometric are the controlling variables in the NO formation process.

Fuel nitrogen is also a source of NO through a different chemical mechanism. Heavy distillate oils that are used in large two-stroke diesel engines contain sufficient nitrogen (about 1%) for this to be a significant additional source. The fraction of the fuel nitrogen converted to NO depends on the relative fuel/air ratio, being much higher (approaching 100%) under lean conditions than under rich conditions.

Nitrogen dioxide (NO_2) concentrations are not significant in premixed spark-ignition engine exhaust gases (NO_2/NO ∼ 1%). However, in diesel engines, NO_2 can be 10–30% of the exhaust NO_x. Equilibrium considerations in high-temperature burned gases indicate there should be negligible NO_2. It is thought that NO formed in the flame zone can be oxidized to NO_2 by radicals in the flame, and that the subsequent conversion back to NO is quenched by mixing with cooler gases.[4]

Nitrous oxide (N_2O) plays a (subsidiary) intermediate role in NO formation within the actual flame reaction "sheet," where oxygen atom concentration exceeds postflame equilibrium levels substantially. However, in engines where combustion occurs in a confined volume, and the pressure rise it produces compresses most of the postflame burned gases to higher temperatures, the amount of flame-front-formed NO is usually much smaller than the amount of NO formed in the postflame gases. N_2O also adds to the NO formation mechanism [reactions (7-1)–(7-3)] in the postflame burned gases under lean mixture conditions, increasing the amount of NO formed by about 20% for $\phi \leq 0.75$, but negligibly for stoichiometric and rich mixtures.

Carbon Monoxide. Carbon monoxide (CO) is always a constituent in hydrocarbon–air combustion products. In fuel-rich mixtures it is present in significant amounts because there is insufficient oxygen for complete oxidation of the fuel carbon to CO_2. It is also present in high-temperature burned gas mixtures of any relative stoichiometry because a fraction of the CO_2 molecules at these conditions will be dissociated into CO and oxygen. CO levels in the burned gases inside the engine cylinder are initially close to equilibrium levels. However, as the gas temperature falls during the expansion process, the CO to CO_2 recombination, which would occur if the burned gases remained in chemical equilibrium, becomes too slow and "freezes" levels of CO in the exhaust gases that are higher than equilibrium calculations predict. This freezing of the CO chemistry occurs at about 1800 K. This freezing of the chemistry leads to moderate (~0.5–1%) levels of CO in stoichiometric exhaust gases and lower levels (~0.2%) in lean exhaust gases.

A secondary mechanism for CO formation is the incomplete oxidation of hydrocarbon compounds that escape burning during the primary combustion process, but which partially burn later during expansion and exhaust. At low enough temperatures, the hydrocarbon to CO to CO_2 oxidation process cannot occur fast enough for it to be completed.

Hydrocarbons. The key question regarding hydrocarbons (HC) in the engine exhaust is: Why didn't the fuel burn completely during the normal combustion process? The answer is fuel can be stored in regions the flame cannot enter (e.g., the crevice regions between the piston, rings, and cylinder liner), the flame can be quenched before the in-cylinder fuel–air charge has been fully burned, some fuel may not mix with enough air to form a combustible mixture (e.g., liquid fuel in the cylinder), fuel may short-circuit the cylinder (especially important with premixed-charge two-stroke SI engines) and have no chance to burn, and (again with two-stroke engines) oil added to the fuel or air stream to lubricate the piston may not fully burn. A much larger fraction of the fuel escapes burning during normal combustion in conventional spark-ignition engines than is usually realized: in four-stroke SI engines it is about 7%[5]; in two-stroke premixed-charge SI engines it is about 20% or more.

Some of this fuel will oxidize within the cylinder as it mixes with the burned gases during expansion and exhaust. Some will remain within the cylinder with the residual gas. Some may oxidize in the exhaust system before the exhaust gas becomes too cool for HC burnup to occur. Thus the engine-out HC emissions are less than the amount of fuel that escapes burning during the normal combustion process.

Other sources of HC emissions are evaporative losses from the fuel system, as described in Section 7-1-1, that vent to the atmosphere and, depending on the two-stroke engine design, the crankcase blow-by gases. A small fraction of the in-cylinder gases (a percent or so) blow by the piston rings into the crankcase. This blow-by gas is a mixture of air, fuel (primarily in premixed-charge engines), and burned gases. In crankcase-scavenged two-stroke engines, the blow-by gases are returned to the cylinder. In engines that are not scavenged via the crankcase, the blow-by gases must be fed back to the engine intake to avoid additional HC emissions.

Note that hydrocarbon emissions comprise many different types of individual organic compounds. Compounds present in the fuel are present in the HC emissions; so also are organic compounds formed by partial reaction of the fuel compounds.

Particulates. Particulate emissions from all engines are of increasing concern due to growing awareness of their impact on human health. The chemical and physical characteristics of engine particulate emissions depend on the fuel composition, the engine concept (carburetted or port-injected SI engine, direct-injection SI engine, diesel) and the cycle (two-stroke or four-stroke). With leaded gasolines, lead compounds dominate the particulate emissions. With unleaded fuels, the particulates comprise a solid insoluble fraction, which is primarily soot particles sometimes with small amounts of inorganic compounds (ash) from the oil, and a (solvent) soluble fraction, which is absorbed hydrocarbons (from the fuel and the oil) and sulfate (from the fuel sulfur). We will examine these various particulate emissions in detail as the emissions from each different type of engine are discussed. However, we will review here the fundamentals of soot particle formation in flames.

Soot forms in fuel-rich mixtures in flames when the fuel molecules pyrolyze—that is, undergo extensive decomposition and rearrangement. The fuel molecules, which in diesel fuel have 12–22 carbon atoms and a molar H/C ratio of about 2, break up into smaller fragments that then aggregate into much larger polycyclic aromatic hydrocarbons (PAH) or polyacetylenes, losing hydrogen in the process, which when large enough condense to form tiny nuclei (< 2 nm) and then grow through surface growth and agglomeration to form larger particles, typically a few hundred nanometers in diameter. These soot particles are loose clusters of many spherules, each about 20 nm in diameter containing some 10^5 carbon atoms and having an H/C ratio of about 0.1.

At lower temperatures (≤ 1700 K), at equivalence ratios richer than about 2–3, aromatic hydrocarbons can produce soot relative rapidly via PAH forming condensation reactions. At higher temperatures (above about 1800 K), aromatic and aliphatic hydrocarbons undergo fragmentation reactions. These fragments then polymerize to form larger polyacetylene molecules, which then condense to form particle nuclei. Dehydrogenation occurs throughout this process. At each stage in these processes of particle generation and growth, oxidation can occur if an oxidizing environment is created through mixing with excess air, and the temperature is high enough. These are the major steps in the process that leads to soot particles in the burned gases that exit the engine cylinder during exhaust.[1]

7-1-3 Measurement Parameters

Engine emissions levels are expressed in several different ways. The basic emissions measurements usually made are the relative concentrations of the pollutants NO, CO, and hydrocarbons in the engine exhaust gases. These are expressed as a mole fraction $(\tilde{x}_i)$, or as parts per million (ppm) by volume (the mole fraction $\times 10^6$) or as mole percent (the mole fraction $\times 10^2$, usually for CO). For example, the mole fraction of CO is

$$\tilde{x}_{CO} = \frac{n_{CO}}{n_e} = \frac{(\dot{m}_{CO}/M_{CO})}{(\dot{m}_e/M_e)} \tag{7-6}$$

where n_{CO} is the number of moles of CO in a given volume of mixture, $\dot{m}_{CO}$ is its mass flow rate through the exhaust, and M_{CO} is its molecular weight; subscript e denotes exhaust. The molecular weight of the exhaust gases is close to that of air; it is about 28.5 and varies only modestly with relative A/F.[1] With NO_x (NO and NO_2, combined) it must be made clear when relating to mass NO_x emissions whether the molecular weight of NO or NO_2 (the more common approach) was used. With HC, the hydrocarbons comprise many different compounds and their concentration in Eq. (7-6) is usually expressed as ppm C_1 (see Section 7-2-4).

It is often more useful to compare the pollutant emission rates with the fuel flow rate or (similarly to bsfc) to engine power. Thus an *emission index* is often used, where, for example,

$$EI_{CO} = \dot{m}_{CO}/\dot{m}_f \tag{7-7a}$$

Here, for convenience, the pollutant emission rate is often expressed in grams per second and the fuel flow rate in kilograms per second, so the units of EI are grams per kilogram. Combining Eqs. (7-6) and (7-7) gives

$$EI_{CO} = \tilde{x}_{CO}[1 + (A/F)](M_{CO}/M_e) \tag{7-7b}$$

A specific emissions parameter is also used: for example,

$$sCO = \dot{m}_{CO}/P \tag{7-8}$$

where P is the engine's power. Both brake and indicated specific emissions quantities are defined, depending on whether brake or indicated power is used. This quantity can be related to the emissions index via the specific fuel consumption: for example,

$$EI_{CO} = (sCO/sfc) \times 10^3 \tag{7-9}$$

7-2 EMISSIONS FROM SPARK-IGNITION ENGINES

7-2-1 Overview

We need to separate two-stroke spark-ignition engine into two categories; those where the fuel is mixed with the scavenging air prior to entry to the cylinder and those where the fuel is injected directly into the engine cylinder sufficiently late in, or after, the scavenging process to avoid any short-circuiting of the fuel. The hydrocarbon emissions characteristics of the second type, the direct-injection SI engine, are obviously substantially different from those of the carburetted or port-fuel-injected version. In this section we discuss the emissions characteristics of gasoline-fueled two-stroke spark-ignition engines, where the air and fuel enter the cylinder essentially premixed. In Section 7-3 we will review emissions from direct-injection two-stroke SI engines.

The spark-ignition engines exhaust gases contain three major pollutants in the following relative amounts: oxides of nitrogen, NO_x (mainly NO and small amounts of NO_2), 500–1000 ppm; carbon monoxide, CO, 0.5–3%; and unburned or partially burned organic compounds, HC, 500–5000 ppm (as C_1). In addition, there are three classes of spark-ignition engine particulate emissions: lead, organic particulates (including soot), and sulfates. For traditional two-stroke cycle engines, where the lubricant is mixed with fuel, higher HC levels of up to 10,000 ppm are expected. The amounts of each pollutant in the exhaust depend on the engine concept, engine design details, and operating conditions. One of the more important variables in determining emission levels in spark-ignition engines is the fuel/air equivalence ratio (ϕ) or the relative air/fuel ratio (λ), which equals $1/\phi$. Figure 7-2 shows how the CO_2, CO, O_2, HC, and NO concentrations in the exhaust gas of a single-cylinder carburetted two-stroke SI engine vary with the air/fuel ratio. The stoichiometric A/F is indicated. The figure shows that leaner mixtures yield lower emissions until the combustion quality becomes poor, when the HC emission would then rise sharply. Rich mixtures yield lower NO emissions, but steadily rising HC and CO emissions.

Several technology changes are being introduced into standard production two-stroke gasoline engines, or developed for possible introduction, to reduce emissions yet retain the simplicity and low cost of this type of engine. The most promising modifications are fuel injection into the intake port or into the crankcase, exhaust port flow control valves, and exhaust catalysts. The first of these provides better control of short-circuiting fuel and therefore can significantly reduce HC emissions. The second area of technology (see Section 5-5-4) improves the engine's trapping efficiency and can be used to control the amount of burned gas residual trapped within the cylinder. These changes reduce HC emissions by reducing the amount of short-circuiting fuel, and can reduce NO_x. Addition of a catalyst to the exhaust system significantly reduces the engine-out emissions, provided the engine's relative air/fuel ratio is adjusted appropriately so the exhaust gas composition and temperature match the catalyst's needs.[7] Thus, interpretation of engine emissions data must be done in the context of the complete engine system, with fuel metering technology and any additional scavenging and/or exhaust flow control devices especially important. Whether or not an exhaust catalyst will be used obviously has a substantial effect on tailpipe emission levels, as well as impacting engine design features and especially how the engine is operated.

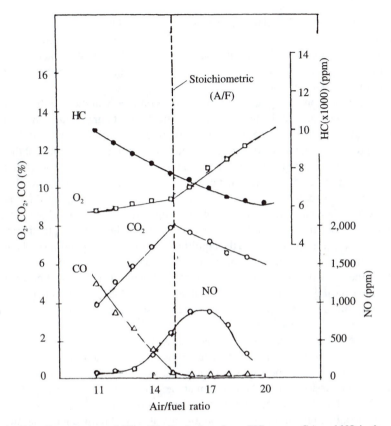

Figure 7-2 Concentrations of CO_2, CO, O_2, hydrocarbons (HC as ppm C_1), and NO in dry exhaust gas from a small two-stroke SI engine, as a function of air/fuel ratio: stoichiometric $A/F = 15.1$; engine speed 2500 rev/min, spark timing 15° BTC (MBT).[6]

7-2-2 Oxides of Nitrogen Emissions

The oxide of nitrogen that is important here is nitric oxide (NO) because the amount of NO_2 in the SI engine's exhaust is of order 1% of the NO level. Extensive data from four-stroke SI engines have established that the critical engine variables that affect NO emissions are the relative air/fuel ratio (λ), the amount of burned gas in the in-cylinder mixture of fuel, air, and residual gas (often expressed in terms of the fraction of the burned exhaust gas recycled to the intake, EGR, although the amount of burned residual that stays inside the cylinder is just as important and must be added to the recycled exhaust to get the total burned gas amount), and spark timing relative to the optimum timing that gives maximum brake torque (MBT timing).[1] These engine variables relate directly to the more fundamental variables that govern the NO formation rate, burned gas temperature, and oxygen concentration, as given in Eq. (7-5). The relative A/F or F/A, λ or ϕ, affects both temperature and oxygen concentration. Burned gas temperatures peak slightly rich of stoichiometric, decrease progressively (approximately as ϕ, the fuel/air equivalence ratio $= 1/\lambda$) on the lean side of stoichiometric, vary only modestly between $\phi = 1$ and 1.2, and then decrease for $\phi \geq 1.2$. Oxygen concentrations are low in high-temperature rich burned gas mixtures, increase close to stoichiometric, and then increase more rapidly as the mixture becomes leaner. Note that cylinder pressure enters as a variable too, because the concentrations of the burned gas species that take part in reactions (7-1)–(7-3) scale with pressure.

Because combustion in engines takes place in a confined volume and the pressure rises steadily through most of the burning process, the highest burned gas temperatures result from both the immediate chemical energy release from combustion, and the subsequent compression process that occurs as combustion continues. Thus most of the NO forms between start of combustion and time of maximum pressure. As the pressure falls from its maximum value 15–20° ATC, the burned gas temperatures decrease, so the NO formation rate and NO equilibrium levels both decrease. During the expansion process, therefore, the NO chemistry effectively ceases as temperatures fall below about 2250 K; the reactions become "frozen." Prior to that, however, when due to falling temperatures the local NO concentrations are above the equilibrium values corresponding to the local temperature, pressure, and composition, some NO decomposition may occur. Thus the NO levels in the burned gases at the end of expansion and during exhaust correspond to what formed during and shortly after combustion, and these levels remain frozen well above equilibrium levels corresponding to exhaust gas temperatures because the relevant chemistry is far too slow.[1]

Nitric oxide concentrations in the exhaust of a simple carburetted Schnürle-type loop-scavenged two-stroke SI engine, as a function of air/fuel ratio and throttle position, are shown in Fig. 7-3. The data show the expected dependences on λ and burned gas fraction. NO emissions show a maximum at each throttle setting close to the stoichiometric A/F of 14.5, in agreement with the predictions in Fig. 7-1 of the effect of relative fuel/air ratio on the initial NO formation rate at typical burned gas conditions.[†] Throttling shows the effect of charge dilution with increasing amounts of trapped burned gas residual. As the throttle is closed, the delivery ratio goes down substantially and, although the increase in trapping efficiency somewhat offsets this, the residual burned fraction in the trapped mixture increases. The relative amount of burned residual can be estimated, approximately, from

$$x_r = 1 - \left(\frac{p_i}{p_0}\right)\eta_{ch} \tag{7-10}$$

where the charging efficiency (the mass of fresh charge retained divided by the reference mass) is given by Eq. (2-8) and (p_i/p_0), the ratio of inlet manifold pressure to atmospheric pressure, provides the appropriate air density scaling. At wide-open throttle, for this engine, the residual fraction would be about 0.5; at the most throttled condition shown it is about 0.7.

---

†Engine data show the maximum NO_x emissions occur with mixtures between stoichiometric and about 10% lean. Exactly where this occurs depends on additional details such as (i) how spark timing was controlled during the tests (adjusted for MBT at each operating point, or held fixed as in the tests shown in Fig. 7-3); (ii) whether the in-cylinder burned gas temperatures were high enough for substantial NO decomposition to occur during expansion (fundamental studies indicate the dependence of formation and decomposition of NO on λ are not the same).

Throttle open angle	Delivery ratio	Trapping efficiency
WOT (80)	0.76	0.62
40/80	0.65	0.66
30/80	0.54	0.71
20/80	0.39	0.76

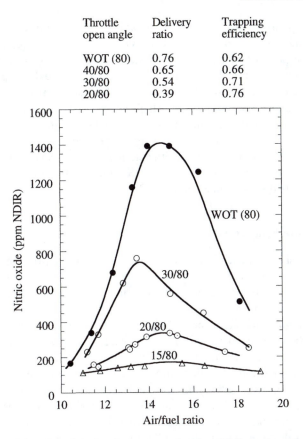

Figure 7-3 NO levels in dry exhaust gas, as a function of air/fuel ratio and throttle opening, from a Schnürle-scavenged carburetted 356-cm^3 displacement two-cylinder two-stroke SI engine: engine speed 4000 rev/min; effective compression ratio 6.5; spark timing 16° BTC; delivery ratio and trapping efficiency at each throttle open angle are given.[8]

Increased dilution with burned gases reduces the burned gas temperatures by effectively increasing the thermal capacity of the unburned charge per unit mass of fuel.† This substantially reduces the NO formation rate, at fixed relative A/F, as expected from Eq. (7-5).

The third major variable, spark timing relative to optimum MBT timing, affects NO formation through its effect on burned gas temperatures. Spark retard from MBT timing results in lower and later peak cylinder pressures, and hence lower peak burned gas temperatures. Typically, 5° of retard from the optimum timing, which results in only about 1% loss in torque, produces a 15–20% reduction in engine-out NO_x.[1]

---

†The temperature rise produced by burning the in-cylinder mixture is given approximately by $\Delta T_{comb} \approx m_f Q_{HV}/(mc_v)$, where Q_{HV} is the fuel heating value, and c_v is the gas specific heat. Higher residual reduces m_f/m, where m is the total mass in the cylinder.

NO$_x$ emissions from premixed two-stroke spark-ignition engines are usually significantly lower, at the same operating conditions, than NO$_x$ emissions from equivalent four-stroke engines. This is primarily due to the two-stroke's scavenging process, which leaves substantially more residual burned gas within the cylinder after the exhaust ports (or valves) close. As explained previously, the higher burned residual fraction in the fuel–air charge reduces the burned gas temperatures. (It also makes fast enough combustion more difficult to achieve.) Controlling the amount of burned gas in the in-cylinder fuel–air–residual mixture is one means of controlling engine NO$_x$ emissions. This can be done with exhaust port flow control valves or with external recycled exhaust (EGR) systems. These techniques are more widely used with direct-injection two-stroke gasoline engines and are therefore discussed in Section 7-3-2.

7-2-3 Carbon Monoxide Emissions

Carbon monoxide (CO) is also a product of the combustion process. In fuel-rich burned gas mixtures it is present in substantial amounts (several percent) because there is insufficient oxygen to burn fully all the carbon in the fuel to carbon dioxide. In high-temperature lean burned gas mixtures, CO is also present but in smaller amounts ($\sim$0.2%) because, at the high temperatures prevailing inside the combustion chamber, partial dissociation of carbon dioxide to carbon monoxide occurs. The measured engine exhaust CO concentrations for fuel-rich mixtures are close to equilibrium concentrations in the exhaust gases. For close-to-stoichiometric and fuel-lean mixtures, measured CO emissions in the exhaust are substantially higher than levels corresponding to equilibrium. The rapid cooling of the gases that occurs during the expansion stroke does not allow the burned gases to remain in chemical equilibrium, and the CO concentrations are left far in excess of levels corresponding to equilibrium at exhaust conditions. Another possible explanation is that only partial oxidation to CO occurs of some of the hydrocarbons that react during expansion and exhaust.[1]

Figure 7-4 shows the CO exhaust emissions of a carburetted two-cylinder 356-cm^3 displacement two-stroke SI engine as a function of A/F and throttle setting.[8] The data are comparable to those in Fig. 7-3. The strongest variable is, as expected, the relative A/F or F/A ratio. For very lean mixtures CO levels are low; for mixtures well rich of stoichiometric, CO levels rise almost linearly as A/F decreases. A transition between these limits occurs around the stoichiometric mixture ($A/F = 14.5$).

Three factors affect this behavior: the overall relative A/F or F/A of the burned gases; the nonuniformity in relative concentrations of fuel and air within the cylinder; and mixture short-circuiting. The effect of overall relative A/F can be explained by Eq. (7-11):

$$(CH_y)_n + \lambda n \left(1 + \frac{y}{4}\right)(O_2 + 3.773N_2)$$
$$= [aCO + bCO_2 + cO_2 + dH_2O + e(CH_y)_n + fH_2 + gN_2]_b$$
$$+ R_{sc}\left[(CH_y)_n + \lambda n \left(1 + \frac{y}{4}\right)(O_2 + 3.773N_2)\right]_{sc} \tag{7-11}$$

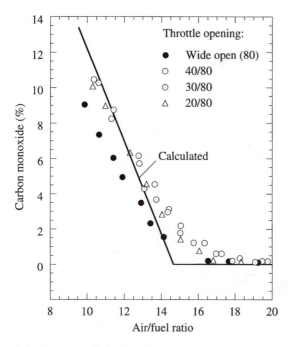

Figure 7-4 Percentage CO in dry exhaust gas, as a function of air/fuel ratio, from carburetted two-stroke SI engine of Fig. 7-3: calculated line corresponds to Eq. (7-11) with no short-circuiting; engine speed 4000 rev/min; spark timing 15° BTC (MBT).[8]

This equation relates the fuel and air entering the engine cylinder to the composition of the exhaust gases leaving. [It is analogous to Eq. (3-19), which was used to relate relative A/F to exhaust gas composition.] The exhaust gases (right-hand side) contain a burned gas component (subscript b) and a short-circuiting fuel–air mixture component (denoted by subscript sc and multiplied by R_{sc}, the short-circuiting ratio). For lean mixtures ($\lambda > 1$), the levels of CO and H_2 are very low ($a \approx 0$, $f \approx 0$) because there is sufficient oxygen for complete fuel oxidation. Relative to the short-circuiting contribution so are the hydrocarbons ($e \approx 0$). For rich mixtures, the oxygen concentration in the burned gases is negligible ($c \approx 0$). Both C and H are incompletely oxidized, and a simple relation between the concentration of CO and H_2 ($f = 0.5a$) accurately matches the exhaust gas composition data (see Section 3-2-2). The calculated line in Fig. 7-4 corresponds to Eq. (7-11), with no short-circuiting, with the simplifying assumptions for lean and rich mixtures explained previously. Although comparable in shape to the CO data it does not match the slope of the data for rich mixtures. The CO mole fraction $\tilde{x}_{CO}$ in dry exhaust gas (the measurements are with water removed, $d = 0$), and with $e \approx 0$, is given by

$$\tilde{x}_{CO} = a / \{a + b + c + f + g + R_{sc}[1 + 4.773\lambda n(1 + y/4)]\} \qquad (7\text{-}12)$$

Equation (7-12) shows how short-circuiting ($R_{sc} > 0$) dilutes the concentration of CO in the burned gases, and therefore reduces the slope of the CO-versus-λ line for rich

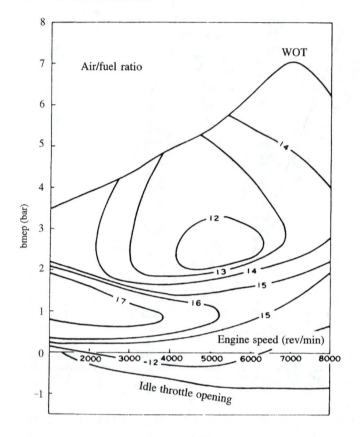

(a)

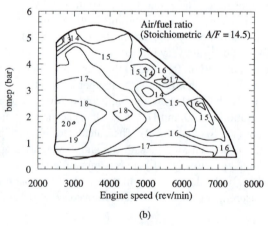

(b)

Figure 7-5 Air/fuel ratio variation over the performance maps of two different carburetted crankcase-scavenged two-stroke SI engines: (a) older Yamaha two-cylinder 347-cm³ displacement motorcycle engine optimized for performance[11]; (b) recent 2-kW single-cylinder moped engine optimized (with an oxidation catalyst) for low emissions.[7]

mixtures. The wide-open throttle data indicate the short-circuiting ratio is about 40%. A more detailed analysis of Eq. (7-12) can be found in Babu et al.[9]

The shape of the curve for close-to-stoichiometric mixtures results from fuel–air nonuniformity. Rich burned gas regions contribute more CO; lean regions do not offset this significantly, as the calculated curve in Fig. 7-4 indicates. The amount of CO in this curved region of the graph can be used to assess the extent of mixture nonuniformity.[10]

Throttle setting affects the concentration of CO (most apparent for rich mixtures where CO concentrations are larger) through the amount of dilution by short-circuiting fresh mixture. As the engine is throttled, the delivery ratio goes down and the trapping efficiency increases (see Fig. 7-3), so short-circuiting goes down. However, relative A/F or F/A is the primary variable governing SI engine CO emissions. Figure 7-5 shows how the air/fuel ratio varies over the operating maps of two different carburetted two-stroke SI engines: the first is an older Yamaha crankcase-scavenged two-cylinder 347-cm^3 displacement motorcycle engine,[11] and the second is a more recent 2-kW (2.7-hp) single-cylinder crankcase-scavenged moped engine design.[7] Contours of constant A/F are plotted on each engine's performance map, a graph of brake mean effective pressure versus engine speed. The rich regions $A/F \leq 14.5$ are at low (or low to medium) speed and at high bmep (torque) to give a good response to a rapid throttle opening from cruise (lighter load) conditions. The more lightly loaded parts of the map run stoichiometric or lean. The more modern engine runs significantly leaner; it has been designed for low engine-out emissions. Carbon monoxide emissions correspond to these A/F distributions as indicated in Fig. 7-4.

7-2-4 Hydrocarbon Emissions

The critical issue here is why all the inducted fuel does not burn during the normal engine combustion process. In the premixed-charge two-stroke SI engine, there are two major reasons. The most important reason is that significant fuel short-circuits each cylinder during its scavenging process. Second, the fuel trapped within the cylinder does not fully burn. There are places where fuel can be stored during the normal combustion process and escape burning and, under lightly loaded operation when conditions for combustion are poor, flame propagation may be incomplete. The first problem dominates HC emissions with standard two-stroke carburetted or port-fuel-injected SI engines. Bulk flame quenching becomes the dominant mechanism for HC emissions with direct-injection two-stroke SI engines (see Section 7-3). For both of these engine concepts, the important HC sources are significantly different from those that are most important in four-stroke cycle SI engines[5]: see Table 7-1. The largest of the *in-cylinder* HC sources in Table 7-1 for premixed two-stroke engines is the crevice regions between the piston, rings, and cylinder liner; fuel–air mixture is compressed into these narrow volumes (especially above and behind the top ring) yet the entrance is too narrow for the flame to enter. An additional potentially significant component of HC emissions with crankcase-scavenged engines is the lubricating oil mixed with the fuel, or with the scavenging air stream, to lubricate the piston rings and skirt, which escapes the cylinder unburned.

Table 7-1 Hydrocarbon sources in SI engines

HC source‡	Relative importance†		
	Premixed 2-S	Premixed 4-S	DI (2-S, 4-S)
Fuel short-circuiting	H	N	N
Fuel–air mixture in crevices	M	H	L/M§
Fuel vapor absorbed in oil film	L	L	L
Fuel vapor absorbed in deposits	L	L	L
Flame quenching at walls	L	L	L
Occasional poor combustion	?	L	M
Bulk flame quenching	?	N	H
Liquid fuel on walls	L	M	?
Leakage past valves or rings	L	L	L
Lubricating oil	L§	N	L/N§

†Relative importance: H = high, M = medium, L = low, N = negligible,? = unclear how important.
§Depends on the engine concept.
‡Substantial post-normal-combustion oxidation of HC occurs in the cylinder and in the exhaust. Also, some HC that escape normal combustion remain in the cylinder with the residual gases.

A significant fraction of the hydrocarbons that escape burning via the aforementioned mechanisms can oxidize within the cylinder and in the exhaust if oxygen is present and the temperature of the mixture of HC, air, and burned gas is high enough for the appropriate reactions to occur. Also, some of the (unoxidized) HC will be retained in the cylinder at the end of the scavenging process with the residual gas. Any HC remaining in the cylinder after the scavenging process is over (the residual HC), will get burned up during the next cycle. We will now examine these processes in more detail. First, however, some comments on how HC emissions are measured.

Hydrocarbon emissions are usually measured with a flame ionization detector (FID), which functions as a carbon atom counter. Thus one butane (C_4H_{10}) molecule, produces four times the signal of a methane molecule (CH_4). Data in the literature are often provided on different parts per million (ppm) scales; it is most convenient to convert all data to ppm C_1 by multiplying by the number of carbon atoms in the reference molecule. For example, ppm propane (C_3H_8), which is often used, should be multiplied by 3 to convert to ppm C_1. Older HC concentration data were taken with nondispersive infrared (NDIR) analyzers where n-hexane was used as the calibrating hydrocarbon. Because the IR absorption characteristics of HC emissions are not the same as n-hexane (and, in fact, vary somewhat with the composition of the HC emissions) a correction factor is required to get the total HC concentration. With four-stroke HC emissions data, a multiplier of 2 is normally used (as well as the factor of 6 to convert ppm hexane to ppm C_1), to convert the NDIR HC concentrations to FID HC concentrations. Piaggio have used a factor of 1.8 for two-stroke engines; Yamaha have used a factor between 1.5 and 1.8, depending on HC concentration (highest factor with lower NDIR HC concentrations).[11] HC concentrations given here are in (FID) ppm C_1 unless explicitly noted otherwise.

The relative amounts of various HC compounds in the engine-out HC emissions from a carburetted crankcase-scavenged two-stroke SI engine, and from a similar-

Table 7-2 Comparison between gas chromatographic analysis of HC emissions from a carburetted crankcase-scavenged two-stroke SI engine, with a similar-capacity four-stroke engine[12]

HC Component (% weight)	Formula	Two-stroke	Four-stroke
Methane	CH_4	1.00	4.21
Acetylene	C_2H_2	0.99	7.81
Isopentane	C_5H_{12}	8.42	5.38
2-Methyl pentane	C_7H_{16}	4.13	2.98
3-Methyl pentane	C_8H_{18}	2.67	2.06
n-Hexane	C_6H_{14}	2.85	2.38
3-Methyl hexane	C_9H_{20}	1.71	1.31

capacity four-stroke SI engine are shown in Table 7-2. Four-stroke engine-out HC emissions are typically about half unreacted fuel (represented in the table by C_5 and higher hydrocarbons) and about half partial reaction products of the fuel molecules (represented by the methane and acetylene). We see in Table 7-2 that the higher molecular weight paraffins that are compounds in the gasoline are present in larger amounts (by 20–60%) than are the nonfuel species, CH_4 and C_2H_2, which are present in much smaller amounts (lower by a factor of 4 and 8, respectively). In-cylinder and exhaust HC composition measurements add further information on the sources of HC.[13] Figure 7-6 shows the relative proportions of HC components, grouped by carbon number (C_1, C_2, etc.), in gas sampled from the cylinder 50° ATC during the expansion process (well before scavenging commences) and gas sampled in the exhaust, near the exhaust port, at 70° ABC when scavenging is essentially complete. The HC concentration in the in-cylinder sampled gas, at both loads, was about 5500 ppm C_1, much lower than the average exhaust levels (39,000 ppm C_1 at full load, 19,000 ppm C_1 at part load). Only a small fraction (12–15%) of these HC had a carbon number higher than C_5. In the gases sampled from the exhaust at full load 78% of the total HC were higher than C_5 compounds, and at part load 49% were higher than C_5. That the short-circuiting fuel dominates is further shown by the time-resolved exhaust HC concentration profiles in Fig. 7-7.[13] Under both full- and part-load conditions, the HC concentration falls rapidly soon after exhaust port opening as burned gas from the cylinder, with relatively low HC levels, mixes with and displaces high-HC-level gas that was stationary in the exhaust duct prior to exhaust port opening. Then, during the scavenging process, as short-circuiting becomes more significant, the exhaust HC levels rise due to the fuel, which increasingly flows straight through the engine. At the higher load, where the delivery ratio is higher, the trapping efficiency lower, and fraction of fuel short-circuiting is larger, the rise in HC occurs earlier and the HC levels in the exhaust at the end of scavenging are three times as high. That the minimum HC levels during blowdown are not lower is probably due to mixing between the low HC burned gas and the stagnant exhaust gas with its high short-circuited fuel content. Note that during the closed-port

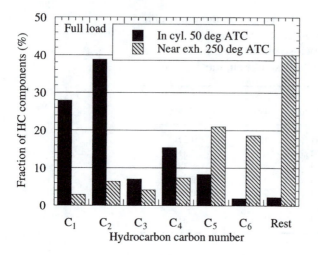

(a)

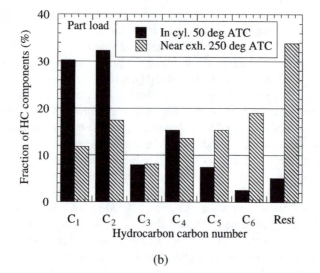

(b)

Figure 7-6 Hydrocarbon composition of gases sampled from the cylinder during expansion at 50° ATC, and in the exhaust gases near the exhaust port at 250° ATC, from a single-cylinder crankcase-scavenged 145-cm^3 air-cooled carburetted SI engine: HC components separated by gas chromatography into C_1–C_6 and higher carbon number hydrocarbons; engine speed 3500 rev/min; full load bmep = 386 kPa; part load bmep = 190 kPa; relative air/fuel ratio = 0.9.[13]

portion of the engine cycle, the HC levels measured in the exhaust fall, presumably due to some oxidation of HC as the burned and short-circuited gases continue to mix and react.

Because HC emissions from carburettd and port-fuel-injected two-stroke SI engines are dominated by the short-circuiting fuel, evaluating the short-circuiting ratio, or fresh

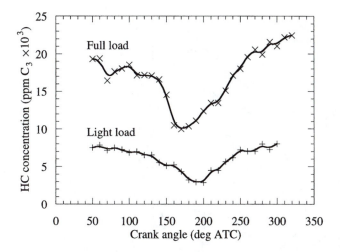

Figure 7-7 Exhaust HC concentration (ppm C_3) as a function of crank angle through the engine cycle, at full and part load, for same engine and operating conditions as in Fig. 7-6.[13]

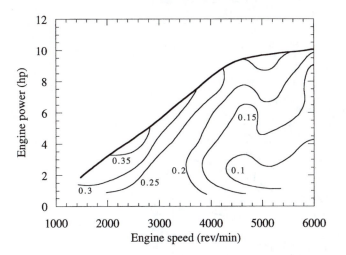

Figure 7-8 Short-circuiting ratio variation across the bmep versus engine speed operating map for a standard small two-stroke SI engine.[12]

charge loss ratio, is useful. This ratio R_{sc} [see Eq. (7-11)] is given by

$$R_{sc} = \frac{m_{lost}}{m_{del}} = 1 - \frac{m_{tr}}{m_{del}} = 1 - \eta_{tr} \tag{7-13}$$

Its magnitude for a typical two-stroke of this type is shown in Fig. 7-8. The short-circuiting ratio is especially high at low to medium speed and wide-open throttle. The dependence of HC emissions on short-circuiting ratio [and hence trapping efficiency via Eq. 7-13)] is shown by the HC emissions and η_{tr} contour maps in Fig. 7-9.[11] The lines of constant HC concentration and η_{tr} are essentially identical in layout, and the scaling

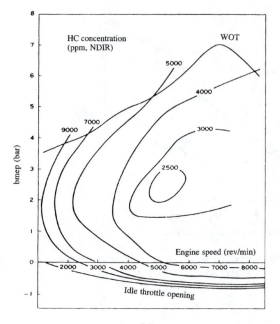

(a)

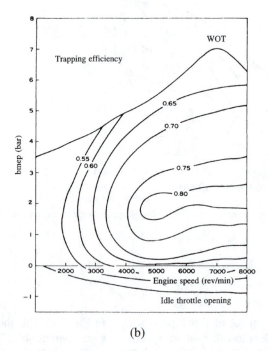

(b)

Figure 7-9 Hydrocarbons emissions and trapping efficiency maps for a two-cylinder 347-cm^3 displacement crankcase-scavenged carburetted two-stroke SI engine[11]: HC concentrations in ppm C_6 (NDIR); multiply by $6 \times 1.8 \approx 11$ to obtain ppm C_1 (FID).

between HC (ppm C_6, NDIR) and R_{sc} ($= 1 - \eta_{tr}$) is almost linear. The importance of direct fuel injection, which effectively prevents fuel short-circuiting, in reducing HC emissions (see Section 7-3) is obvious.

The other sources of HC listed in Table 7-1, which are the major sources in four-stroke cycle SI engines, have been extensively reviewed elsewhere.[1,5] A short summary is given below. During compression and combustion, some of the fresh mixture inside the cylinder is forced by the increasing cylinder pressure into small volumes or crevices connected to the combustion chamber. These crevices include the narrow volumes between the piston, rings, and liner and small volumes in and around the spark plug, head gasket, and any cylinder-head valves. Most of this gas is unburned fuel–air mixture that remains unburned because the entrance to these crevices is too narrow for the flame to enter. The mixture that leaves these crevices later during expansion, and during the gas exchange blowdown process, is one significant source of HC emissions.

Another source for the HC is the quench layer on the combustion chamber walls. This thin layer (~ 0.1-mm thick) of mixture left on the wall as the approaching flame front extinguishes, contains unburned and partially burned fuel–air mixture. However, these layers do usually burn up during expansion. Other sources of HC are the porous deposits on the piston and cylinder head, and the thin oil film left by the piston rings on the cylinder liner, which can absorb fuel hydrocarbon components during compression and desorb those components after combustion, thus contributing to HC emissions.

Also, any liquid fuel impinging on the piston, liner, and cylinder head may not vaporize and mix with air rapidly enough to burn and may be exhausted unburned, or may flow into the piston, ring, liner crevices, or the combustion chamber deposits, or mix with the oil and become HC emissions. This source would be more significant when the engine is cold than when it is fully warmed up.

Note that the unburned and partially burned fuel–air mixture that escape burning during the primary combustion process can then mix with the products of combustion and fully or partially oxidize during the expansion and gas exchange processes. The amount of oxidation depends on the temperature and oxygen concentration time histories of these hydrocarbons as they mix with the bulk gases within the cylinder and in the exhaust. Also a fraction of the in-cylinder HC will remain inside the cylinder as part of the residual gas and will burn during the next cycle.

Relative air/fuel ratio λ and ignition timing affect hydrocarbon emissions from premixed two-stroke SI engines. Figure 7-10 shows the effect of varying A/F at fixed speed and load, at a light load, and at a high load, for a 347-cm^3 displacement, two-cylinder, carburetted crankcase-scavenged two-stroke engine. Both mass and concentration (ppm C_6, NDIR) HC emissions are shown. Mass hydrocarbon emissions show a minimum slightly lean of stoichiometric, $A/F \approx 14.5$, and increase for richer and leaner mixtures. At constant load, the delivery ratio Λ and trapping efficiency η_{tr} will change as λ changes, so the loss ratio R_{sc} will change. For stoichiometric and rich mixtures, where combustion is fast and repeatable, the output is airflow limited so Λ and η_{tr} will change little and the mass of HC in the short-circuiting mixture will vary as $1/\lambda$. For slightly lean mixtures where combustion is also good, Λ must increase (and hence R_{sc} will increase) to maintain constant load. Since HC mass emissions scale as R_{sc}/λ, they will change little (this explains the minimum). For leaner mixtures where combus-

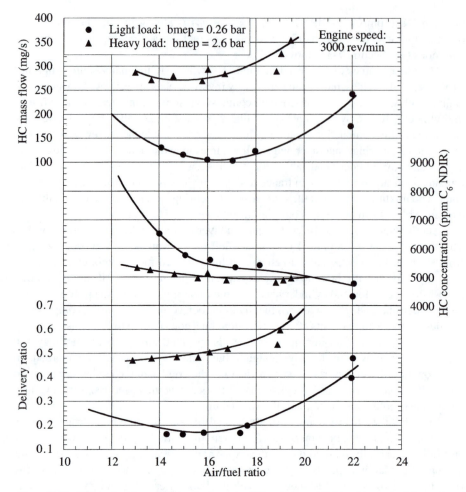

Figure 7-10 Effect of varying air/fuel ratio on engine HC emissions at light and heavy load for engine of Fig. 7-9: HC concentrations in ppm C_6 (NDIR).[11]

tion increasingly deteriorates (and hence requires a more rapid increase in Λ, and will add HC emissions), mass HC emissions will rise increasingly more rapidly than R_{sc}/λ. This is especially apparent in the light-load data.

The effect of spark timing on HC emissions is more modest. Retarding the spark from MBT timing, at fixed load, requires an increase in Λ to maintain constant torque. However, gas temperatures during expansion and exhaust will increase as timing is retarded, which will increase the amount of HC afterburning, provided oxygen is available. These trends affect HC emissions in opposite directions; thus they (partially) offset. With very advanced or retarded spark timing, combustion quality deteriorates, which will cause HC emissions to increase due to occasional partial burns or misfires (see Section 6-2-5).

7-2-5 Particulate Emissions

Particulate emissions from spark-ignition engines may be classified into three categories; lead, organic particulate (including soot), and sulfates.

For spark-ignition engines operated with leaded gasoline, the particulate emissions are primarily lead compounds. In four-stroke gasoline engines, the particle size is typically on the order of 1 μm diameter. The particles are formed and grow in the exhaust system due to vapor-phase condensation enhanced by coagulation, and therefore the exhaust temperature has a significant effect on particulate emission levels. Not all the lead in the fuel that is burned inside the cylinder is exhausted. Some of it deposits in the cylinder and exhaust system. With premixed two-stroke gasoline engines there is additional lead in the lead alkyls in the gasoline in the short-circuiting fuel–air mixture. Cold engines emit higher levels of particulate with leaded gasolines. Use of unleaded gasoline significantly reduces particulate emissions. The continuing removal of lead antiknock additives from gasoline will eventually eliminate this source of particulates.

With unleaded gasoline, organic-based particulate emissions are most important. One component of particulate emissions in many two-stroke engines is the blue opaque smoke that comes from the oil added to the fuel or sprayed into the scavenging air, in crankcase-scavenged engines. The usual oil to gasoline ratio is about 1:50. This ratio is dictated by low cost and ensuring low risk of failure; the most sensitive lubrication area affected by this ratio is the big-end bearing of the connecting rod. Lower oil/fuel ratios can be used with higher quality oils to reduce this component of emissions.[7]

The particulate collected from the exhaust with a filter consists of a solid (insoluble) component, which is primarily soot formed in the combustion chamber from fuel-rich mixture regions, with hydrocarbons and sulfate (the soluble component) absorbed onto the solid (submicron) particles. See Section 7-1-2 for additional details. Particulate emission levels of small carburetted two-stroke SI engines are substantially higher than comparable four-stroke SI engines (e.g., engine-out particulate emissions for various small two-stroke moped and motorcycle engines are between 0.05 and 0.5 g/km in simulated driving emissions-measurement cycles compared with about 0.006 g/km from a four-stroke motorcycle engine[7]).

The sulfur in the gasoline contributes the sulfate portion of these particulate emissions. Unleaded gasoline contains 100–600 ppm by weight sulfur (S). Sulfur is a yellow nonmetallic reactive element that combines with most other elements, including oxygen. Sulfur is, therefore, oxidized within the engine cylinder to sulfur dioxide (SO_2) and, in the presence of an appropriate catalyst (some vehicle catalyst types are among these), this SO_2 can be oxidized to SO_3 under normal spark-ignition engine operating conditions. At close to ambient temperatures, SO_3 combines with water to form a sulfuric acid (H_2SO_4) aerosol, which is a strong dibasic acid. Levels of sulfate emissions depend on the fuel sulfur content, the operating conditions of the engine, and the details of the catalyst system used. A high sulfur content in the gasoline, with rich operation with engines with catalysts, can lead to formation of hydrogen sulfide with its annoying odor.

7-2-6 Trace Emission Components

Specific trace emissions compounds of concern are polycyclic aromatic hydrocarbons (PAH), and carbonyls (adlehydes, ketones) and phenols. PAHs are compounds of four to six benzene rings and can be gaseous, liquid, or absorbed onto particulate. The most hazardous PAHs are benzo(a)pyrene and 1-nitropyrene. The limited data available indicate that high-performance two-stroke motorcycle engines have substantially higher emissions (in micrograms per kilometer) of PAHs than do smaller moped two-stroke engines and four-stroke motorcycle and car engines (in the same units, μg/km) in appropriate driving cycles. The quantity of PAH in the gasoline is an important factor: gasolines with high relative amounts of PAH result in higher engine-out PAH emissions. Rich engine operation also increases PAH emissions, relative to lean operation. An oxidation catalyst reduces the PAH levels.[7]

Aldehyde and phenol engine-out emissions, which are toxic and highly reactive in the photochemical smog atmospheric reactions, are also of concern in small premixed two-stroke SI engines, as in four-stroke SI engines. An oxidation catalyst substantially reduces these emissions. See Laimböck[7] for additional details.

7-2-7 Emissions Reduction Potential

Due to its simple construction, compactness, high power-to-weight ratio and low cost, the standard two-stroke SI engine is widely used in small sizes for handheld equipment, mopeds, motorcycles, and outboard marine applications. Its major emissions problem is the high hydrocarbon emissions that result from fuel–air mixture short-circuiting the engine, and reduction of these emissions is essential if this type of two-stroke is to maintain or expand its market.

Direct fuel-injection into the cylinder is one promising approach (see Section 7-3), but it is expensive. Here, we review other approaches that help resolve this fuel short-circuiting problem. Three areas are promising: (i) better control of fuel metering to avoid fuel-rich operation; (ii) better control of the scavenging flow to minimize short-circuiting; (iii) use of a catalyst to clean up the exhaust stream (see Section 7-5).

Figure 7-11 illustrates the more important effects that premixed two-stroke SI engine design and operating variables have on bsfc and engine-out emissions. The engine variables are listed in the center of the diagram. The more fundamental variables or factors are whether the mixture is rich, stoichiometric, or lean, whether the maximum burned gas temperatures (and pressure) increase or decrease, and whether combustion becomes significantly faster ($\Delta\theta_b$ decreases) or slower ($\Delta\theta_b$ increases). The most effective means of reducing emissions are the following. CO and HC emissions can be reduced by running the engine leaner (see Figs. 7-4 and 7-10), especially at partial load, by avoiding rich excursions with more precise control of mixture enrichment for cold engine starting, by acceleration enrichment during transients that compensates for the lag in increased fuel vapor delivery to the cylinder on throttle opening due to the delay in fuel vaporization, and by using fuel cutoff during highly throttled engine decelerations. Because richer mixtures are traditionally used to compensate for potential combustion problems, such leaner engine operation requires a good combustion system. So combustion improvements may well be required. Excessively lean excursions that lead to

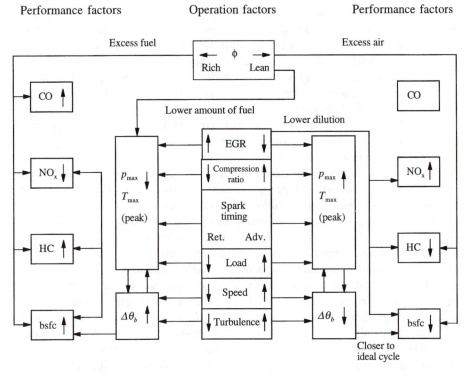

Figure 7-11 Diagram showing the directional effects of major two-stroke SI engine design and operating parameters on engine-out CO, NO_x, and HC emissions, and bsfc.

partial burns and misfires must be avoided; a higher energy spark discharge and better control of scavenging (see below) will help resolve such combustion problems.[7]

A more efficient scavenging process would reduce the amount of short-circuiting fuel. More precise control of the scavenging flow can be achieved with an exhaust port flow control valve (see Section 5-5-4). By varying the exhaust port opening timing and effective port flow area, more of the scavenging flow can be retained within the cylinder (especially at partial load). Exhaust port control valves are increasingly being considered for larger two-stroke SI engines. Stratified scavenging techniques have also been developed, where the earlier part of the scavenging flow, which contributes much of the short-circuiting flow, is largely air and the latter part of the scavenging flow contains most of the fuel. One proposed approach is to stratify the crankcase mixture by injecting the gasoline through the crankcase wall onto the underside of the piston, through an elongated hole in the piston skirt, with a low-cost solenoid-controlled fuel injector. The piston crown is thereby cooled, and rapid fuel vaporization is achieved. A dividing wall at the bottom end of the piston skirt partly separates the space inside the piston from the crankcase and by arranging for the later-opening scavenging ports to be fed rich mixture through this elongated hole from the inside of the piston, a stratified scavenging can be obtained.[7] An alternative approach, called air-head stratified scavenging, also achieves significant HC and CO emissions reductions (see Blair[14] for a brief history of this approach). During intake the airflow into the engine is divided: part flows through the carburetor into the crankcase as in a conventional engine; part flows into the scavenging or transfer ducts without mixing with fuel. The throttle and reed valves control these flows. Then, when scavenging commences, the initial phase of the scavenging process is done by air alone, and fuel enters the cylinder later. Thus much of the short-circuiting flow is air alone. An additional claimed benefit is improved (faster) combustion at partial load, achieved by containing the fuel within the cylinder and preventing excessive dilution. Reductions in HC emissions by a factor of 3–4, and CO emissions by about a factor of 2, were demonstrated.[15]

Addition of a catalyst to the exhaust of these simple two-stroke SI engines, combined with appropriate control of the relative air/fuel ratio of the exhaust gas entering the catalyst, achieves significant reductions in the HC and CO emissions. Some reduction in NO_x can also be obtained if the engine can be operated close to stoichiometric, with a three-way catalyst. Catalysts are discussed more fully in Section 7-5. The critical engine issues are the engine's relative air/fuel ratio and then whether additional (secondary) air needs to be added to the exhaust to achieve an oxidizing environment. The catalyst conversion efficiency (the fraction of entering emissions that are removed in the catalyst) depends strongly on temperature and relative air/fuel ratio. Catalyst temperatures above about 300°C are required for effective removal of HC and CO. A catalyst problem, however, with premixed two-strokes, with significant short-circuiting of fuel-air mixture, is overheating the catalyst at high engine loads due to the combination of high exhaust gas temperatures and substantial energy release from the short-circuiting fuel within the catalyst. High HC and CO catalyst conversion efficiencies in small-engine applications, along with good catalyst durability (which requires avoiding catalyst overheating), can be achieved by running the engine slightly lean of stoichiometric over much of its operating map, and holding scavenging mixture losses to a minimum by keeping port widths and open times as small as feasible while meeting engine performance requirements.[16] Alternatively, with engines using richer mixtures, use of secondary air injection into the exhaust with a dual catalyst system (one upstream, one downstream, to split the energy release as HC and CO are oxidized) or with a single catalyst that has excellent thermal stability, is an effective strategy.

7-3 EMISSIONS FROM DIRECT-INJECTION SI ENGINES

7-3-1 Benefits of Direct Injection

Direct-injection (DI) gasoline two-stroke SI engines have been extensively developed and examined over the past 15 or so years. One obvious reason for this interest is that direct injection solves the standard two-stroke's problem of short-circuiting fuel. A major reason that so much effort has been invested in DI two-stroke engines is that the technology for direct gasoline injection has been improved significantly and is now capable of reliably preparing the in-cylinder fuel–air mixture distribution this approach requires. Although there are several gasoline direct-injection SI engine concepts, and several fuel injection approaches, the overall objectives are to confine the bulk of the fuel–air mixture within the cylinder at the lighter loads to a region around the spark plug (so the engine can operate with excess air or substantial burned residual and improve its efficiency), and at high load to fill the cylinder as effectively as possible with stoichiometric or slightly rich mixture to obtain the maximum bmep. To accomplish these objectives, injection is usually later during compression at the lighter loads, and is earlier at high loads, but always after the point in the cycle where any fuel becomes involved in the scavenging process to avoid any fuel short-circuiting.

The fuel injection technologies employed are either high-pressure ($\sim$3 MPa) liquid injection or air-assist fuel injection into the cylinder, or injection into a cavity or chamber connected to the cylinder via an auxiliary valve or port that admits (pressurized) rich fuel–air mixture to the cylinder at the appropriate time. Examples of these technologies and additional discussion can be found in Sections 1-6, 6-3-2, 9-2-2, and 9-3-4 and Figs. 1-24–1-26, 2-25, 9-3, 9-7, and 9-10. Here, we will compare the emissions characteristics of these DI two-stroke SI engine approaches with those of equivalent carburetted two-stroke engines. In the next two sections, we will examine NO_x and HC emissions control strategies with DI two-stroke SI engines in more detail because these are the two pollutants it is most important to control.

Figure 7-12 shows a photo of a DI crankcase-scavenged scooter engine and the layout of the direct-fuel-injection system.[17] Figure 7-13 shows the brake specific fuel consumption (bsfc), CO and HC emissions contours on the bmep versus speed performance maps for the stock carburetted engine and the direct-injection version. The CO data are in mole % and HC data in ppm hexane measured with NDIR (to obtain ppm C_1 HC equivalent to FID data, multiply by 11). The fuel consumption of the DI engine is about 30–50% better, depending on the region of the performance map, due to elimination of the fuel short-circuiting and operation with less-rich mixtures (as evidenced by the CO contours: Fig. 7-4 can be used to assess mixture enrichment). The HC emissions are reduced by about an order of magnitude, except at very light loads, where the DI benefit is significantly less (see Section 7-3-3).

Nitric oxide emissions data from DI two-stroke engines can be higher or lower than standard carburetted two-stroke engine emissions, depending on the details of how the engines are calibrated. Relative to four-stroke cycle engines, NO emissions from two-stroke SI engines are significantly lower because (i) for comparable power output, the two-stroke operates at lower bmep due to its two times higher firing frequency and (ii) it retains a higher fraction of the burned gas in the cylinder as residual. DI two-strokes typically run leaner than carburetted two-strokes, which increases NO emissions; however, they often control the amount of trapped residual more precisely and increase that amount, which would decrease NO emissions (see Fig. 7-3).

The details of the direct-injection fuel spray, the in-cylinder air motion, and the combustion chamber shape (both cylinder head and piston crown) all have an impact on the performance and fuel consumption of DI two-stroke engines, and especially on the emissions characteristics of the engine. At lighter loads, when the amount of fuel injected and delivery rates are low relative to full-load values, the spray cone angle, mean drop size, droplet velocity or spray penetration, and start and end of injection are especially important parameters. A contained "cloud" of combustible fuel-vapor–air–residual mixture (relatively close to stoichiometric) adjacent to the spark plug location

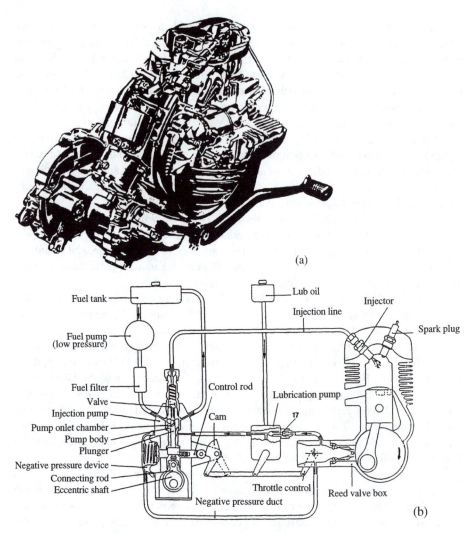

Figure 7-12 (a) Direct in-cylinder injection crankcase-scavenged two-stroke SI scooter engine; (b) schematic of system arrangement. (Courtesy of Piaggio Co. S.p.A.[17])

must be produced at time of spark discharge. The in-cylinder airflow pattern and the piston crown and cylinder head shapes often assist in forming this contained mixture cloud. It takes substantial effort to develop an injection and combustion system that provides good performance and low emissions over the important regions of the engine's performance map. One challenge, therefore, in interpreting data from DI two-stroke SI engines is whether the injection and combustion system have been developed sufficiently for the performance and emissions data to be truly illustrative of the potential of that concept and technology. As we compare the characteristics of high-pressure liquid fuel injection systems, air-assist fuel injection systems, and externally prepared direct

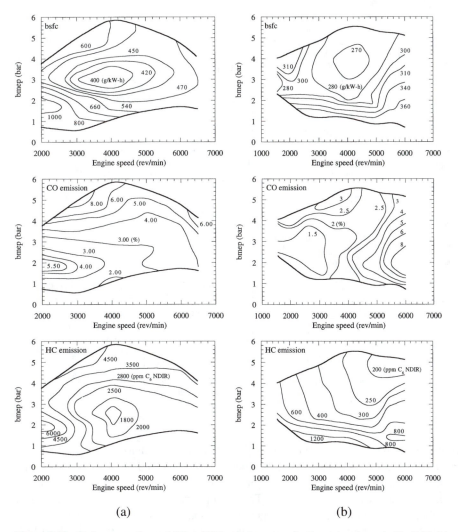

Figure 7-13 Fuel consumption and CO and HC emissions maps for the engine shown in Fig. 7-12: (a) carburetted engine; (b) direct in-cylinder fuel injection engine[17]; bsfc in grams per kilowatt-hour; CO in percent HC in ppm C_6 (NDIR); multiply by 11 to obtain ppm C_1 (FID).

mixture-injection systems (like IFP's IAPAC concept; see Sections 6-3-2 and 9-2-2 and Fig. 2-25), this caution should be kept in mind.

The data in Fig. 7-14 illustrate the phenomena that variations in injection timing, with high-pressure liquid fuel direct-injection (HPFI) and air-assist direct-injection systems (AAFI), can produce. The effect on bsfc, bsHC, and bsCO at a fixed light-load low-speed operating point (1200 rev/min, 2 bar bmep) of varying the start of injection is shown. The HPFI spray has a cone angle of 55°, the Sauter mean droplet diameter was 15 μm, and the injection pulse duration was 0.3 ms at a fuel injection pressure of

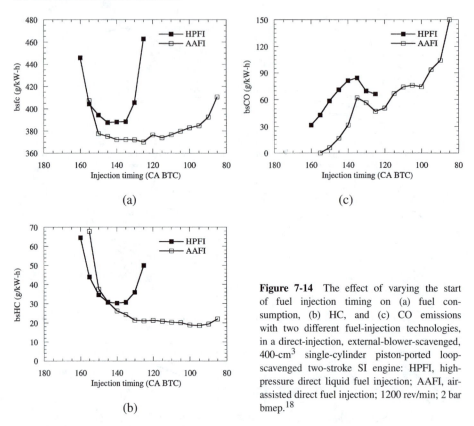

Figure 7-14 The effect of varying the start of fuel injection timing on (a) fuel consumption, (b) HC, and (c) CO emissions with two different fuel-injection technologies, in a direct-injection, external-blower-scavenged, 400-cm^3 single-cylinder piston-ported loop-scavenged two-stroke SI engine: HPFI, high-pressure direct liquid fuel injection; AAFI, air-assisted direct fuel injection; 1200 rev/min; 2 bar bmep.[18]

about 70 bar. The AAFI system injects fuel at 7.5 bar pressure into a cavity in the injector containing 5 bar compressed air, the open period for injecting the air–fuel mixture into the cylinder was 3.5 ms, the spray cone angle was about 70°, and the Sauter mean droplet diameter was less than 6 μm. These drop sizes are sufficiently small to give rapid fuel vaporization. Results at higher speeds and light load were similar to those shown.[18] The injection timing sweep was done at the air/fuel ratio that gave best bsfc and MBT spark timing. The data show that the "softer" longer duration spray of the air-assist injector is more tolerant of variations in injection timing. The exhaust port closes at 87° ABC (93° BTC), and overly early injection shows a rise in bsfc and bsHC due first to overmixing and then to increasing amounts of fuel short-circuiting. Note that the injector is located in the cylinder head, so there is a substantial transit time (about 7 ms) for the spray to reach the exhaust port at this engine speed, which corresponds to a spray penetration rate of about 10 m/s. The HPFI system also shows a rise in bsfc and bsHC as injection moves closer to TC. This is most likely due to increasingly poor positioning of the fuel-vapor–air cloud relative to the spark plug location, leading to partial burning and even occasional misfires. Note that the general trend with CO is for bsCO to rise as injection moves closer to TC due to the increasingly rich regions in the spray that result from less time available for mixing. Generally, the smaller drop sizes, broader spreading, and some premixing of fuel and air, with air-assist fuel injection technology result in less sensitivity to injection parameters and better combustion and emissions characteristics than can be achieved with high-pressure liquid-fuel injection systems. There are however, an increase in engine cost and a performance penalty (associated with compressing the injector air) with the air-assist injection technology.

An alternative to liquid "direct fuel injection" is direct *mixture* injection after scavenging is complete, as exemplified by the Institut Français du Petrole's (IFP) IAPAC concept—Injection Assistée Par Air Comprimé. Here (see Sections 2-3-6 and 9-2-2 and Figs. 2-25 and 9-7), the fuel is injected just ahead of the intake valve by a single injector of the standard intake-port-injection type, into a surge tank into which compressed air from the crankcase has been admitted through a reed valve. This tank connects with the cylinder via a small intake poppet valve, which is opened as or after scavenging ends in these crankcase-compressed cross-scavenged engine designs. Rich "low-pressure" fuel–air mixture is thus delivered to the cylinder with high spray quality, small fuel droplets, and low penetration. Examples of IAPAC technology two-stroke engines have been developed for several applications: outboard marine,[19] scooters,[20] and automotive.[21,22] Figure 7-15 illustrates the performance and emissions characteristics of this concept with bsfc, bsHC, $bsNO_x$, and standard deviation in imep, data as a function of load (bmep) at 2000 rev/min. Three engine configurations of this three-cylinder 1230-cm^3 displacement cross-scavenged outboard-marine engine were tested: the standard engine with conventional port fuel injection; the same engine with IAPAC direct mixture injection; the same engine with IAPAC mixture injection, with additional control of the transfer flow at light load (Fluid Dynamically Controlled Combustion Process). Relative to the standard (premixed) engine, the bsfc and bsHC are reduced substantially by elimination of any significant fuel short-circuiting, but $bsNO_x$ increases significantly at medium and high loads because the engine runs leaner than the standard engine. The combustion cycle-by-cycle variability, which causes the cyclic variability in imep (see Section 6-2-5) is improved, but still deteriorates unacceptably at light loads due to the fact that the use of crankcase compressed air and low-pressure (2–3 bar) fuel injection dictates early opening of the intake valve, resulting in overmixing and partial burns and some misfires. The light-load combustion performance can be significantly improved by controlling the scavenging process with flow control valves either on the exhaust (see Section 5-5-4) or transfer port. The data shown are with control of the crankcase to cylinder transfer passages. The slowing down of the main scavenging flow reduces the mixing between the fresh air charge and residual gas, better stratifying the air. Throttling the transfer passages also increases the fraction of the air that enters the surge tank which then enters the cylinder later, with the fuel, through the auxiliary intake valve, improving mixture preparation, and air trapping. Along with substantially improved combustion go reduced bsfc and bsHC.[21] Substantial additional information is available on the performance and emissions characteristics of this concept.[19–22]

A low-pressure pneumatic direct-injection concept using a mechanically driven rotary injector in the transfer passage (analogous to the IAPAC approach) has been combined with the active radical concept (see Section 6-3-1) to produce a low-emissions motorcycle two-stroke engine concept.[23] This combination solves both the fuel short-

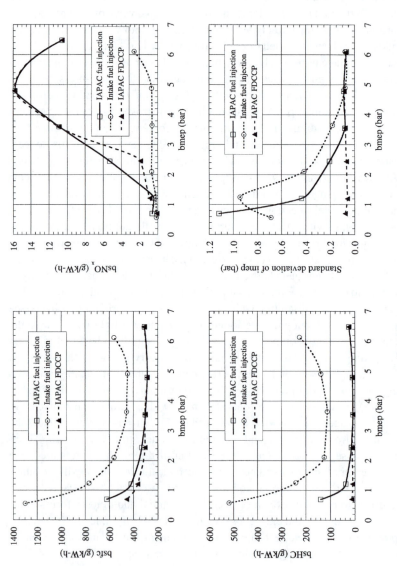

Figure 7-15 Performance and emissions characteristics of IFP's IAPAC air-assisted direct mixture injection two-stroke SI concept: bsfc, bsHC, bsNO$_x$, and standard deviation in imep are plotted versus bmep at 2000 rev/min. The three engine configurations are standard intake-system fuel injection, IAPAC air-assist direct mixture injection into the cylinder, and IAPAC with additional control of the transfer air flow at light load (FDCCP); for a three-cylinder, 1230-cm^3 displacement, cross-scavenged outboard marine two-stroke engine.[21]

circuiting problem, via direct mixture injection, and the problem of poor combustion at light load, through use of the ARC approach.

7-3-2 NO$_x$ Emission Control

As explained previously, the two-stroke SI engine inherently produces low NO$_x$ emissions due to the engine operating at lower bmep (relative to the four stroke) for comparable shaft output because the firing frequency is doubled, and due to its retention of a substantial amount of burned residual gas within the cylinder. However, because the direct-injection two-stroke SI engine usually operates with a lean-of-stoichiometric mixture at light to medium load, the standard three-way catalyst technology of the four-stroke engine, which requires stoichiometric operation, cannot be used, and additional NO$_x$ control must be achieved within the cylinder. This two-stroke engine NO$_x$ emissions advantage can be enhanced by more precise control over charge stratification and increased exhaust gas retention. Additional improvements in NO$_x$ emissions can be achieved by exhaust gas recirculation (EGR)—recycling a fraction of the burned exhaust gases into the scavenging or intake flow—to supplement or replace internally retained burned gases. Use of EGR is a well-established four-stroke SI engine NO$_x$ control strategy.[1]

Figure 7-16 shows an Orbital Engine Corp. design of piston-ported EGR system feeding directly into the crankcase under each cylinder, which can be used with crankcase-scavenged two-strokes. The timing and dimensions of port opening need to be carefully optimized. A limitation of this system is the sensitivity to mean crankcase pressure, so the EGR flow potential decreases at higher loads as its driving pressure difference (exhaust to crankcase) is reduced. Figure 7-16 also shows an alternative, with higher EGR flow potential; the EGR is introduced into the intake manifold where the av-

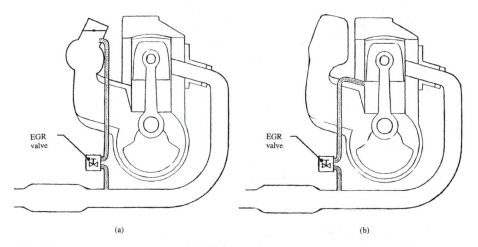

(a)   (b)

Figure 7-16 Two approaches to exhaust gas recycling (EGR) in direct-injection two-stroke SI engines: (a) EGR flow into the intake manifold; (b) EGR flow via a piston-controlled port into the crankcase. (Courtesy Orbital Engine Corp.)

erage pressure (vacuum) is lower than in the crankcase. Figure 7-17a shows the effects of piston-ported and manifold EGR on the bsfc/bsNO$_x$ characteristics of an Orbital Engine Corp. air-assist DI two-stroke OCP-X 1.2-liter three-cylinder engine, at fixed speed and load (bmep = 288 kPa).[24] Moving from A to B in the figure shows the effect of leaning out the in-cylinder fuel–air mixture, with B representing the lean relative A/F ratio

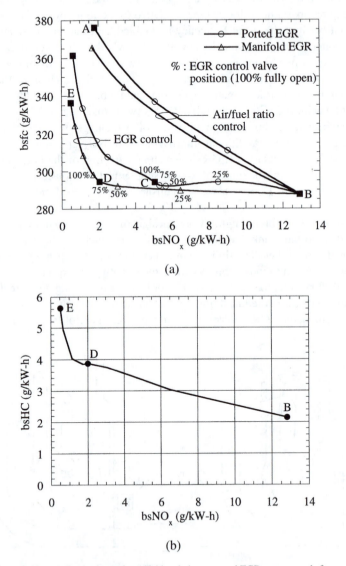

(a)

(b)

Figure 7-17 (a) Comparison of manifold and piston-ported EGR systems on bsfc versus bsNO$_x$ plot, at 2600 rev/min and 55 N-m torque, for direct-injection Orbital combustion process 1.2-liter three-cylinder two-stroke SI engine. A–B shows effect of enleanment only, with B at best torque conditions. B–C shows ported EGR; B–D shows intake manifold EGR, both at fixed relative A/F; D–E shows effect of enrichment. (b) Effect of manifold EGR on HC emissions.[24]

that gives the highest efficiency. B to C shows how the engine responds to EGR with the air/fuel ratio held roughly constant. The percentages show the relative open area of the EGR valve. Beyond 100% open, D to E, the graph then shows the effect of mixture enrichment. Note that substantial reductions of NO_x can be achieved with (up to 30%) EGR, without any deterioration in bsfc, with only moderate impact on combustion stability and HC emissions (see Fig. 7-17b). This is because Orbital's air-assist injection and combustion system produces a confined and stratified fuel-vapor–air mixture cloud, and combustion of the richer regions of this cloud only deteriorates modestly as EGR is added to (and effectively dilutes) the intake air. Additional benefits at higher loads have been obtained by cooling the EGR flow with a heat exchanger prior to mixing with air in the inlet manifold. This reduces the mixture temperature, and hence burned gas temperatures and NO_x emissions, further. With manifold EGR, it is important to distribute the EGR flow equally between the different cylinders.[24]

7-3-3 HC Emission Mechanisms and Control

Recent detailed studies of fuel distribution, combustion characteristics and hydrocarbon emissions in air-assisted direct-injection two-stroke engines have identified the primary mechanisms for HC emissions.[25–27] Speciation of the hydrocarbon emission from different types of direct-injection two-stroke SI engines has added additional insights.[28] We will now summarize these findings and relate them to the possible HC source mechanisms listed in Table 7-1, focusing on lighter loads where the direct-injection engine operation is characterized by later fuel injection and stratified combustion, rather than heavier loads (characterized by earlier injection and more homogeneous combustion).

Figure 7-18 shows the combustion and HC emissions profiles through a typical cycle with an air-assist direct-injected gasoline-fueled cross-scavenged two-stroke engine, at 900 rev/min and an imep of 100 kPa. Measured cylinder and exhaust pressure data were used to calculate the mass-burning rate and the exhaust mass flow rate. A fast-response FID system was used to determine the time-varying HC concentrations in the exhaust, close to the port. Then the HC mass flow rate out of the engine cylinder was calculated from the relation

$$\dot{m}_{HC} = \tilde{x}_{HC}(M_{HC}/M_e)\dot{m}_e \tag{7-14}$$

where M is the molecular weight and subscript e denotes exhaust gas values. Use of Eq. (7-14) through each engine cycle gave the mass of HC emitted from that cycle. The HC concentration in the exhaust varies significantly during the total gas exchange process, rising to a maximum at about the time the scavenging flow rate peaks. During the period when the exhaust port is closed, the gas in the port is essentially stagnant and the HC concentration is almost constant. The cylinder blowdown process (EPO to SPO) carries about 20–30% of that cycle's HC into the exhaust, at the light-load conditions tested. The bulk of the HC (60–70%) are exhausted during the primary scavenging flow (IPO to about BC). A backflow at this light load into the scavenge ports occurs from about BC to scavenge port closing. Then, additional gas of composition close to the residual burned gas leaves through the exhaust port before it closes, due to piston motion. About 10% of that cycle's HC leave during this last phase (SPC to EPC).[27]

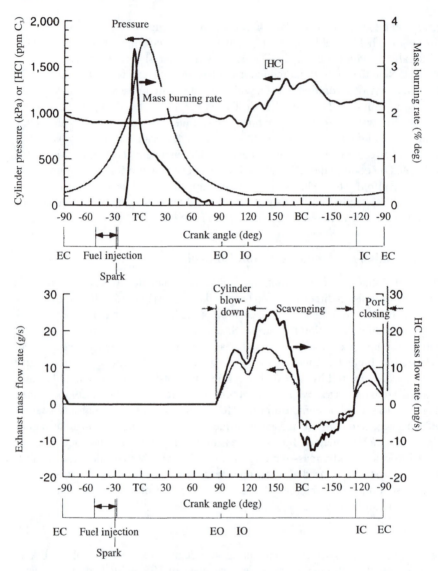

Figure 7-18 Combustion and HC measurements through a typical individual cycle with a direct-injection two-stroke SI engine at light load (900 rev/min, imep = 100 kPa). Upper graph: cylinder pressure, mass-burning rate (from p_{cyl}), and HC concentration from fast FID, versus crank angle. Lower graph: exhaust port mass flow rate and HC mass flow rate (HC mass fraction × mass flow rate) versus crank angle.[27]

Two types of air-assist injectors, a wide spray and a narrow spray that had different combustion stability characteristics and different light-load HC emissions were tested. Their different spray shapes 40° BTC are indicated by the Mie-scattering images produced by the fuel droplets in the spray in Fig. 7-19. Compressed air was supplied to both injectors at 540 kPa, and the liquid fuel at 620 kPa 3 ms before injection; the

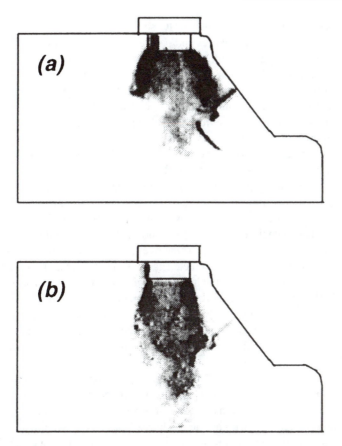

Figure 7-19 Planar Mie-scattering images of direct-injection air-assist fuel sprays: (a) wide-spray and (b) narrow spray, at about 40° BTC.[27]

masses of fuel and air injected each cycle were comparable. The narrow spray required significantly more spark advance (some 10°) than the wide-spray injector for best HC emissions, and the level of HC emissions was about half. Apparently minimum injection to ignition time leads to lower HC emissions, which is consistent with our understanding that overmixing of the fuel with air is a major source of HC emissions in direct-injection gasoline engines.[27]

The extensive data and analysis of Fansler et al.[25-27] lead to the following description of the light-load HC emissions mechanisms in this type of direct-injection engine. We will relate this discussion to Table 7-1. Direct fuel short-circuiting is eliminated as a source, through suitable choice of injection technology and injection timing. Fuel trapped in the piston–top-ring-land crevice is relatively unimportant at light load because, due to the limited spray penetration under these conditions, little fuel will reach the entrance to this crevice region. At higher loads, where the spray penetrates deeply into the combustion chamber to achieve as full air utilization as possible, this crevice will become more important.

Misfires and partial burns are a significant HC source. At light load, where the charge is highly stratified, occasional misfires and partial burns can occur due to cyclic variations in the fuel/air equivalence ratio and residual burned fraction in the mixture near the spark gap. The excess HC emissions resulting from a single misfiring cycle are spread out over a few (about four) engine cycles due to the high residual fraction. Consequently, they amount to only about one-third to one-half of the average mass of fuel injected per cycle with the remaining misfire-cycle fuel mass being burned within the cylinder in the following few cycles.

Bulk quenching of the combustion process is the major HC source. Extinction occurs as the flame propagates into mixture regions that are too lean or dilute to burn, around the periphery of the spray-produced cloud of fuel vapor, air, and residual gas. This mechanism is primarily responsible for the increase in hydrocarbon emissions as engine loads become lighter and as the delay between injection and ignition increases.

Another form of bulk quenching occurs during the final mixing-limited phase of the direct-injection combustion process, when some over-rich or slow-burning mixture regions fail to burn completely. The last fuel injected has less time to mix and, depending on details of injector design, may be injected at low velocity with poor atomization due to the low-pressure differential between injector and cylinder at the end of injection. It therefore mixes more slowly, leaving overly rich pockets of mixture near the injector. The release of fuel trapped in the injector-nozzle exit crevice can be an appreciable HC source too, for similar reasons: any fuel in this region emerges slowly from this crevice and remains largely unmixed and unburned.[27]

Some of the hydrocarbons that escape burning during the normal DI engine combustion process due to these mechanisms will partially or fully oxidize within the cylinder and the exhaust, during the expansion and exhaust processes, if they achieve a high enough temperature through mixing with previously burned gas. Also, a significant fraction will remain in the cylinder with the residual gases (the residual HC concentration from Fig. 7-18 is about 3500 ppm C_1) and will largely burn up during the next cycle. Speciated hydrocarbon emissions data indicate that while the bulk of the hydrocarbons emitted are unreacted fuel compounds, some are secondary oxidation products. Figure 7-20 shows the weight percent of 24 C_1 through C_8 hydrocarbons in the exhaust of a direct-injection gasoline-fueled two-stroke SI engine, identified through gas chromatography.[28] Both speciated exhaust HC and fuel HC distributions are shown. The hydrocarbons have been classified into two groups:

1. secondary oxidation products, the C_1–C_4 compounds (except butane and isobutane), which are not present in the fuel;
2. unburned fuel compounds such as isopentane, isooctane, and toluene.

By summing the weight percentages shown in the figure, the fraction of exhaust HC corresponding to secondary oxidation products is found to be about 20%.†

---

†The precise value depends on how the other hydrocarbons are apportioned. These are most probably dominated by higher carbon-number fuel compounds, which is what we have assumed here.

Weight (%)

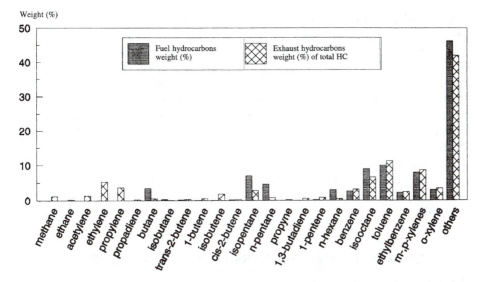

Figure 7-20 Comparison between speciated exhaust HC and fuel HC in a direct-injection gasoline-fueled two-stroke SI engine at 2475 rev/min and bmep = 1.6 bar.[28]

Variations of injection timing, spark timing, and fuel injector type on these distributions produced changes expected from the previously described HC mechanisms. Earlier injection increases the fuel hydrocarbon emissions (in grams per kilowatt-hour) much more than the secondary oxidation hydrocarbons, due to increased overmixing of fuel vapor, and eventually with very early injection, due to fuel short-circuiting. Retarded spark timing, at fixed injection timing, increases the fuel hydrocarbons emissions due to increased overmixing with significantly less increase in secondary oxidation HC. Direct high-pressure liquid-fuel injection increases the speciated HC emission amounts significantly, by a factor of 2, above pneumatic injection amounts, with both fuel and secondary oxidation HC compounds increasing by comparable ratios. With direct fuel injection and active radical combustion which can be made to occur at light load (see Section 6-3), comparable amounts of secondary oxidation HC products were produced, the more reactive fuel HC were absent, only the less reactive aromatic HC were present and these in much lesser amounts (by a factor of about 3 for benzene and factor of about 10 for toluene) than with direct injection and normal combustion. Only when injection was sufficiently advanced for fuel short-circuiting to occur were significant amounts of fuel hydrocarbons present.[28]

Control of engine-out HC emissions from direct-fuel-injected SI engines is effected by the following strategies. The appropriate combination of fuel injection system, combustion chamber shape, and scavenging and compression in-cylinder flows must be developed, that produces a confined cloud of fuel vapor, air, and residual gas that is appropriately stratified and located relative to the spark plug electrode gap for a robust (fast) start to combustion. Overall lean operation at light load is desirable and feasible, but overmixing of fuel must be minimized. Direct-injection engines have the benefit that relatively little enrichment is required with a cold engine, relative to premixed engine operation. An exhaust port flow control valve can be used to increase the amount of residual retained and therefore increase the temperature of the retained in-cylinder gases. This enhances in-cylinder HC oxidation after normal combustion ends. There is, however, a trade-off between in-cylinder residual temperature levels, and exhaust stream temperature levels, which affect the performance of an exhaust oxidation catalyst. Earlier effective exhaust port opening, giving less in-cylinder HC oxidation and more catalyst HC oxidation, may be a better overall system solution. Another strategy for both reducing engine-out HC, increasing exhaust gas temperature, and improving catalyst performance at idle and light load is cylinder cutout. Under these conditions, cutting out a cylinder by not injecting fuel increases the bmep in the firing cylinders, reduces the light load overmixing problem and increases in-cylinder HC burnup, thereby decreasing engine-out HC, and increases catalyst temperature, which increases catalyst HC oxidation.[24]

7-4 EMISSIONS FROM COMPRESSION-IGNITION ENGINES

7-4-1 Formation Processes

Diesel engine NO_x emission levels are comparable to those from spark-ignition engines. Carbon monoxide emission levels are typically lower due to the lean fuel/air ratios used and the efficient diesel combustion process (especially in large bore engines). Diesel hydrocarbon emissions are significant although exhaust concentrations are lower by about a factor of 5 than a typical SI engine level. Hydrocarbons are responsible for the white smoke that may occur during engine starting and warm-up due to condensation in the exhaust. Specific hydrocarbon compounds in the exhaust gases are the source of diesel odor. Particulate emissions, which mostly result from incomplete combustion of fuel hydrocarbons, are an especially important component of the diesel emissions problem. Between 0.2 and 0.5% of the fuel mass is emitted as small particles (typical particle size is less than 1 μm), which consist primarily of soot with additional absorbed hydrocarbon and sulfate material. Figure 7-21 illustrates a typical breakdown of the major components of air, fuel, lubricating oil, and exhaust gas for a large low-speed two-stroke diesel engine.[29]

In the diesel engine, the liquid fuel is injected into the cylinder as one or more jets, just before combustion starts. The injected fuel atomizes, and the droplets vaporize while they penetrate and mix with the air contained within the cylinder. Spontaneous ignition occurs as soon as the fuel-vapor–air mixture comes within the flammability limits. Local regions of fuel-rich, stoichiometric, and lean mixture strength are formed; thus different portions of the fuel burn under different conditions. See Section 6-4 for a more detailed discussion of the diesel combustion process. The pollutant formation processes are strongly dependent on the fuel distribution and how that distribution changes with time, which depends on droplet vaporization rates and fuel–air mixing rates. Nitric oxide forms most rapidly in those regions where the burned gas is slightly lean of stoichiometric and the local temperature is high. Soot forms in the fuel-rich regions of the spray where the fuel vapor, heated by mixing first with air and then with hot burned gases within the flame region, pyrolyzes to form polycyclic aromatic hydrocar-

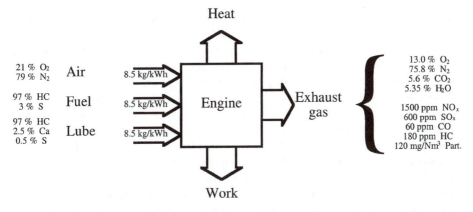

Figure 7-21   Typical breakdown of the major components of the air, fuel, lubricating oil, and exhaust gases, for a large low-speed two-stroke diesel (courtesy of MAN B&W Diesel A/S).[29]

bons which then condense to form soot particles, which then grow and agglomerate (see Section 7-1-2). When soot contacts unburned oxygen at sufficiently high temperatures, it oxidizes giving rise to the yellow luminous character of the diesel flame. The hydrocarbons components and aldehydes originate in regions where the flame quenches on the walls or where excessive dilution with air prevents the completion of the combustion process. We now examine these mechanisms in more detail.

NO_x emissions.   The kinetic mechanisms for NO and NO_2 formation were described in Section 7-1-2. At the high prevailing temperatures in the burned gases during and shortly after combustion, oxygen and nitrogen will inevitably react to form oxides of nitrogen. The primary process is the formation of nitric oxide (NO), and the assumptions made regarding equilibration of species in the C–O–H burned gas system apply to diesels as well as to spark-ignition engines. During combustion and expansion, and in the exhaust system, some nitrogen dioxide (NO_2) and some nitrous oxide (N_2O), typically 5–10% and 1%, respectively, of the NO amount, will form. The conversion of NO to NO_2 also occurs in the atmosphere. Oxides of nitrogen emitted to the atmosphere contribute to photochemical smog formation. Because NO_2 is soluble in water, it is washed out by rain, which increases the acidity of the soil. The formation of NO_x during diesel combustion (for convenience, the various species are grouped and termed NO_x) is controlled by the local conditions within the enflamed region that surrounds each diesel fuel spray. Local burned gas temperature and oxygen concentration are the controlling parameters, and a useful rule of thumb is that a temperature change of 100 K changes the NO_x emissions by a factor of about 3.[29]

CO emissions.   Carbon monoxide emissions from internal combustion engines are controlled primarily by the relative air/fuel or fuel/air ratio in the combustion zone. Spark-ignition engines operate close to stoichiometric, and therefore CO emissions are significant. Diesels, however, operate well on the lean side of stoichiometric, and CO emissions are low enough to be unimportant.

HC emissions.   During combustion, a small fraction of the fuel hydrocarbons remain unburned and other hydrocarbons may be formed. Most of these, with some HC from the lubricating oil, leave the cylinder still unburned in the exhaust. Hydrocarbon emissions are normally expressed as an equivalent C_1 amount. The exhaust gas from large-bore diesel engines can contain up to 300 ppm of HC, but the emission level depends on the fuel and the engine design and operating details. Some of the higher molecular weight HC species may condense to form tiny particles, and some may be absorbed onto the surface of the soot particles formed from the fuel inside the cylinder during combustion. The two major causes of diesel HC emissions are (i) fuel that has mixed too rapidly with air and, during combustion, is too lean to burn (called overleaning) and (ii) fuel that mixes too slowly during combustion, is too rich to burn. Although some of this fuel if mixed with sufficiently hot burned gases early enough in the expansion process may then oxidize, some will not and will be emitted. Note that fuel injected at the end of the injection process as the injector is closing, and any fuel that subsequently leaks out from the injector, contribute to HC emissions via the second mechanism.[1]

Particulate emissions.   Diesel particulates consist principally of combustion-generated carbonaceous material (soot) that forms from the fuel during combustion, on which some higher molecular weight organic compounds in the exhaust and sulfate from the sulfur in the fuel have become absorbed. Most particulate material results from incomplete combustion of fuel hydrocarbons, although some is contributed by the lubricating oil. Although largely formed by agglomeration of and surface deposition on very small particles originating from hydrocarbons in the diesel fuel, ash from the fuel and cylinder lubrication oil, and the peeling-off of deposits from the combustion chamber and exhaust system walls, are also mechanisms that contribute to exhaust particulates. Particulates filtered from the exhaust consist mainly of collections of primary spherules about 30 nm in diameter agglomerated into aggregates. Individual particles range in appearance from clusters of spherules to chains of spherules. Clusters may contain as many as 4000 spherules.[1]

The composition of diesel exhaust particulate material depends on many parameters. These include the fuel composition, lubricating oil composition and consumption, the local conditions under which combustion takes place, conditions in the engine exhaust system, and the particulate collection system. Particularly important are the fuel injection system parameters that determine the fuel injection rate, the fuel droplet size distribution and penetration, fuel vaporization and especially fuel–air mixing rates, air temperature, and air motion. It is, therefore, difficult to state typical particulate emission rates for two-stroke diesels, but with heavy fuel oil values are about 1 g/kW-h.[29] Except for flakes of deposits that have peeled off the combustion chamber or exhaust system walls, the particles are small: over 90% are less than 1 μm in size.

Most particulate measurement techniques attempt to determine the amount of particulate emitted to the atmosphere. Techniques for particulate measurement and characterization range from straightforward smoke meter opacity readings to analyses of particulate in exhaust dilution tunnels. It is, however, important to note that the exhaust gases leaving the exhaust pipe may be visible for several reasons: their particulate matter; nitrogen dioxide (NO_2) level (NO_2 is a yellow brown color); and condensing water vapor. Opacity measurements (e.g., Hartridge smoke value with a scale of percent Hartridge, and Ringelman number with a scale of 0–5), generally express the percentage of light absorbed when a beam is passed through a given path length of exhaust gas. Then particulate matter, as well as nitrogen dioxide, contributes to the measured value. Other methods of particulate measurement are the Bosch smoke number (with a scale of 1–10 BSN), and Bacharach smoke number (scale of 0–9 BSN). Both of these measure the degree of blackening of a white filter paper, through which a certain volume of exhaust gas has been sucked. Because the various methods are so different in principle, and because the emitted particulate may be altered due to coalescence, through loss to surfaces, and through chemical reaction with other species in the exhaust, it is not possible to convert the results of smoke measurements taken by one method into values that can be compared with results obtained from another method.

Sulfur oxide emissions. The commonly used diesel fuels contain sulfur. The sulfur is oxidized in the engine's combustion process to yield SO_2 and some SO_3, typically in a ratio of about 15:1. Under normal combustion conditions, the oxidation of the sulfur in the engine's combustion chamber is unavoidable, and the emissions of SO_x (SO_2 and SO_3) are, essentially, determined by the sulfur content of the fuel. SO_x emissions can be controlled by either removing the sulfur from the fuel or by removing the SO_x from the exhaust gas by cleaning (see Section 7-5). To protect the engine from corrosion by these sulfur oxides, the sulfur that reaches the cylinder walls (a minor part of the total) is neutralized to form calcium compounds by using alkaline lubricants. The SO_x emitted to the atmosphere will eventually be washed from the atmosphere by rain and will increase the acidity of the soil.

Quantifying emission levels. In the standard mass-emission measurement procedure, dilution tunnels are used to simulate the physical and chemical processes the engine's exhaust emissions undergo as they enter the atmosphere. The exhaust gases are mixed in a dilution tunnel with a substantial amount of ambient air to a temperature of 52°C or less. The total dilution tunnel flow is maintained constant, and gas is sampled at a constant flow rate from the mixture of exhaust gas and ambient air in the tunnel, is analyzed to determine the engine's gaseous emissions, and is filtered to remove and then analyze the particulate emissions. Table 7-3 shows various ways of expressing the exhaust gas composition. In this example, the exhaust gas was analyzed without any dilution, and the data were then adjusted to represent different levels of dilution with ambient air. The concentration measurements in the top half of the table have been expressed as emission factors, where the pollutant emission rate is normalized by an engine operating or performance parameter (such as fuel flow rate, power, etc.).[29]

7-4-2 Engine Emissions Data and Control

The available emissions data for two-stroke diesel engines is much more limited than for four-stroke diesels. Also, the different applications of two-stroke diesels lead to substantially different engine configurations. There is a long history of using large low-speed long-stroke two-stroke diesels for marine propulsion and power generation. Two-stroke diesels have competed with four-stroke engines in the locomotive and urban bus arenas. Recently, prototype smaller IDI and DI two-stroke diesels based on current four-stroke diesel technology have been developed and tested as potential automotive engines.

In the large compression-ignition two-stroke engines used today for marine and power plant applications, the heavy fuel oil used customarily contains metals and sulfur. Thus engine particulates and SO_x emissions are relatively high, and these contaminants cause standard exhaust catalysts to deteriorate after a short period of operation. However, an oxidation catalyst could be used successfully to reduce CO and HC emissions if a particulate trap is employed. However, the high stroke-to-bore ratios employed in these engines and their low operating speed make more complete combustion possible and CO and HC emissions are lower, compared with smaller engines. Specific NO_x emissions are relatively high (10–20 gNO_2/kW-h, see Table 7-3) for these engines. An extensive review of the emissions characteristics of these large diesel engines and appropriate control approaches, written by Henningsen, can be found in Sher.[2] Here, we provide a summary.

Table 7-3 Different ways of expressing the emissions from a 18.9-MW large two-stroke diesel

A. Concentrations

	NO_x (ppm)	CO (%)	HC (ppm C_1)	SO_x (ppm)	O_2	CO_2	H_2O	Parti-culates
Wet analysis	1570	57	284	516	13.0	5.2	5.4	—
Dry analysis	1660	60	300	545	13.7	5.6	0	—
Dilution to 15% O_2	1362	49	246	447	15.0	4.6	0	—
Dilution to 13% O_2	1820	66	329	598	13.0	6.1	0	—
Dilution to 5% O_x	3650	132	660	1,200	5.0	12.3	0	—
Dilution to 0% O_2	4770	172	861	1,570	0	16	0	—

B. Emission factors

	NO_x	CO	HC	SO_x	O_2	CO_2	H_2O	Parti-culates
g/m^3 (STP)	3.41	0.08	0.22	1.56	196	108	0	0.12
g/bhp-h	13.7	0.30	0.86	6.26	790	440	180	0.48
g/kW-h	18.6	0.41	1.17	8.52	1,070	590	250	0.66
g/MJ input	2.49	0.05	0.16	1.14	142.9	78.9	33.6	0.09
g/kg fuel	99.9	2.2	6.3	45.7	5,740	3,170	1350	3.52
kg/h	352	7.7	22.2	161.0	20,200	11,200	4750	12.1

In principle there are two ways to reduce NO_x emissions from diesel engines: primary methods, aimed at reducing the amount of NO_x formed during combustion; and secondary methods, aimed at removing NO_x from the exhaust gas.[29] The primary methods include the following: reducing the scavenging air temperature by fully cooling the air after compression; reducing the peak cycle temperature and pressure by retarding fuel injection and, hence, combustion; increasing the amount of residuals either by recirculating part of the exhaust gas (EGR) or by reducing the delivery ratio; injection of water into the combustion chamber or using water emulsified into the fuel. Although a combination of these methods will reduce NO_x emission by 20–30%, the cycle efficiency deteriorates and brake specific fuel consumption is increased by some 2–4%.

The scavenging air temperature should be reduced as much as is practical, prior to entry to the cylinder, to hold the burned gas temperatures as low as possible. Performance and efficiency benefits as well as NO_x benefits result. The temperature of the available coolant sets the lower limit. The effect of retarding injection timing is to retard combustion from its optimum phasing relative to piston motion, and hence to retard and reduce peak cylinder pressures and temperatures. Obviously bsfc increases, as does the Bosch smoke number; 5° of injection retard gives about 20% reduction in NO_x and 4% increase in bsfc. (For each 10 bar reduction in peak pressure, a 10% NO_x reduction is achieved.) Recycling exhaust gas is used effectively in four-stroke engines to reduce NO_x emissions. In large two-stroke diesels, the exhaust gas needs to be cooled and cleaned for both sulfur and particulates before mixing with the scavenging air. Thus although difficult to implement in practice in these engines, EGR is an effective NO_x control strategy, with 10% EGR producing about 30% reduction in NO. Exhaust gas has a higher heat capacity than the air it effectively replaces; so for a given fuel chemical energy release, the maximum burned gas temperatures are reduced.[29,2]

Another way to reduce peak burned gas temperatures is to use water as the thermal-capacity-adding diluent. Water can be sprayed directly into the combustion chamber or added, emulsified, with the fuel. Figure 7-22 presents data showing how water-emulsified fuel can be used to reduce NO_x emissions in a 20-MW two-stroke low-speed diesel engine. The NO_x emissions have been corrected to 15% O_2. An uncontrolled NO_x level of about 1100 ppm has been reduced to 770 ppm with a fuel containing 22% emulsified water. To apply this method in practice, these modifications in the engine design were made: a larger fuel pump was used to accommodate the higher fuel and water flow rates; the low-pressure part of the fuel system was replaced by a closed pressurized system to avoid cavitation and flashing of the water; a dosage control system and homogenizer were installed. It is interesting to note that general engine cleanliness and the condition of the exhaust valve and the fuel pumps were very satisfactory.[29] When low-sulfur fuel is used, however, achieving satisfactory water emulsification is a major problem. Further, due to its limited reduction potential (1% reduction per 1% water addition), other means of NO_x reduction, such as exhaust gas after treatment (see Section 7-5-3), should be considered.

Medium-duty two-stroke diesels have been popular as engines for urban buses due to their high power density. Conversion of these engines to use natural gas as fuel is

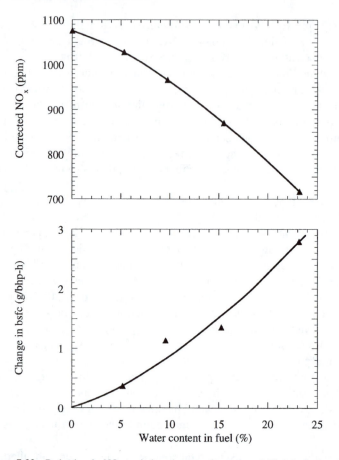

Figure 7-22 Reduction in NO$_x$ emissions by use of water emulsified in fuel, and increase in bsfc in a 20-MW low-speed two-stroke cycle diesel.[29]

of interest because urban bus emissions are a major public concern. Figure 7-23 shows the bsNO$_x$-versus-load and bsNO$_x$-versus-bsfc characteristics of a Detroit Diesel six-cylinder 9-liter displacement two-stroke diesel under standard diesel operating conditions† and when operated with natural gas. With natural-gas fueling, a high-pressure (185 bar) direct in-cylinder gas injection system was used and pilot injection of diesel fuel was used to initiate combustion. The amount of pilot diesel fuel injected was held roughly constant and was 6% of the full load amount. The diesel bsNO$_x$ emissions (Fig. 7-23a) show modest variation with load (until the highest load is reached at 1260 rev/min).[30] As load is increased, the amount of fuel injected per cycle increases so the overall engine fuel/air equivalence ratio increases. Thus the amount of high-temperature close-to-stoichiometric burned gas produced also increases and NO$_x$

---

†Note that this diesel engine, as configured and tested, does not meet the current (1998) U.S., European, or Japanese heavy-duty diesel NO$_x$ and particulate standards.

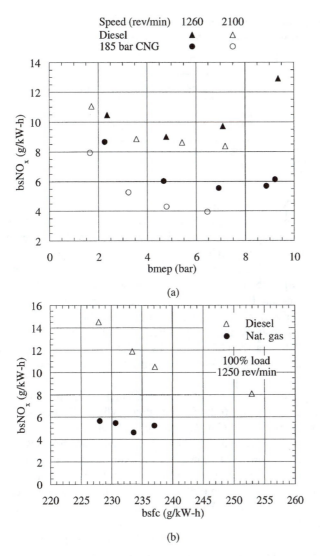

Figure 7-23 (a) BsNO$_x$ versus load and (b) bsNO$_x$ versus bsfc, for a Detroit Diesel 9-liter six-cylinder two-stroke medium-duty diesel under standard diesel fueled operating conditions, and with direct in-cylinder injection of natural gas (at 185 bar) with pilot diesel fuel injection to initiate combustion.[30]

emission expressed as ppm NO$_2$ in the exhaust will steadily rise.[1] However, because bmep is also increasing, bsNO$_x$ at fixed speed will only change moderately because the stoichiometric adiabatic flame temperature in the engine is a weak function of load. Figure 7-23b shows the effect of injection retard on bsNO$_x$ emissions. Injection retard, and hence combustion retard, lowers peak cylinder pressures and temperatures and is widely used as an NO$_x$ control strategy. It results, of course, in a fuel consumption penalty, as shown.

Table 7-4 Brake specific fuel consumption and emissions from Detroit Diesel medium-duty two-stroke engine† with diesel and natural-gas fueling[30]

g/kW-h	bsfc	$bsNO_x$	$bsCH_4$	bsnmHC‡	bsCO
Diesel	274	11.5	0.4	0.3	0.7
Natural gas	280	6.6	1.9	0.3	3.1

†9-liter six-cylinder 123-mm bore 127-mm stroke 6V-92TA turbocharged and aftercooled, electronically controlled two-stroke diesel engine.
‡nmHC = nonmethane hydrocarbon emissions.

The natural-gas-fueled NO_x emissions show similar trends, but are lower. The engine's brake fuel conversion efficiencies with the two fuels are essentially the same over the full load range (reaching a maximum of 37% at full load and 1260 rev/min). The lower $bsNO_x$ emissions are due primarily to the lower flame temperature of natural gas (here 95% methane, 3% ethane, and 0.5% propane) relative to diesel, and perhaps due to differences in the fuel–air mixing processes between the two fuels. The emissions characteristics and bsfc with the two fuels, averaged over the 13-mode steady-state emissions test procedure [measurements made at idle (weighted 20%), 2%, 25%, 50%, 75%, and 100% load (weighted 8% each, at medium and rated speed)], are shown in Table 7-4. Note that the CO_2 emissions with the natural gas and about equal fuel consumption are 25% lower due to the different C/H ratio of the two fuels.[30]

Examples of recent prototype small two-stroke diesels include the Toyota 2.5-liter four-cylinder S-2 IDI engine (see Fig. 6-25),[31] two single-cylinder Daimler-Benz DI engines,[32] and a 1-liter three-cylinder AVL DI engine.[33] Because these were developed as potential high-efficiency car engines, their emissions characteristics at partial load are most important. Figure 7-24 shows how emissions and fuel consumption vary with indicated mean effective pressure at 1600 rev/min for two single-cylinder DI diesels: a ported loop-scavenged engine ($350-cm^3$ displacement), and a uniflow engine (four exhaust valves in the head, $493-cm^3$ displacement).[32] These engines used a common-rail fuel injection system. Because the loop-scavenged engine did not (and could not) generate in-cylinder swirl whereas the uniflow design achieved a swirl at BC of between 2 and 3 times crankshaft speed, fuel–air mixing rates were lower and a higher injection pressure and smaller fuel nozzle hole diameters were required to achieve comparable smoke and particulate levels. Even with higher injection-rail pressures, the data in the figure show the loop-scavenged engine has higher smoke levels and worse indicated specific fuel consumption and HC emissions, and the maximum load achievable at a Bosch smoke number of 2.5 is reduced. This is because the gas exchange process with loop-scavenging arrangements is not as effective as with uniflow scavenging, and combustion quality is poorer because, without swirl in these small engine sizes, satisfactory mixture preparation is not fully achieved. The NO_x emissions are comparable, and at $isNO_x$ levels of 4–6 gNO_2/kW-h are significantly less than the high-load levels typical of large-bore two-stroke diesels due to the high residual gas fraction trapped in the cylinder at the low delivery ratios

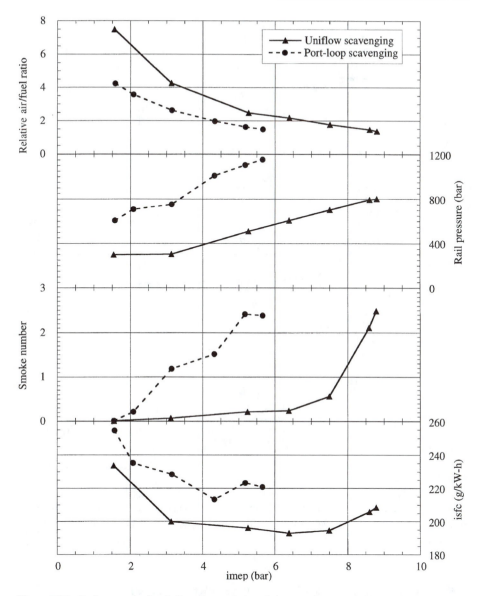

Figure 7-24 Performance and emissions characteristics of two prototype automobile two-stroke DI diesel engines, as a function of load (imep), at 1600 rev/min: (i) ported loop-scavenged crankcase-compression 350-cm^3 single-cylinder engine; (ii) uniflow-scavenged external-blower 493-cm^3 single-cylinder engine.[32]

used at light load. Engine size and speed also impact NO formation: larger size reduces the impact of heat losses on burned gas temperatures, and lower speed allows NO levels to approach closer to equilibrium values. It is suggested that the good combustion characteristics of the uniflow prototype engine would allow external exhaust gas recycle (see Section 7-3-2) to further reduce engine-out NO_x emissions.

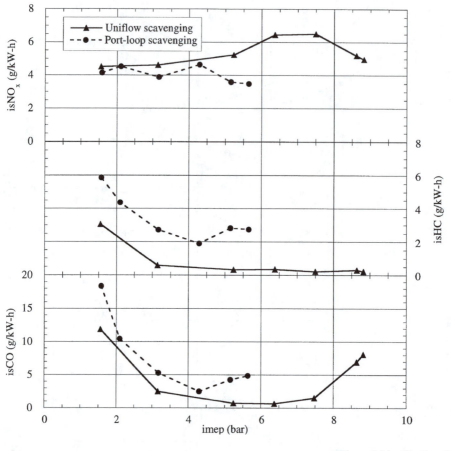

Figure 7-24 (Continued).

These indicated values of specific power and emissions are comparable to those of modern four-stroke DI diesel engines with four valves per cylinder and common-rail injection.[32]

7-5 CATALYSTS AND EXHAUST GAS TREATMENT

A significant reduction in emissions can be obtained by removing pollutants from the exhaust gases by using catalytic converters (this includes oxidizing catalysts for HC and CO, storage and reducing catalysts for NO_x, and three-way catalysts for all three pollutants). Thermal reactors for HC and CO burnup and filters or traps for removing particulate emissions have also been developed and tested. Although some oxidation of HC undoubtedly occurs in the exhaust port of two-stroke gasoline engines (i.e., the exhaust does act as a "thermal reactor"), it has proven more effective to use catalysts instead of specially designed exhaust thermal reactors for additional HC and CO removal. In this section we discuss catalyst systems for two-stroke SI engine emissions reduction, and exhaust NO_x catalysts and particulate control for two-stroke diesels.

7-5-1 Catalyst Fundamentals

A catalyst is a material that increases the rate of a chemical reaction while not itself undergoing any permanent change. In engine applications, the catalyst is a solid surface. For a molecule A to be catalytically converted to a product molecule B, the following steps must occur. Molecule A diffuses through the thin layer of gas adjacent to the catalyst surface, toward the active catalyst sites. It is then chemisorbed onto an active site and converted to product B. The product then desorbs from the site and diffuses through the gas layer adjacent to the surface into the bulk gas stream, where it is carried away. The catalyst increases the reaction rate for a reaction (reactant A producing product B) by reducing the energy barrier to the reaction (the so-called activation energy) thus enabling a larger fraction of the molecules to react at a given temperature. Catalysts influence the rate at which a reaction proceeds and not the final equilibrium state. Through appropriate selection of the catalytic material, it can be made *selective*: that is, preferentially lower the activation energy of a particular reaction to increase the yield of the product of that reaction.[34]

A catalytic converter is a device through which the exhaust gases are directed that converts specified reactant species in the exhaust into the desired products. The catalytic converter is usually installed in the exhaust system between the engine and the exhaust muffler or silencer. It consists of a metal housing, a substrate (usually a ceramic or metallic honeycomb), a carrier (porous oxide materials such as Al_2O_3, SiO_2, and TiO_2), and an active catalytic material (noble metals such as Pt, Rh, Pd, or base metal oxides). Figure 7-25 shows the design of a typical automotive catalytic converter. The specially designed metal casing directs the exhaust gas flow through the catalyst bed. Because the number of molecules converted per unit time is directly related to the number of catalytic sites available, maximizing the surface area of the active catalyst material is an important design task. To accomplish this, the active catalytic components are dispersed over a high-surface-area carrier. The carrier is not catalytically active but plays an important role in dispersing the active catalyst material and maintaining its stability and durability. The highly porous carrier has typical pore sizes between 20 and 50Å and a surface area of 50–$200 \, m^2/g$. The carrier plus catalyst is thinly spread over a supporting structure or *substrate*. Usually the substrate is in the form of a monolith or honeycomb as shown in Fig. 7-25. This form of substrate provides a large surface area for exhaust gas contact with the catalyst, and, most important, a low pressure drop across the catalytic converter as the exhaust flows through it. When the carrier, impregnated with the catalytic components, is bonded to a monolithic support it is called a *washcoat*.[34]

Three types of substrates are currently used: ceramic pellets, ceramic monoliths or honeycombs, and metallic monoliths or honeycombs. The ceramic pellet type uses layers of spherical pellets or beads resting on top of each other to form the catalyst bed. The pellets are made of a ceramic such as high-temperature-resistant magnesium aluminum silicate. The spaces surrounding the bead contact points provide a large exposed surface area. The pellet stack is packed in a perforated container, one half of

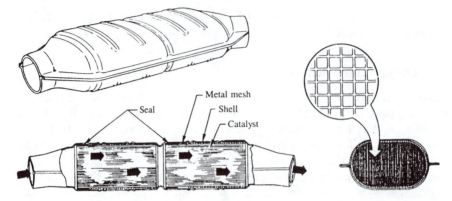

Figure 7-25 Monolithic design of catalytic converter.[1] Reproduced with permission of The McGraw-Hill Companies.

which is exposed to the incoming exhaust gases while the other is the exit leading to the exhaust silencer. The pellet type of substrate is steadily loosing ground in favor of ceramic and metal monoliths. A typical straight-channel monolithic substrate or support is made of extruded ceramic material with about 60 cells per square centimeter (400 cells per square inch). The monolith has a large open frontal area that minimizes flow resistance. The channel's shape can be triangular, hexagonal, square, or any other convenient shape. Ceramic monoliths are usually surrounded by a resilient matting material to absorb mechanical vibration and thermal expansion, and to provide thermal insulation. The package is mounted in a metal housing. Metallic monoliths[35] are more durable but also more expensive. They are employed where higher cell densities with low-pressure drops are important, or in catalysts mounted close to the engine as a supplement to the main catalytic converter to provide more rapid catalyst warm-up following a cold start and better high-temperature performance because the higher conductivity of the metal results in a more uniform temperature. Thus metal-supported catalysts are attractive for two-stroke cycle engines, especially in applications like chain saws where mechanical vibrations are substantial. The metal substrate is obtained by winding together a flat and a corrugated steel sheet and brazing the joins, as illustrated in Fig. 7-26.[7]

Oxidizing catalysts transform hydrocarbons and carbon monoxide into water vapor and carbon dioxide through the following overall reaction steps[34]:

$$C_n H_m + \left(n + \frac{m}{4}\right) O_2 \rightarrow n CO_2 + \frac{m}{2} H_2 O \qquad (7\text{-}15)$$

$$CO + \frac{1}{2} O_2 \rightarrow CO_2 \qquad (7\text{-}16)$$

$$CO + H_2 O \rightarrow CO_2 + H_2 \qquad (7\text{-}17)$$

To accomplish this, they require an environment with excess oxygen. The required oxygen is provided either by a lean engine mixture calibration or by a secondary air stream supplied to the exhaust upstream of the converter. See Fig. 7-27a. Self-inducting air

Standard

S-type

Brazing

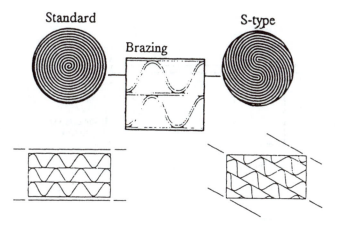

Figure 7-26 Metal substrate design for catalytic converters that is more robust at higher temperatures.[7]

valves or a centrifugal pump driven by the engine are commonly employed. In oxidizing catalysts, oxides of nitrogen (NO_x) levels are essentially unchanged.

To remove NO_x (largely nitric oxide, NO), it must be reduced. Nitric oxide is converted into nitrogen (N_2) through reduction by CO, H_2, or hydrocarbons:

$$NO + CO \rightarrow \frac{1}{2}N_2 + CO_2 \qquad (7\text{-}18)$$

$$NO + H_2 \rightarrow \frac{1}{2}N_2 + H_2O \qquad (7\text{-}19)$$

$$\left(2n + \frac{m}{2}\right)NO + C_nH_m \rightarrow \left(n + \frac{m}{4}\right)N_2 + nCO_2 + \frac{m}{2}H_2O \qquad (7\text{-}20)$$

To achieve reduction of NO and oxidation of CO and HC, several different catalysts and converter arrangements have been used (see Fig. 7-27b and c). When the engine is operated with an air/fuel ratio close to stoichiometric, then both NO reduction and CO and HC oxidation can be done in a single catalyst bed. Such a catalyst is termed a *three-way catalyst*. Use of three-way catalysts requires that the exhaust gases be at the stoichiometric air/fuel ratio. This usually requires the use of an exhaust gas oxygen sensor (a lambda sensor) and closed-loop air/fuel ratio control. The three-way catalyst is the most effective pollutant reduction system presently available. To achieve rapid catalyst light-off while the engine and the catalyst in its normal location some distance from the exhaust port are still warming up, a fast-light-off catalyst, close-coupled to the engine, is sometimes used; see Fig. 7-27c. Its primary purpose is HC and CO oxidation, but, depending on the relative air/fuel ratio of the engine and the catalyst details, some NO_x reduction can be achieved as well.

The attractiveness of direct-injection engines, which operate lean at lighter loads, is prompting the development of NO_x removal catalyst systems in net oxygen-rich exhaust streams. Two approaches are shown in Fig. 7-27d and e. Direct NO_x decomposition is the most desirable approach, but currently available catalysts are not practical for engine use. Catalysts that selectively reduce NO_x by direct reaction with hydrocarbons under

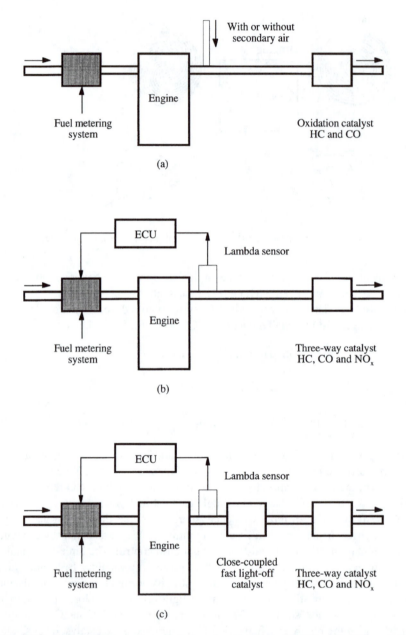

Figure 7-27 Various catalyst arrangements in the exhaust to remove CO, HC, and/or NO_x, and effect rapid catalyst light-off with adequate durability. For premixed charge engines: (a) oxidation catalyst for HC and CO removal; (b) three-way catalyst with exhaust gas oxygen sensor and fuel metering feedback control, for HC, CO, and NO_x removal; (c) similar to (b) with a close-coupled fast-light-off catalyst for HC and CO oxidation (and some NO_x reduction) during engine warm-up. For direct-injection engines: (d) NO_x storage/reduction catalyst plus three-way catalyst system, with fast-light-off warm-up catalyst upstream; (e) lean NO_x selective reduction catalyst, with three-way catalyst downstream. ECU: engine control unit.

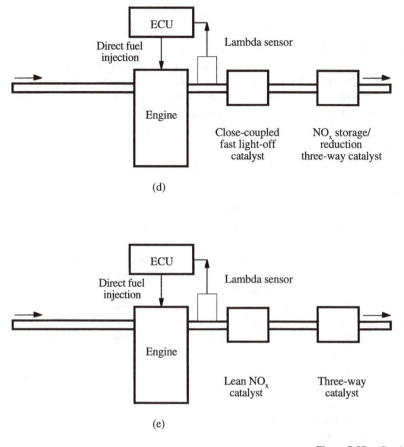

(d)

(e)

Figure 7-27   (Continued).

lean conditions [reaction (7-20)] are developing adequate durability but are significantly less effective than three-way catalyst technology.[36] An alternative approach, the NO_x trap technology, is also being developed and used. NO is oxidized to NO_2 and stored in the catalyst during lean engine operation as a nitrated species. Then, after one or more minutes of engine operation under these (lean) conditions, the engine is operated fuel-rich for a short period (seconds), the NO_2 is released and is reduced by HC, CO, and H_2 to N_2. Promising performance is being achieved, although sulfur poisoning of the active catalyst sites requires use of low-sulfur gasolines.[37]

The effectiveness of a catalyst is defined by its *catalyst (conversion) efficiency*,

$$\eta_{cat} = 1 - \frac{\dot{m}_{i,out}}{\dot{m}_{i,in}} \qquad (7\text{-}21)$$

where $\dot{m}_i$ is the mass flow rate of pollutant i into (in) or out of (out) the catalyst. It is the amount (or percentage) of entering emissions that the catalyst removes. Catalyst efficiency depends on the amount of the catalyst, the catalyst composition and surface area, the catalyst operating temperature, the exhaust gas composition (usually defined by the relative air/fuel ratio), and the gas residence time in the converter (defined by its reciprocal, the *space velocity*—the volume flow rate of exhaust through the catalyst, divided by the volume of the catalytic converter). Especially important is the extent to which use (or aging) has deteriorated the catalyst through poisoning some of the active catalyst sites, through coating some of these sites, or by reducing the number of active sites through coalescence produced by overheating, or by loss of washcoat porosity (and hence catalyst surface area) through operation at still higher temperatures.

Figure 7-28 shows the effect of catalyst temperature on the efficiency of an oxidation catalyst. The temperature at which 50% conversion efficiency is achieved is referred to as the catalyst *light-off temperature*. Typical HC light-off temperatures for aged modern Pt/Rh and Pd catalysts are in the 300–400°C range depending on catalyst noble metal loading. Ideal operating conditions for high conversion efficiencies and long life are between about 400 and 800°C. Above this range, the thermal aging process accelerates by sintering of the noble metals and the carrier, thus reducing the effective active surface area. Present NO catalysts operate effectively in the temperature range between about 350 and 600°C. Figure 7-29 shows how the conversion efficiency of a three-way catalyst for NO, CO, and HC depends on the air/fuel ratio. It illustrates how narrow is the range near stoichiometric conditions in which high conversion efficiencies for all three pollutants are achieved. To keep high conversion efficiencies at all times, a closed-loop air/fuel ratio control must be employed. Modern three-way catalysts include ceria (CeO_2) in the washcoat to store oxygen when the exhaust gases are slightly lean so oxidation at the catalyst surface can continue when the exhaust is rich. Also, the relative air/fuel ratio is deliberately oscillated about the set point (usually slightly rich) at about 1 Hz frequency and 0.5 air/fuel ratio amplitude to broaden the operating window.

Figure 7-30 shows the effect of increasing space velocity on catalyst efficiency. Too high a flow rate (i.e., too short a residence time in the catalyst) results in a fall-

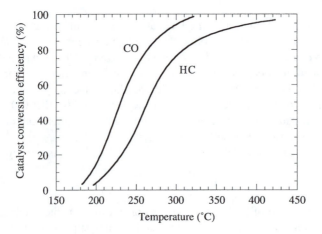

Figure 7-28 Conversion efficiency for CO and HC as a function of temperature for typical oxidizing catalytic converter.[1] Reproduced with permission of The McGraw-Hill Companies.

off in catalyst performance. Typically, catalysts are sized for a space velocity of about 50,000–100,000 hr^{-1} (a residence time of about 0.05 s).

Especially important for good catalyst durability is avoiding excessive catalyst temperatures. In four-stroke SI engines, substantial enrichment of the mixture at high engine power is often done to reduce the exhaust gas temperature to avoid catalyst over-temperature conditions when the exhaust flow rate is high. In two-stroke cycle SI engines, the higher potential for misfires and partial burns, and the short-circuiting of fuel–air mixture during scavenging, creates additional catalyst durability problems. Any HC, CO, and H$_2$ in the exhaust, *if sufficient oxygen is available*, can release chemical energy within the catalyst. The concept of (releasable) exhaust chemical energy, or combustion inefficiency,[1] is often used to define this potential. The rate at which the

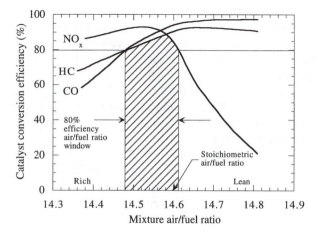

Figure 7-29 Conversion efficiency for CO, HC, and NO$_x$ as a function of air/fuel ratio for typical three-way Pt/Rh catalyst.[1] Reproduced with permission of The McGraw-Hill Companies.

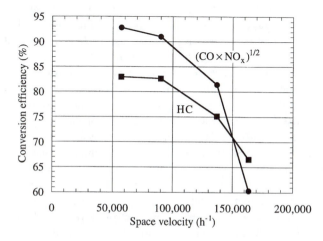

Figure 7-30 Effect of space velocity on conversion efficiencies of an aged automotive three-way catalyst.[38]

chemical energy in the engine's exhaust can be released is given by

$$\dot{E}_{chem,e} = \dot{m}_{HC} Q_{HV,HC} + \dot{m}_{CO} Q_{HV,CO} + \dot{m}_{H_2} Q_{HV,H_2} \qquad (7\text{-}22)$$

where $Q_{HV,i}$ is the heating value of each combustible emission (44, 10.1, and 120 MJ/kg for HC, CO, and H_2, respectively). Note that whether all this energy is released depends on whether sufficient oxygen is available. In fuel-rich exhaust gases, the molar concentration of H_2 is about half the molar concentration of CO (see Section 3-2-2).

We can apply Eq. (7-22) to the data in Fig. 7-13. For the carburetted stock two-stroke SI engine, the emissions at 6000 rev/min, 5 bar bmep, are 3500 ppm (NDIR) C_6 for HC ($\approx 3500 \times 1.8 \times 6 = 38,000$ ppm C_1 FID) and 5.6% CO (which would give $H_2 \approx 2.8\%$). These are dry exhaust molar fractions; the H_2O mole fraction at this A/F of 12 would be about 12%. Correcting to wet emission concentrations, Eq. (7-22) then gives $\dot{E}_{chem,e}$ equal to 40% of the chemical energy flowing into the engine with the fuel. The relative contributions are HC 47% (dominated by short-circuiting fuel and air), CO 37%, and H_2 16%. For the direct-injection SI engine data in Fig. 7-13(b), the emissions at the same operating point are 160 ppm (NDIR) C_6 HC (≈ 1700 ppm C_1 FID) and 3.5% CO (which gives 1.8% H_2) at an $A/F \approx 13$. Again, correcting to wet concentrations, Eq. (7-22) gives $\dot{E}_{chem,e}$ equal to about 15% of the incoming fuel energy, with HC:CO:H_2 percentages 6%, 65%, and 28%, respectively. The benefit of direct injection in reducing the short-circuiting fuel is clear. Note that the release of this exhaust chemical energy in the catalyst depends on the availability of oxygen in the exhaust. With short-circuiting fuel–air mixture, even when fuel-rich, a substantial fraction can be released. Thus good control of mixture air/fuel ratio, and short-circuiting with premixed two-stroke SI engines, is important in preventing possible catalyst damage. With repeated partially successful or unsuccessful ignition, extremely high temperatures (above 1400°C) can result, and catalyst destruction is then inevitable.

7-5-2 Catalyst Systems for SI Engines

Combining a catalyst with a two-stroke spark-ignition engine raises several integration issues largely related to exhaust gas composition and temperature. Premixed small two-stroke SI engines normally run fuel-rich to obtain good combustion characteristics. With an oxidation catalyst, either the engine must be leaned out (to stoichiometric or leaner), which makes combustion slower and more irregular, or secondary air must be added to the exhaust stream. At light loads, the exhaust temperature is relatively low; at high loads it is much higher and the energy in the short-circuited fuel–air mixture causes a substantial temperature rise in the catalyst. So the placement of the catalyst in the exhaust system relative to the exhaust port is an important design issue.

In larger automotive two-stroke SI engine sizes, direct fuel injection has become essential. These DI engines run lean at light to medium load, so secondary air for an oxidation catalyst is not required. However, at the lighter engine loads, the exhaust temperature may be too low for effective catalyst operation, and means of increasing exhaust temperature (e.g., use of an exhaust port flow control valve, cylinder cutout in multicylinder engines) may need to be considered. In automotive applications, fast light-off of the catalyst is essential, and means of achieving this are being developed (e.g., use of substantial spark retard while the engine is cold to increase exhaust gas temperature, use of low-thermal-capacity and insulated exhaust manifolds to reduce heat losses during warm-up, use of a close-coupled catalyst with good high-temperature durability). Here we review different examples of engine–catalyst systems that have been developed, to illustrate these issues and the overall system performance that can be achieved.

We now review catalyst systems that have been developed for small carburetted (or port-injected) two-stroke SI engines. A simple example, used for both a 0.9-kW (1.2-hp) and 2.0-kW (2.7-hp) single-cylinder loop-scavenged moped engine, is shown in Fig. 7-31 and additional details are given in Table 7-5.[16] The exhaust system consists of a circular cross-section pipe of length 505 mm connected to the cylinder exhaust port,

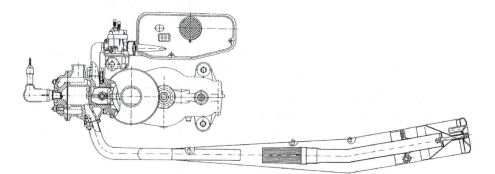

Figure 7-31   Engine, exhaust system, and catalyst assembly of 2-kW (2.7-hp) moped engine.[16]

Table 7-5   Details of the 1.2 and 2.7 hp single-cylinder loop-scavenged two-stroke moped engines

Characteristic	Units	1.2 hp	2.7 hp
Max. power/speed	kW/rev/min	0.9/3500	2.0/4500
Max. torque/speed	N-m/rev/min	2.95/3000	4.44/3500
Max. bmep	kPa	390	550
Displacement	cm^3	49.0	49.9
Bore × stroke	mm × mm	40 × 39	38 × 44
Carburetor venturi diam.	mm	10	14
Compression ratio	—	6.7	7.0
Compression ratio from SC	—	4.96	5.38
Crankcase compression ratio	—	1.38	1.28
Catalyst diameter × length	mm × mm	33.5 × 90.5	33.5 × 90.5
Washcoat coating	g/cm^3	4.3	4.3
Precious metal (Pt/Rh = 5:1)	g/liter	2.74	2.74
Number of cells	cpsi†	400	400

†cpsi stands for cells per square inch (400 cpsi = 60 cells per square centimeter).

which connects to a diffuser, followed by a cylindrical section with a deflecting wall.†
To avoid any deterioration in the gas dynamic behavior of the exhaust, the catalyst is
held in position in the diffuser by the carrier pipe from the rear, welded to the deflecting wall. A faster light-off temperature is achieved by shortening the catalyst outside
shell length, thus exposing the substrate to the hot exhaust gas. Figure 7-32 shows the
temperatures of the exhaust gases 10 mm upstream of the frontal area of the catalyst
and at the catalyst exit, under steady-state conditions for the larger engine. The exhaust
gas temperature increases by some 150–250°C over the catalyst due to the exothermic
conversion of CO and HC to CO_2 and H_2O.

The steady-state postcatalyst emissions of HC and CO, and the respective catalyst efficiencies (for the smaller engine with the same catalyst system) are shown in
Fig. 7-33. Under steady-state conditions, high catalyst efficiencies between 92 and 98%
have been achieved through operating at a low space velocity (values between 40,000
and 100,000 hr^{-1} in the larger engine, lower by about a factor of 2 in the smaller engine) and by adjusting the air/fuel ratio delivered by the carburetor so that the engine
operates close to stoichiometric at wide-open throttle and increasingly leaner as the
throttle is closed (to about 20:1 at 200 kPa bmep and 3000 rev/min). Thus engine-out
NO_x emissions increase due to this leaner calibration, average about 2 g/kW-h $bsNO_x$
for the smaller engine with higher high-load numbers for the larger engine, and are
unaffected by the catalyst. Even under these stoichiometric and lean conditions, the releasable chemical energy or exothermicity in the exhaust due to the HC, CO, and H_2
emissions is substantial and at wide-open throttle is nearly as large as the power output
from the engine.[7]

Some alternative catalyst configurations are shown in Fig. 7-34. Figure 7-34a shows
a two-catalyst exhaust system tested with the standard 2-kW (2.7-hp) moped engine
described previously. Average emission levels of 0.15 gHC/kW-h and 0.015 gCO/kW-h
were achieved, about one-tenth of the levels the one-catalyst system achieved. Note that
sufficient oxygen must be present in the exhaust for additional oxidation to be feasible.
(Oxygen levels in the exhaust after one catalyst were 1–5%, because the engine was
operated on the lean side of stoichiometric.)[16]

Figure 7-34b shows a catalyst system for a 3-kW (4-hp) chainsaw. A typical duty
cycle for professional woodcutters is a few 15–25-s wide-open-throttle cutting operations where engine speed drops from about 11,000 to 9000 rev/min, followed by de-
branching operations where the engine operates at WOT, lightly loaded, at speeds between 12,000 and 15,000 rev/min for a few seconds with short ($\sim$ 1 s) idling periods
in between. The chainsaw application for catalysts is especially demanding as this duty
cycle can produce very high catalyst temperatures (950–1100°C) at WOT conditions
followed by a much lower temperature at idle but substantial mass flow rates due to the
still high engine speeds. This thermal cycling combined with high levels of mechanical vibration have required special catalyst substrate technology to be developed using
increased metal monolith sheet thicknesses, a different wrapping technology, and an

---

†When we discuss engine exhaust noise, in Section 7-6-3, the logic behind this exhaust system configuration is explained.

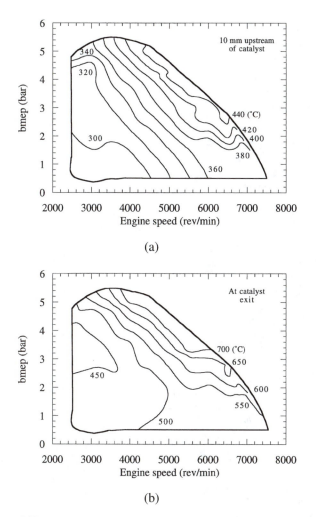

Figure 7-32 Exhaust gas temperature map of the 2-kW (2.7-hp) moped engine in Fig. 7-31: (a) 10 mm upstream of the front of the catalyst; (b) at the catalyst exit.[16]

austenitic steel outside tube to allow for increased thermal expansion. The catalyst efficiencies achieved (in a 3.5-hp chainsaw engine) under steady-state operation are shown in Fig. 7-35, along with the relative air/fuel ratio map. The engine runs 10–15% fuel rich over most of its operating map, reaching stoichiometric at WOT. This achieves good throttle response, as well as reducing the bsfc and unreleased chemical energy in the exhaust at WOT. Due to this slightly rich to stoichiometric operation and the rhodium content of the catalyst, all three pollutants are reduced by the catalyst (see Fig. 7-29). The NO reductions are greatest at partial load, with slightly rich-of-stoichiometric operation. The HC and CO reductions greatest at stoichiometric conditions (about 70 and 60%, respectively) where sufficient oxygen for substantial burnup is available.[7]

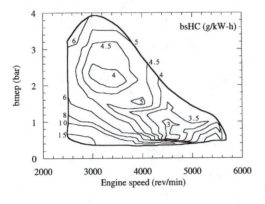

(a)

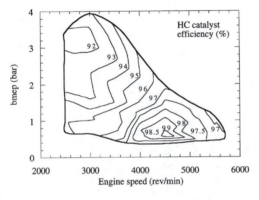

(b)

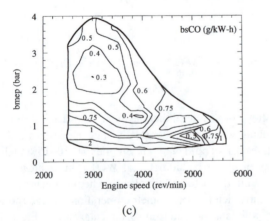

(c)

Figure 7-33 Postcatalyst HC and CO emissions (a) and (b), and catalyst conversion efficiency maps (c) and (d), for 0.9-kW (1.2-hp) moped engine with a similar exhaust system layout and the same oxidation catalyst shown in Fig. 7-31.[16]

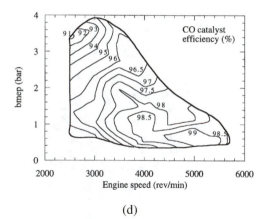

(d)

Figure 7-33 (Continued).

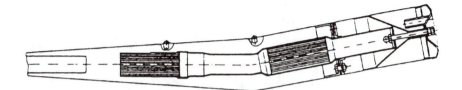

(a)

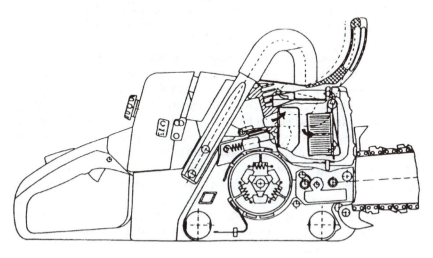

(b)

Figure 7-34 Various two-stroke SI engine exhaust and catalyst systems: (a) two-catalyst system[16]; (b) production chainsaw engine and catalyst system[16]; (c) 125-cm^3 motorcycle engine with fuel injection into the crankcase, exhaust port emission control valve and auxiliary exhaust volume, precatalyst, and main catalyst.[7]

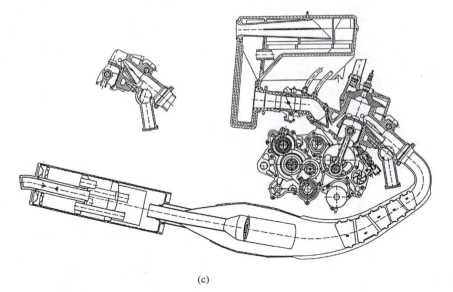

(c)

Figure 7-34   (Continued).

Figure 7-34c shows a more sophisticated engine catalyst system, for a 125-cm^3 displacement high-performance motorcycle engine that achieves 142 kW/liter and a maximum bmep of 890 kPa. It includes indirect fuel injection, an exhaust emission control valve and auxiliary exhaust volume, and a precatalyst. The injector position shown (beneath the exhaust port) sprays fuel into the crankcase onto the inside of the piston for rapid fuel vaporization. Other possible injector locations are into the intake upstream, or just downstream, of the reed valve. The exhaust emission control valve, shown in the closed position in the insert, throttles the exhaust port by raising a gate from the bottom of the port that closes the side exhaust ports and reduces the open area of the main exhaust port. The side exhaust ports increase the bmep above 8500 rev/min but also increase HC emissions at low speeds due to higher scavenging losses. As the gate is closed, the auxiliary exhaust volume is opened, which increases the bmep at medium speeds. To raise the exhaust temperature to high enough levels to achieve rapid light-off of the main catalyst, a precatalyst is used. This consists of five steel catalyst-coated plates, held in position in the diffuser section of the exhaust system by the seam between the two welded diffuser halves as shown. An electronic control module actuates ignition, fuel injection, and the emission control valve in response to parameters such as throttle position, engine speed, and coolant temperature. A higher energy ignition system is used to ensure reliable ignition with the close-to-stoichiometric and higher residual gas content in-cylinder mixture.[7]

Two important impacts that use of exhaust catalysts have on engine operation are the increased back pressure due to the increased pressure losses in the exhaust system, which increase with increasing exhaust gas temperature, and the increased speed of sound. Both of these affect the engine's power. The gas temperature after the catalyst ranges between about 400 and 1000°C, whereas without a catalyst it is about half

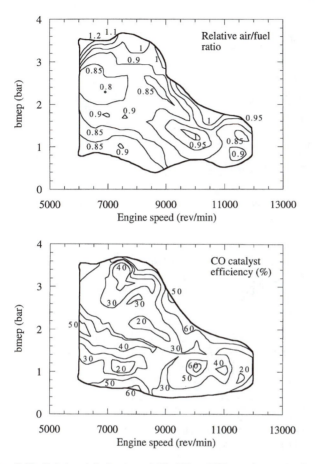

Figure 7-35 Relative air/fuel ratio, and CO, HC, and NO$_x$ catalyst conversion efficiency maps from a carburetted 2.6-kW (3.5-hp) chainsaw two-stroke SI engine.[7]

these values. The exhaust cross-sectional areas after the main catalyst were accordingly increased by 60–90% in this high-performance motorcycle engine.

An alternative emissions control approach with small premixed two-stroke SI engines is to maintain the fuel-rich operation, which gives good combustion characteristics, and add secondary air to the exhaust upstream of the catalyst to provide an oxidizing environment for CO and HC removal. An example of such a secondary air system is shown in Fig. 7-36.[39] It consists of a sponge to reduce noise, a filter to clean the air and also reduce noise, and a reed valve to prevent reverse flow. Air is drawn into the exhaust when the exhaust system pressure at the air entry point falls sufficiently below atmospheric pressure (at the end of exhaust blowdown). A precatalyst is used—a conical diffuser section with a catalytic coating on the inside wall—to achieve adequate CO and HC burnup when the engine is lightly loaded and exhaust gas temperatures are low, and to spread out the exothermic energy release between the precatalyst and main catalyst. Conversion of CO in the pre-catalyst is substantial (70–90% of total CO conversion);

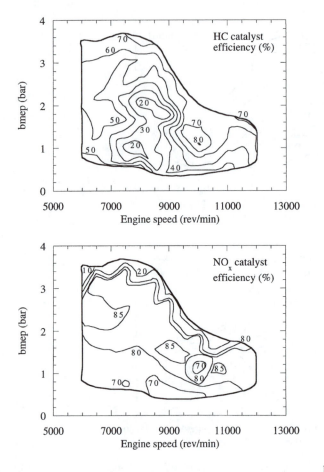

Figure 7-35 (Continued).

conversion of HC is about half these values. A secondary air flow of about 20% of the engine airflow can be achieved with this system; although a higher airflow would give higher catalyst conversion efficiencies, it is difficult to achieve. The substantial effect of adding the catalyst system and then the secondary air system, on CO and HC emissions from a 50-cm^3 displacement motorcycle engine in the ECE-R40 driving-cycle test is shown in Fig. 7-37.[39]

Exhaust gas treatment requirements for direct-injection (DI) two-stroke gasoline engines are different from those of premixed engines. These DI engines operate stratified and lean at partial load, have inherently low NO_x emissions (further reduced by use of EGR, see Section 7-3-2), and have much lower HC emissions than their premixed equivalents due to the elimination of fuel short-circuiting. However, the combination of lower exhaust gas temperatures, high exhaust oxygen levels, and high mass flow rates is such that the light-off characteristics of the catalyst system play a bigger role than they do with conventional four-stroke stoichiometric-operating SI engines. For these

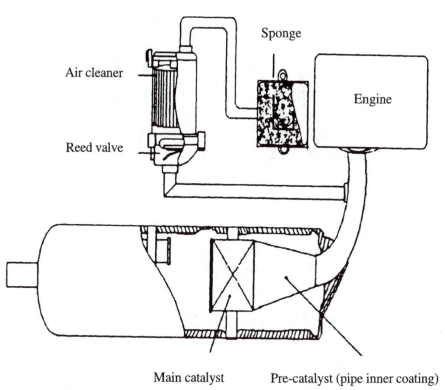

Sponge

Air cleaner

Engine

Reed valve

Main catalyst   Pre-catalyst (pipe inner coating)

Figure 7-36 Schematic of exhaust system of a carburetted small-displacement (50–125 cm^3) two-stroke motorcycle SI engine, with a reed valve controlled secondary air system and muffler containing a precatalyst and main catalyst.[39]

engines, the focus is on platinum or palladium oxidation catalysts. The light-off characteristics of Pt and Pd oxidation catalysts are shown in the test rig results in Fig. 7-38. The light-off test rig fed engine exhaust gas at a controlled temperature into the catalyst. For each catalyst shown, light-off occurs at the higher temperature branch of each curve; then, as exhaust gas temperature is decreased, HC conversion continues at a high level to a lower temperature before it falls off rapidly. Platinum-based catalysts have higher initial light-off temperatures, but retain high conversion efficiencies for longer; Pd and Pt/Pd mix catalysts have lower light-off temperatures but conversion efficiency falls off faster.[24]

Two catalyst-system approaches, developed by the Orbital Engine Corp. for the DI engine in Figs. 1-24 and 1-25, are shown in Fig. 7-39: a twin catalyst system with one catalyst close-coupled to the exhaust manifold, and a (vehicle) underfloor single-catalyst system with a low-thermal-capacity insulated exhaust system. The close-coupled catalyst system in Fig. 7-39 starts to convert HC and CO earlier after engine start-up and maintains a higher efficiency during the light-load portions of the vehicle emissions test cycle. However, the underbody system has a number of inherent advantages: reduced HC and CO loading (and hence catalyst exothermicity) during high-load operation due

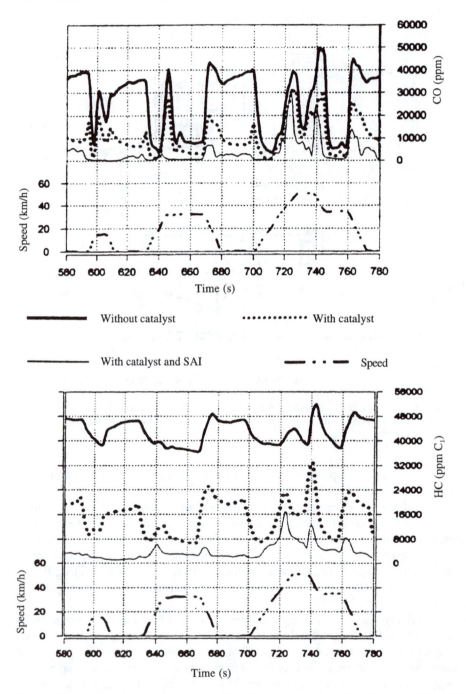

Figure 7-37 CO and HC concentrations in the exhaust gases of two-stroke motorcycle engine during ECE-R40 mode driving-cycle emissions test without catalyst, with catalyst, and with catalyst and secondary air (see Fig. 7-36).[39]

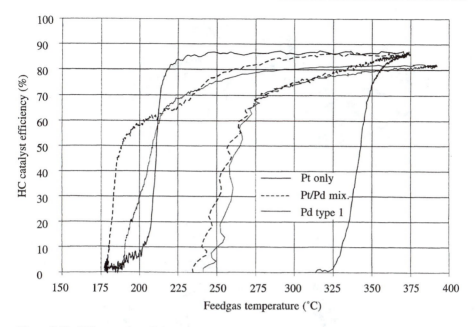

Figure 7-38 HC conversion efficiency hysteresis curves for different Pt and Pd catalyst washcoat compositions as a function of simulated two-stroke exhaust gas temperature measured in a light-off test rig. Once light-off occurs, high conversion efficiencies are maintained to temperatures significantly cooler than the light-off temperature.[24]

to oxidation occurring in the manifold and downpipe, and hence improved durability; improved exhaust system dynamics (better tuning), and therefore improved power; improved vehicle packaging flexibility. The performance of the underbody system can be improved by the following measures: minimizing heat losses from the exhaust gas by reducing heat transfer to the coolant and by thermal insulation of the exhaust system; reducing the thermal capacity of the exhaust system; optimizing the space velocity across the catalyst bricks to improve light-off and light-load conversion efficiency; improving the exhaust gas distribution across the face of each catalyst brick to improve durability and aged-catalyst light-off characteristics; refining the position of the catalyst bricks and selecting appropriate washcoats.[24]

Additionally, a number of techniques can be used to raise the engine-out exhaust gas temperature and achieve faster catalyst light-off: (i) optimization of the exhaust port flow control valve to balance lower engine-out HC and CO with higher exhaust gas temperatures; (ii) cutting out one cylinder on the three-cylinder Orbital DI two-stroke engine by stopping fuel injection (this increases in-cylinder charge temperature, and exhaust gas temperature from the firing cylinders—both desirable); (iii) use substantially retarded spark timing (up to 30° ATC) at idle to increase the exhaust gas temperature. The first two of these can improve minimum engine-out exhaust temperatures by up to 100°C. By optimizing (iii), initial exhaust gas temperatures can be increased by up to 300°C.[24]

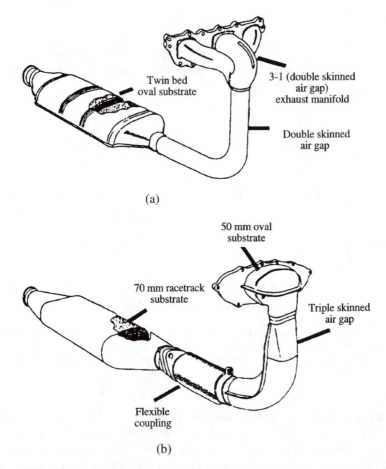

Figure 7-39 Fast light-off exhaust and catalyst systems for automotive two-stroke direct-injection SI engines: (a) single-catalyst system with insulated exhaust manifold and pipe; (b) twin-catalyst system with close-coupled catalyst, insulated header pipe, and under-floor main catalyst.[40] (Courtesy Orbital Engine Corp. Ltd.)

7-5-3 Exhaust Gas Treatment for Diesels

Exhaust after treatment technologies for diesel emissions reduction consist of the following[41]:

1. selective catalytic reduction (SCR) of NO_x;
2. wet or dry desulfurization;
3. traps or filters for particulate matter;
4. oxidation catalysts for HC.

Note that exhaust after-treatment approaches for the large diesel engines that use heavy fuel oil, and small diesels that use standard light diesel oil, are significantly different.

The former use fuels that have much higher sulfur content, fuel-bound nitrogen, and significant ash content.

The diesel operates with a large excess air ratio, so standard SI engine NO_x-reducing catalysts cannot be applied. It is, however, possible to reduce NO_x by using a selective catalytic reduction (SCR) system, in which NH_3 (or urea) is introduced into the exhaust gas, mixes, and in the presence of a catalytic material reacts selectively with NO and NO_2 to yield N_2 and H_2O through the following two main reaction steps:

$$4NH_3 + 4NO + O_2 \rightarrow 4N_2 + 6H_2O \tag{7-23}$$

$$4NH_3 + 2NO_2 + O_2 \rightarrow 3N_2 + 6H_2O \tag{7-24}$$

When urea is used, it must first decompose to NH_3. The exhaust gas should have a temperature between 300 and 400°C. At higher temperatures, NH_3 would burn rather than react with NO and NO_2. At lower temperatures, the rate of reaction is too slow, and condensation of a mixture of water and sulfur would destroy the catalyst. The lower temperature limit is mainly determined by the fuel sulfur content. To reduce the catalyst sensitivity to fuel sulfur content and increase its efficiency at high temperatures, a non-sulfating TiO_2 carrier and V_2O_5 catalyst pair has been developed. The low operating temperatures of 150–200°C with a Pt catalyst have been increased to 350–400°C at high conversion efficiency with this catalyst pair.[34] More than 90% NO_x reduction can be obtained by using the SCR method; in addition, some of the soot and HC in the exhaust are removed by oxidation in the SCR reactor.[29] A schematic layout of an SCR system for large low-speed two-stroke diesel engine is shown in Fig. 7-40.

The ammonia feed can be either liquid water-free ammonia under pressure, or a 25% aqueous solution at atmospheric pressure. When liquid, the ammonia is heated, and evaporation takes place at around 70°C. Ammonia is a combustible gas, so a double-walled pipe system with an appropriate venting and gas leak detector should be installed. After evaporation, the ammonia is diluted with pressurized air from the scavenge air receiver and subsequently mixed in a static mixer. The mixture is then injected into the exhaust pipe and mixes with the exhaust gas. It is important to ensure uniform mixing with the exhaust gas to minimize ammonia loss. The SCR reactor contains several layers of catalyst. The amount of catalyst required can be determined from the desired "space velocity," which is defined as the number of cubic meters of exhaust gas per hour (under standard conditions) that are treated per cubic meter of catalyst volume.

The degree of NO_x removal depends on the amount of ammonia added. At high NH_3/NO_x ratios, effective NO_x removal can be obtained but the amount of unused ammonia, termed NH_3 slip, in the cleaned flue gas will increase. Figure 7-41 shows an example of how the NO_x reduction and ammonia slip vary with the ammonia-added ratio and catalyst volume. It seems that both the NO_x reduction and ammonia slip increase with an increase in the ammonia-added ratio. It is possible to obtain the same NO_x reduction by using a much smaller catalyst volume, by slightly increasing the ammonia added ratio. At the same time, however, the ammonia slip increases considerably. Ammonia slip of 5–10 ppm is normally an accepted emission level. Alternatively, a higher degree of NO_x reduction, while keeping low ammonia slip, can be achieved with higher catalyst volume.

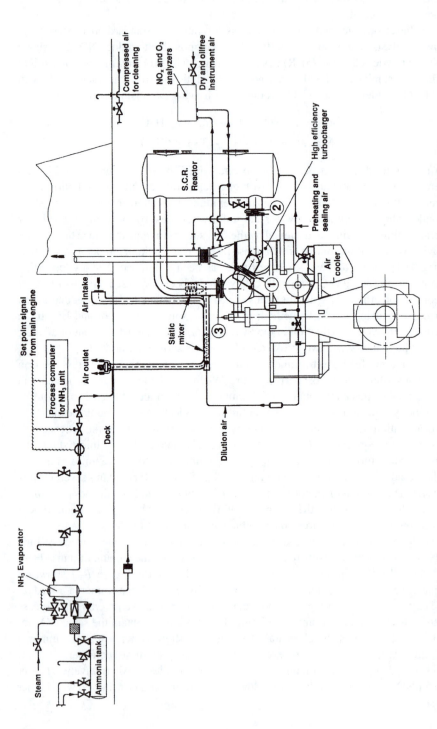

Figure 7-40 Layout of a selective catalytic reduction (SCR) system for reducing NO_x, with NH_3 addition to the exhaust, for a large low-speed two-stroke diesel engine.[29] (Courtesy MAN and B&W Diesel A/S.)

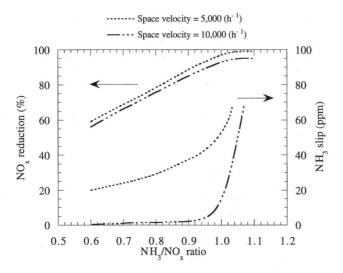

Figure 7-41 Effect of NH_3/NO_x ratio on percentage of NO_x removed, and NH_3 loss or "slip," in an SCR diesel exhaust treatment system. Space velocity: cubic meters of exhaust gas per hour, under standard atmospheric conditions, which are treated per cubic meter of catalyst.[29]

A different catalytic NO_x reduction technology is being developed for light-duty as well as heavy-duty vehicle diesel engines, due to the lower exhaust gas temperatures and practical issues related to lack of space, complexity, maintenance, and cost. Hydrocarbons can be used to reduce NO_x under excess oxygen conditions, on a platinum or base metal catalyst. The method requires sufficient hydrocarbon concentrations in the exhaust gas that the HC must be added either by postcombustion in-cylinder injection or by secondary injection into the exhaust. There is, therefore, a fuel consumption penalty of a few percent. This technology is still developing, and NO_x conversion efficiencies of up to about 30% have been achieved. Particulate, HC, and CO emissions reductions are also achieved with low levels of hydrocarbon addition.[42,43]

For exhaust gas desulfurization in large two-stroke diesels, either wet (scrubber) or dry techniques are available. A chemical compound that reacts with SO_2 and SO_3 to produce sulfates (e.g., calcium carbonate or sodium bicarbonate) is mixed with or sprayed into the exhaust gases. The sulfate formed is removed by washing or filtering (e.g., using baghouse or ceramic filters).[41] Light diesel fuel is relatively low in sulfur, and regulations are decreasing the maximum sulfur levels allowed. So in small two-stroke diesels, sulfur oxide emissions are not an issue except for their contribution to particulates. Fuel sulfur level does affect NO_x catalyst efficiency.

The particulates emitted from a diesel engine typically cover a size range of 0.01 to 10 μm in diameter. Because the average particle size, on a mass basis, is far below 1 μm, filtration (or trapping) and electrical separation are the only feasible technologies for reducing solid particulate emissions. These technologies are not specifically designed for two-stroke cycle engines and are being developed primarily for four-stroke diesel engines.

With this particulate trap technology, the particles are collected by passing the exhaust gases through an appropriate filter mesh. The collected particulate (soot and absorbed hydrocarbons) can be burned off with the excess air in the exhaust gases when the exhaust gas temperature is high enough (typically 550°C). However, local peak temperatures of up to 1200°C can be reached during burn-off, which requires high-temperature-resistant filter materials such as ceramics. An extruded ceramic honeycomb substrate element, similar in design to the support for a monolithic SI engine catalyst, is usually employed. The details of the gas flow passages are different to allow particle collection: half of them are blocked at one end, half are blocked at the other end. To prevent an excessive back pressure and possible blockage, trap regeneration aids must be employed. Providing external energy via a fuel burner, and introducing metal–organic substances to reduce ignition temperatures, resulting in forced regeneration of the filter, are two techniques employed in large-bore engines.

The electrical separator is an efficient method for cleaning particles from the exhaust gases. Unlike the particulate burn-off filter, the pressure drop across the electrical separator is independent of the particulate quantity collected, and blockage risk is minimal. In the electrical separator, a high voltage is applied between short edge electrodes (discharge electrodes), which are situated along the gas flow passage, and blunt electrodes, which are situated downstream. The electrodes generate free charge carriers, which attach themselves to the soot particles. When the charged particles flow downstream, they are separated out by the opposing polarity of the blunt electrodes. The particles are then collected and may be removed by a cyclone, and disposed of by burn-off techniques.

Oxidation catalysts are used in truck and car diesel engines, with distillate diesel fuel, to reduce the exhaust hydrocarbon concentration, and hence the amount of higher molecular weight hydrocarbons that are absorbed onto the particulates. About a one-third reduction in particulate mass results.

Diesel exhaust treatment technology is being extensively explored and developed, largely for four-stroke engine application. As these catalyst technologies improve and become practical, they will obviously be available for two-stroke diesels as well.

7-6 ENGINE NOISE

Noise is often defined as "unwanted sound." It is a significant problem with internal combustion engines, and especially with the small two-stroke gasoline engines used in mopeds, motor scooters, motorcycles, snowmobiles, powerboats, and hand-held devices such as chainsaws and weed whackers. It is a difficult problem to resolve because effective noise suppression usually reduces engine power, increases engine bulk size and weight, and increases cost, all of which run counter to the competitive pressures in these low-cost engine markets. Noise is an extensive and well-defined field, with its own vocabulary and literature, which often separate it from those aspects of engine operation that we cover in this book. Increasingly, however, as noise limits on engines become more stringent, the reduction of engine noise must be integrated into a systems approach to engine design. An extensive treatment of engine noise is beyond the scope of this book. So in this section, we develop enough of the fundamentals of the subject for the reader to be able to understand the sources of noise in two-stroke cycle internal combustion engines and the means available to reduce them. Additional details and more extensive reviews can be found in Annand and Roe,[44] Blair,[14] and the Bosch Automotive Handbook.[45]

7-6-1 Fundamentals and Definitions†

Sound propagates through a medium as pressure fluctuations. The frequency of these fluctuations determines the *pitch* of the sound. The human ear can detect frequencies from about 20 Hz to 20,000 Hz. A simple sinusoidal pressure oscillation produces a sound wave that is called a pure tone. Usually sound is composed of a fundamental frequency with smaller amplitude harmonics—simple multiples of this fundamental frequency. This mixture of frequencies gives the sound its "tone"; significant high-frequency components make sounds especially objectionable.[44]

The pressure fluctuations generated by the noise source transfer energy outward through molecular collisions. The more energy transferred, the higher the sound level or volume. The *intensity of sound*, I (W/m^2), is defined as the energy transfer per unit time by the sound wave, through unit area perpendicular to the wave propagation direction. It quantifies the actual energy flux associated with the sound wave. Usually, the effective sound pressure—the root-mean-square (rms) fluctuating pressure δ_p—is what is measured. For plane or spherical waves,

$$I = (\delta_p)^2/(\rho c) \tag{7-25}$$

where ρ is the medium density [for air, $\rho = 1.184$ kg/m^3 (0.0739 lbm/ft^3)] and c is the sound speed [given by $(\gamma RT)^{1/2}$, 346 m/s for air at 25°C]. With a localized noise source, the sound intensity I and the square of the rms fluctuating pressure $(\delta p)^2$ both fall off as the square of the distance from the source.

The range of sound intensities that humans can hear and tolerate is very large, from 10^{-12} W/m^2 to 1 W/m^2, so a logarithmic scale is used. The *sound intensity level L_I* is therefore defined as

$$L_I = 10\log_{10}(I/I_0) \tag{7-26}$$

where I_0 is a chosen reference level. The unit of this relative intensity is the decibel (dB). Equation (7-26) can be expressed in terms of the sound pressure level L_p,

$$L_p = 10\log_{10}\left[(\delta p)^2/(\delta_{p0})^2\right] = 20\log_{10}(\delta_p/\delta_{p0}) \text{ dB} \tag{7-27}$$

where δ_{p0} is the reference sound pressure level. The standard reference level corresponds to the threshold of hearing, where $I_0 = 10^{-12}$ W/m^2 and $\delta_{p0} = 5 \times 10^{-5}$ Pa.[44] Using Eqs. (7-25) to (7-27), a sound intensity of 1 W/m^2, which is the sound level at which the ear begins to feel pain, corresponds to a mean effective sound pressure of 20 Pa, and a sound pressure level of 120 dB.

In the frequency range 500–5000 Hz, the ear responds almost logarithmically to sound intensity, so this logarithmic decibel scale is appropriate. However, sounds of

†The following summary follows the development of noise fundamentals in Annand and Roe.[44]

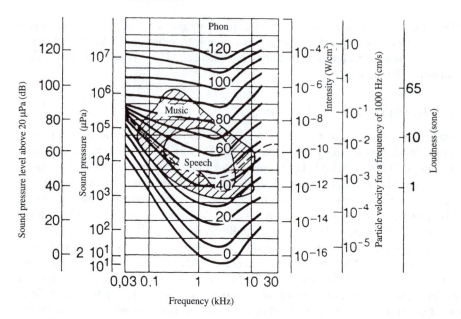

Figure 7-42 Objective (sound pressure level, in decibels) and subjective (A-weighted sound-meter curve, in phons) scales for noise curves of equal loudness level, as a function of frequency.[45]

different pitch or frequency perceived as equally loud do not have the same intensity. To provide a quantitative measure of subjective *loudness*, sound is said to have a loudness in *phons* equal to the intensity level of a pure tone of frequency 1 kHz that appears equally loud.[44] Figure 7-42 shows how the actual intensity required to produce a certain sensation of loudness for an average listener varies with frequency. The weighting shown is known as the *A-weighting* and corresponds to the ear's response at 40 phon. An A-weighting filter is normally applied automatically to noisemeter (a microphone that converts variations in pressure to variations in voltage) measurements and is designated dB(A). Note that the most irritating frequencies are between about 1 and 5 kHz, with increasingly higher sound pressure levels sounding equally loud as frequencies decrease and increase below and above this band, respectively.

With more than one source of sound, in nonreflecting surroundings and sufficiently distant from the sources, the combined effect is obtained by adding the absolute source intensities (the energy fluxes per unit area), as they are each planar sound waves. For example, with two sources of intensity I_1 and I_2, the combined intensity I_{12} is

$$I_{12} = I_1 + I_2 \tag{7-28}$$

where, from Eq. (7-26),

$$I_1 = I_0 \, \text{antilog}_{10}(L_{I1}/10) \quad \text{and} \quad I_2 = I_0 \, \text{antilog}_{10}(L_{I2}/10)$$

The resultant sound intensity level is given by

$$L_{I12} = 10 \log_{10}(I_{12}/I_0)$$
$$= \log_{10}\left[\text{antilog}_{10}(L_{I1}/10) + \text{antilog}_{10}(L_{I2}/10)\right] \tag{7-29}$$

If the two sources (1 and 2) are equal, Eq. (7-29) gives

$$L_{12} = 10 \log_{10} \left[2 \operatorname{antilog}_{10}(L_{I1}/10) \right]$$
$$= 10 \left[\log_{10} 2 + (L_{I1}/10) \right] = L_{I1} + 3.01 \text{ dB}$$

that is, doubling the intensity of sound increases the sound intensity level by 3 dB. If one source is substantially larger than the other, for example, the two sources are 100 dB and 90 dB, then the two sources together give

$$L_{I12} = 10 \log_{10} \left[\operatorname{antilog}_{10} 10 + \operatorname{antilog}_{10} 9 \right] = 100.4 \text{ dB}$$

Here, the smaller source contributes little to the combined sound level.

Two important consequences for reducing noise levels in engines follow from this review of noise fundamentals[14]:

1. The loudest of the several sources of noise in an engine will be much the most important in determining the total sound pressure level. To reduce the noise, it must first be dealt with effectively.
2. The higher frequency range (1–5 kHz range) is much more objectionable to the human ear, at the same sound intensity level, than are lower frequencies. So treatment should focus first on this higher frequency range.
3. When a major noise varies with time and produces short intense pulses (e.g., a poorly muffled engine exhaust system at low speed), these are highly disagreeable and must be heavily damped by the noise treatment system.[44]

7-6-2 Sources of Engine Noise

There are three major sources of engine noise: the exhaust, the intake, and mechanical impacts and vibration. Exhaust noise is easiest to understand; its origin is the pressure waves in the exhaust system set up by each cylinder's exhaust process, and noise is generated as these waves are released to the atmosphere. Intake noise is generated by the pressure waves set up by the periodic induction processes into each cylinder, and by the vibrating reeds when reed valves are used to control the flow into and out of the crankcase in simple two-stroke cycle engines. Mechanical noise is generally lower than exhaust- and intake-generated noise. There are several sources. One is the combustion-produced pressure rise that is transmitted through the block and cylinder-head walls to the atmosphere. The piston slaps against the cylinder walls, causing them to vibrate, when, at certain positions in its travel, the forces acting on the piston cause it to rock from one side of the liner to the other. Mechanical noise also originates from the oscillating forces on the crankshaft bearings. The roller bearings usually used in simple crankcase-scavenged two-stroke engines are noisier than standard lubricated plain bearings where the oil film damps out the mechanical vibration. In air-cooled engines, the cylinder and head cooling fins vibrate and produce noise. (Liquid cooling of the engine damps out some of the vibrations the block and head structure would otherwise transmit.) Gears and chain drives are also significant sources of mechanical noise.[14,44]

In large two-stroke diesels, additional important noise sources are the turbocharger, the exhaust valves, and the fuel injection system. Constant-pressure turbocharger systems, where a gas receiver upstream of the turbine is used to dampen the exhaust blow-down pressure pulses to improve overall turbine performance, are significantly quieter than the alternative impulse systems where the turbine is designed to take advantage of the exhaust pressure pulsations.[46]

We now show some examples of engine noise levels and spectra. Noise level meters provide such information in several ways. One is to weight the sound pressure levels at different frequencies and then integrate to produce an overall sound intensity reading. The A-weighting mentioned earlier, which corresponds to the average ear's response at the 40-phon level, is usually used in engine and vehicle noise tests. However, knowledge of the noise spectral distribution is helpful when designing noise suppression systems. Depending on the bandwidth of the filter used to determine the intensity frequency distribution, the noise spectrum can be given in octave bands (one octave wide, plotted at the center frequency, usually with the points joined by straight lines) on a log scale, or with much smaller frequency bands to give the complete spectrum. Figure 7-43 shows the noise spectra for the exhaust, intake, and major mechanical sources from an unsilenced (unmuffled) 380-cm^3 single-cylinder competition motorcycle two-stroke carburetted SI engine. Table 7-6 gives the overall sound intensity levels, in dB(A), for each source. In this unsilenced engine, the exhaust and intake noise would need substantial reduction for road use. The exhaust expansion is an excessive noise source over a wide frequency range, a common problem with two-stroke engines. The noise radiated from the engine's cooling fins is serious, and the gearbox and chain drive noise also need to be reduced.[47]

These data relate to wide open throttle engine operation. Noise can also be a problem at idle due to irregular combustion behavior and piston slap, which produce an annoying periodic impulsive sound. Piston slap occurs when the piston moves rapidly from one side of the liner to the other (usually around TC and BC crank positions as the connecting rod moves through the cylinder axis). Piston rotation or rocking at other crank angle positions can also cause liner impact, and hence vibration, depending on the details of the piston geometry and cylinder gas pressures. Figure 7-44 shows the relationship between cylinder pressure, piston vibration, and engine noise in an air-cooled

Table 7-6 Noise sources from an unsilenced 380-cm^3 single-cylinder competition motorcycle two-stroke SI engine[47]

Source	Sound intensity level [dB(A)]
Open exhaust (15 cm from open end)	142
Carburetor intake (15 cm from airbox filter)	125
Air filter removed	129
Box also removed	133
Exhaust expansion chamber (2.5 cm from wall)	126
Engine cooling fins (2.5 cm)	124
Primary chain drive (2.5 cm from casing)	121
Gearbox (2.5 cm from casing)	118
Final drive chain (2.5 cm from moving chain)	117

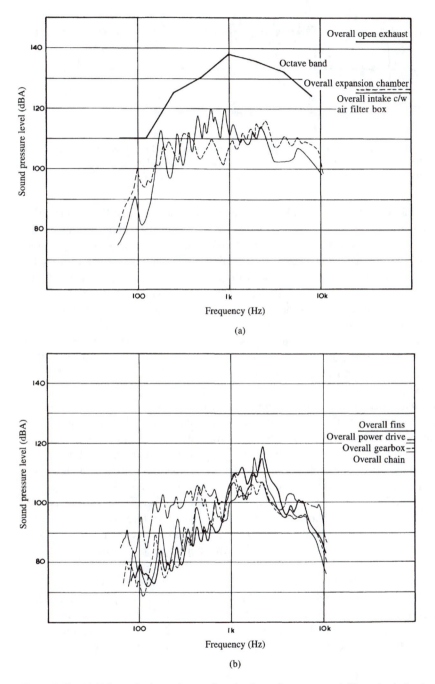

Figure 7-43 (a) Exhaust, intake, and expansion chamber noise spectra and (b) mechanical noise spectra; both for 380-cm^3 displacement single-cylinder unsilenced cross-country motorcycle Greeves Griffon two-stroke engine.[47]

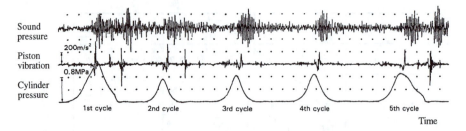

Figure 7-44 Relationship between cylinder pressure, piston vibration, and engine noise over five consecutive cycles for a 96-cm^3 displacement single-cylinder air-cooled two-stroke SI engine, idling at 1300 rev/min, air/fuel ratio of 13.[48]

96-cm^3 displacement single-cylinder two-stroke crankcase-scavenged SI engine, idling at 1300 rev/min and A/F of 13.[48] The first cycle shows normal combustion and a large positive piston vibration is evident at 20–30° ATC due to the piston motion from the minor thrust to the major thrust side of the cylinder.† The next vibration is at 80–85° ATC, and a final large vibration at about 130° ATC after exhaust blowdown. The second, third, and fourth cycles are misfiring cycles. The second cycle shows a vibration at 20–30° ATC similar to that of the first cycle, and also a significant vibration at 100° ATC. In the third and fourth cycles only the 20–30° ATC vibration is evident and it is smaller. In the fifth cycle, a late-burning cycle, the vibration pattern of the first cycle repeats, but with smaller amplitude. The sound pressure level increases substantially at each impulsive piston movement at around 20° and 130° ATC (but not at 80° ATC), and then decays.

Detailed study of the piston motion showed that piston slap—the piston leaving the minor thrust side at 10° ATC and impacting the major thrust side at 20–30° ATC—is the cause of the first sound pressure pulse. At 130° ATC, the piston then slaps the minor thrust side of the liner. These piston motions are affected by the axial piston skirt profile, and the offset of the piston pin. Usually the piston pin is offset from the piston axis toward the major thrust side by a distance equal to 1–2% of the bore, to initiate the close-to-TC transition minor to major thrust sides before TC when the cylinder pressures are lower. Suitable design of the piston, and also retarded spark timing, and use of a pressure release hole above the exhaust port, can reduce the magnitude of the impact and the noise it produces. These last two approaches of course increase fuel consumption.

Gotoh and Maeda[48] propose a weighted sum of the overall loudness (rms 1-s average, A-weighted sound pressure level), the impulsiveness (rms fluctuation, 2-ms average, A-weighted sound pressure level), and the high-frequency loudness (rms 1-s average, A-weighted 2–4-kHz sound pressure level) as an appropriate quantitative measure of the disagreeableness of periodic impulsive engine noise. We will use this

---

†With the connecting rod at an angle to the cylinder axis, the piston is pushed to one side of the liner. Because expansion-stroke pressures are higher than compression-stroke pressures, these are usually called *major* and *minor* thrust sides, respectively (and sometimes "thrust" and "antithrust" sides).

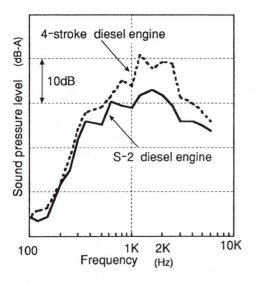

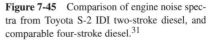

Figure 7-45 Comparison of engine noise spectra from Toyota S-2 IDI two-stroke diesel, and comparable four-stroke diesel.[31]

data in the following section to explain these sources of noise in more detail and how they can be reduced.

Figure 7-45 shows the noise spectrum of the Toyota two-stroke 2.5-liter four-cylinder prototype diesel (see Fig. 9-14 and associated discussion) relative to that of an equivalent four-stroke diesel. The sound pressure levels in the 1–5 kHz range are substantially lower due to the shorter two-stroke ignition delay (and hence slower initial combustion-generated rate of pressure rise, see Section 6-4-2), and the lower two-stroke engine peak pressures. The two-times-higher firing frequency of the two-stroke, at the same engine speed, also reduces mechanical vibration substantially, and the mechanical noise it generates.[31]

In large marine two-stroke diesels, the major noise challenge is the overall airborne engine room noise, which must be kept below 100 dB(A), measured at several locations, at 2–3-m heights, about 1 m from the engine outer surfaces. It is primarily generated by individual engine components and their surfaces. Exhaust gas noise can be brought down by about 25 dB(A) to acceptable levels with large absorption silencers, which are especially effective at damping the higher frequencies (see the following section).[46]

In general, the two-stroke cycle engine, relative to equivalent four-stroke engines, has the following inherent noise characteristics[14]:

1. Because the firing frequency is doubled, noise generated by gas pressure waves is a greater problem because the human ear is more sensitive.
2. The ports in two-stroke engines open more rapidly than poppet valves, so the pressure wave fronts are steeper, which increases the higher frequency components of the sound.
3. Simple two-stroke cycle engines are designed to be light and compact, so the engine surfaces radiate noise more easily, and the space available for noise suppression

devices is limited.

4. Two-stroke engines usually employ a tuned exhaust, with a choked final outlet, to maximize the power output. This exhaust system feature eases the task of designing a silencer–muffler with minimum impact on power.

5. A crankcase scavenging air pump produces lower intake air velocities than does the four-stroke engine's induction of air directly into each cylinder.

6. Due to its less effective filling of each cylinder with air, the two-stroke engine has lower peak cylinder pressures than the four-stroke. The resulting oscillating forces in major engine components are therefore lower, and so is the mechanical noise they generate.

7-6-3 Engine Noise Reduction

Exhaust noise. Exhaust noise is the result of release to the atmosphere of pressure waves in the exhaust system through its open end, and from vibration of the exhaust system surfaces caused by these pressure waves. Various muffler (silencer) technologies and exhaust system configurations have been developed to reduce exhaust noise. The selection of the appropriate approach and its integration into the total exhaust system depend on the engine application (and space and weight constraints) and the type of exhaust system needed to achieve the desired engine power levels.[7]

Mufflers take many forms, and more than one type may be used in a given exhaust system. Two basic principles are involved, absorption and reflection, and mufflers are divided into two types according to which principle is involved, although both absorption and reflection usually occur. Mufflers, together with the exhaust system, form an oscillator with natural resonance frequencies, so the position of the mufflers in the exhaust is critical. To avoid structure-borne noise (and provide thermal insulation) mufflers often have double walls and an insulating outer layer.[45]

Absorption mufflers (see Fig. 7-46a) are constructed with one chamber, filled with sound-deadening material, through which the perforated exhaust pipe is passed. The absorbing material is usually long-fiber mineral wool with a bulk density of 120–150 g/liter. The level of muffling depends on the bulk density and sound-absorbing characteristics of the absorbing material, the muffler length, and material thickness. Damping takes place across a wide frequency band but only starts at higher frequencies (above about 1 kHz). Absorption mufflers are normally used as rear mufflers[45] and are the most frequently used design in large two-stroke diesels.[46]

Reflection mufflers consist of chambers of varying dimensions connected together by pipes. The differences in cross sections of the pipes and the chambers, the dispersion of the exhaust gas flow, and the resonators formed by connecting pipes to chambers, all reduce the proportion of the incident pressure fluctuations transmitted. Diffusing mufflers or silencers (see Fig. 7-46b) absorb noise frequencies other than those at which the expansion box will resonate. Side resonators (Fig. 7-46c), where the flow connects with a separate closed chamber through holes or slits in the pipe wall, absorb noise at a specific frequency, such as the fundamental exhaust pulse frequency. The more chambers used, the more effective the muffler.[14,45]

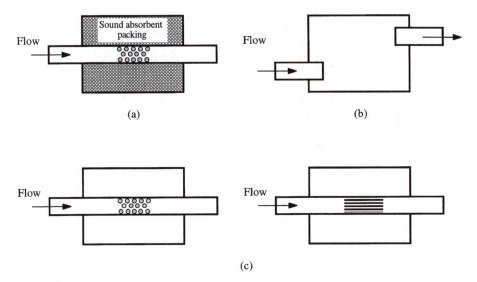

Figure 7-46 (a) Absorption muffler element; (b) diffusing muffler element; (c) side-resonator muffler element.

Reflection mufflers remove annoying components of the noise emitted by the exhaust through several different principles. The operation of the expansion box muffler in Fig. 7-46b can be explained as follows. The performance of an isolated muffler or silencer is often expressed as a *transmission loss*. Provided the relative amplitude of the pressure fluctuations is not too large, and the sound source is not affected by reflections from the muffler, the sound intensity at a given frequency is reduced by a fixed fraction. Thus, the intensity level is reduced by a fixed number of decibels. This reduction in decibels is the transmission loss. The transmission loss of a single pass through expansion box is a function only of the expansion area ratio $m = A_2/A_1$, and the ratio of the box length L to the wave length λ of the sound. The transmission loss T_{loss} is given by[44]

$$T_{\text{loss}} = 10 \log_{10}\left[1 + \frac{1}{4}\left(m - \frac{1}{m}\right)^2 \sin^2\left(\frac{2\pi L}{\lambda}\right)\right] \text{dB} \qquad (7\text{-}30)$$

The box resonates whenever an integral number of half wavelengths just fits the length L [i.e., where $\lambda = 2L, L, 2L/3, L/2$, etc., at frequencies of $f = c(2L), c/L, 3c/(2L), 2c/L$, etc.]; the transmission loss then goes to zero. Between, it rises to a maximum value of

$$T_{\text{loss,max}} = 10 \log_{10}\left[\frac{1}{4}\left(m + \frac{1}{m}\right)^2\right] \text{dB} \qquad (7\text{-}31)$$

Experimental measurements show good agreement with theory until frequencies are reached where the box can resonate in a transverse or radial mode (see Fig. 7-47). For a box with circular cross-section, the first radial resonance occurs, as shown, at a frequency of $1.22c/D$.[44]

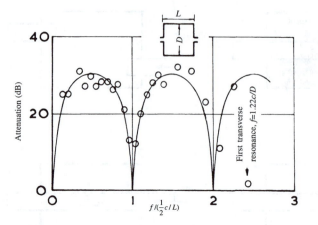

Figure 7-47 Transmission loss of an expansion box of circular cross section showing pass bands at the first two longitudinal resonance frequencies, and the first transverse resonance.[44]

Pipes perforated with holes or slits connected to boxes, as used in side-resonator mufflers (see Fig. 7-46c), absorb strongly at the natural frequency of the box or side cavity. The transmission loss of a side-resonator element of volume V connected to a pipe of cross-sectional area A_p through an opening of conductivity K is given by

$$T_{\text{loss}} = 10 \log_{10} \left[1 + K V \bigg/ \left\{ 2A_p \left(\frac{f}{f_n} - \frac{f_n}{f} \right) \right\}^2 \right] \text{dB} \qquad (7\text{-}32)$$

The conductivity of the opening (which has the dimensions of length) is given approximately by

$$K = \pi r^2 / (l + (\pi/2)r) \qquad (7\text{-}33)$$

where r is the radius of the opening and l is the length of the pipe leading to the cavity (zero for a simple hole, when $K = 2r$). The resonant frequency f_n is

$$f_n = \frac{c}{2\pi} \sqrt{\frac{K}{V}} \qquad (7\text{-}34)$$

The transmission loss is very high at and near the resonance frequency of the cavity. Away from this frequency, the loss depends on $\sqrt{KV/A_p}$, as shown in Fig. 7-48. With high values of $\sqrt{KV/A_p}$, a substantial transmission loss is achieved over a wide frequency range.[44]

For several holes, the conductivity is the sum of the individual values. For a rectangular slit of length l and width b, in a sheet of thickness t, K is

$$K = bl / (t + 0.92k\sqrt{bl}) \qquad (7\text{-}35)$$

where k decreases from 1.0 for $l = b$ to 0.6 at $l = 20b$. The attraction of side-cavity resonators is that the flow through the main pipe is virtually unimpeded.[44]

Practical mufflers are often combinations of more than one expansion box and/or side resonators so their operation is more complex than these simpler illustrations. The

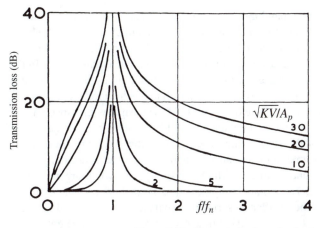

Figure 7-48 Transmission loss of a simple side-resonator muffler element.[44]

preceding transmission loss expressions are for a reflection-free muffler outlet. The open end of the tailpipe is not reflection free; a pressure wave is reflected at an open pipe end as an equally strong rarefaction wave. Muffler location in the exhaust system is also important. The exhaust pipe upstream can be thought of as an organ pipe with one end (the engine end) closed and one end open; the exhaust pipe downstream has both ends open. Their resonant behavior will be different as a result. Also, high-performance two-stroke engines use tuned exhaust system with expanding and contracting sections designed to enhance the scavenging process by varying the gas pressure at the exhaust port appropriately so it aids the scavenging process (see below). The design of exhaust systems, including their noise reducing features, is now done with unsteady one-dimensional gas dynamic computer codes (see Section 5-5-2) as well as based on extensive empirical data. Blair's text[14] reviews the design of exhaust systems in substantial detail.

Figures 7-31 and 7-34 show examples of exhaust systems for small two-stroke SI engines, designed to accommodate a catalytic converter and mufflers to reduce noise. Figure 7-34b shows a chainsaw system, which must be compact and light, and as a consequence is untuned and must be designed to meet environmental regulations while holding the loss in power to a minimum. Figures 7-31 and 7-34c show high-performance engine exhaust systems that are tuned so the unsteady gas dynamic behavior in the total exhaust system enhances the scavenging process.

In compact exhaust systems, an expansion box (or pair of boxes with a short connecting pipe) is normally used to reduce exhaust noise; it can also hold a catalytic converter. The total box volume (one or both together) should be 10 or more times the swept volume of the engine, the cross-sectional area of the exhaust duct leading into the box is usually 15–20% greater than the full exhaust port area. The exit pipe diameter is significantly smaller, depending on performance and silencing requirements.[14] These systems are suitable for moderate bmep levels of 2.7–5 bar and maximum power between 35 and 60 kW/liter.[7]

High-performance tuned two-stroke exhaust systems (Figs. 7-31 and 7-34c) incorporate an expansion chamber that works as follows. When the exhaust blowdown pressure pulse passes through the expanding section of the exhaust, it induces expansion waves that propagate back toward the exhaust port. At the engine speed for which the exhaust is tuned, the maximum rarefaction arrives at the exhaust port at about BC and effectively pulls additional burned gas out of the cylinder to enhance scavenging. In the high-performance exhaust systems used in motorcycles, snowmobiles, and jetskis, this is followed by a subsequent converging section of the exhaust, which reflects a series of positive pressure waves back toward the exhaust port. At the design engine speed, these arrive at the port while it is closing, and enhance cylinder charging by preventing loss of fresh charge at the end of the scavenging process. Both these actions improve the maximum engine power. The maximum expansion chamber diameter is normally about three times the exhaust port duct diameter, which is usually somewhat larger in cross section than the exhaust port. The exhaust tailpipe diameter is about half the diameter of the exhaust port duct.[48] The muffler (a series of expansion boxes) is added at the downstream end, with modest impact in the exhaust system's performance enhancing behavior because the tailpipe is already restricted.[14] Smaller engines with maximum power in the 18,060 kW/liter range use simpler exhaust and muffler configurations like that shown in Fig. 7-31. Systems like that in Fig. 7-34c can achieve bmep values above 10 bar and power levels of 275 kW/liter in sports and racing applications.[7]

Intake noise.   Relative to exhaust noise, is surprisingly high, especially near wide-open throttle (see Table 7-6). Intake noise originates from the pulsating intake flow corresponding to crankcase and/or cylinder filling events. Intake noise reduction is challenging because any flow restriction will reduce engine output. The basic induction pulse frequency is

$$f = n_c N(\text{rev/min})/60 \qquad (7\text{-}36)$$

where n_c is the number of cylinders. Intake noise reduction on small engines is usually accomplished with an expansion box silencer, which incorporates the air filter in its downstream end. It may be fed by a diverging intake duct. The expansion box or cavity has a resonant frequency given by Eq. (7-34), where, because the entry duct is long in relation to its radius, the conductivity K is A_p/l [see Eq. (7-33)]. Intake flow losses limit the minimum value of A_p. The desired deep-toned sound requires the effective attenuation of the higher frequencies. This requires a low natural frequency for the resonator (where no attenuation will occur), requiring a high value for $V/K = lV$; it also requires a high value for $\sqrt{KV}/A_p$, see Eq. (7-32), or a large expansion box cavity volume. In practice, the natural frequency is set at or below the frequency corresponding to the lowest engine speed at which high loads are produced [using Eq. (7-36)] and using a box volume V between 10 and 20 $V_d\sqrt{n_c}$, where V_d is the displaced volume of each cylinder and n_c is the number of cylinders. In some applications (such as hand-held tools such as chainsaws), this volume may be larger than can be practically accommodated.[14] The air filter material in the expansion box acts as a sound absorber, especially at the higher noise frequencies. The intake noise damping that such a system produces is zero at the resonant frequency of the cavity, and rises steadily to about 15–20 dB as frequency increases, before additional wave phenomena effectively limit further improvements.[45]

An engine operating with a partly closed throttle generates high-frequency "hiss" due to eddy shedding at the throttle plate edges.[44] Combustion and piston slap noise can be reduced by reducing piston–liner clearance and using piston designs with reduced top land clearance to reduce the tilting of the piston. Appropriately profiled piston skirts can effectively eliminate any slapping as the piston rotates during the expansion stroke.[7,48] Radiation of noise from the block and cylinder head cooling fins can be reduced by inserting rubber blocks or discs between the fins at critical locations where high amplitude vibrations occur.[47] Water-cooling the engine, instead of air-cooling, significantly reduces the block and head sound radiation.

REFERENCES

1. Heywood, J.B., *Internal Combustion Engine Fundamentals*, McGraw-Hill, New York, 1988.
2. Sher, E., ed., *Handbook of Air Pollution from Internal Combustion Engines*, Academic Press, New York, 1998.
3. Heywood, J.B., Fay, J.A., and Linden, L., "Jet Aircraft Air Pollutant Production and Dispersion," *AIAA J.*, vol. 9, no. 5, pp. 841–850, 1971.
4. Miller, J.A., and Bowman, C.T., "Mechanism and Modeling of Nitrogen Chemistry in Combustion," *Prog. Energy Combust. Sci.*, vol. 15, pp. 287–338, 1989.
5. Cheng, W.K., Hamrin, D., Heywood, J.B., Hochgreb, S., Min, K., and Norris, M.G., "An Overview of Hydrocarbon Emissions Mechanisms in Spark-Ignition Engines," SAE Paper 932708, *SAE Trans.*, vol. 102, 1993.
6. Sato, K., Ukawa, H., and Nakano, M., "Effective Energy Utilization and Emission Reduction of the Exhaust Gas in a Two-Stroke Cycle Engine," SAE Paper 911848, *SAE Trans.*, vol. 100, 1991.
7. Laimböck, F.J., "The Potential of Small Loop-Scavenged Spark-Ignition Single Cylinder Two-Stroke Engines," SAE Paper 910675, *SAE Trans.*, vol. 100, 1991.
8. Yamagishi, G., Sato, T., and Iwasa, H., "A Study of Two-Stroke Cycle Fuel Injection Engines for Exhaust Gas Purification," SAE Paper 720195, *SAE Trans.*, vol. 81, 1972. (Also in *Two-Stroke Cycle Spark-Ignition Engines*, PT-26, SAE, Warrendale, PA, 1982.)
9. Babu, P.R., Nagalingam, B., and Mehta, P.S., "Hydrocarbon Modeling for Two-Stroke SI Engine," SAE Paper 940403, 1994.
10. Eltinge, L., "Fuel–Air Ratio and Distribution from Exhaust Gas Composition," SAE Paper 680114, *SAE Trans.*, vol. 77, 1968.
11. Tsuchiya, K., and Hirano, S., "Characteristics of 2-Stroke Motorcycle Exhaust HC Emission and Effects of Air–Fuel Ratio and Ignition Timing," SAE Paper 750908, 1975. (In *Two-Stroke Cycle Spark-Ignition Engines*, PT-26, SAE, Warrendale, PA, 1982.)
12. Nuti, M., and Martorano, L., "Short-Circuit Ratio Evaluation in the Scavenging of Two-Stroke S.I. Engines," SAE Paper 850177, 1985.
13. Maji, S., Saxena, M., Mathur, H.B., and Bhagat, S.D., "Analysis of Total and Individual Hydrocarbon Components in the Exhaust and Inside the Cylinder of a Two-Stroke Engine," SAE Paper 920728, 1992.
14. Blair, G.P., *Design and Simulation of Two-Stroke Engines*, SAE, Warrendale, PA, 1996.
15. Sawada, T., Wada, M., Noguchi, M., and Kobayashi, B., "Development of a Low Emission Two-Stroke Cycle Engine," SAE Paper 980761, 1998.
16. Laimböck, F.J., and Landerl, C.J., "50 cc Two-Stroke Engines for Mopeds, Chainsaws, and Motorcycles with Catalysts," SAE Paper 901598, *SAE Trans.*, vol. 99, 1990.
17. Nuti, M., "Direct Fuel Injection: Piaggio Approach to Small 2T SI Engines," SAE Paper 880172, 1988.
18. Yoon, K-J., Kim, W-T., Shim, H-S., and Moon, G-W., "An Experimental Comparison between Air-Assisted Injection System and High Pressure Injection System in 2-Stroke Engine," SAE Paper 950270, 1995.
19. Monnier, G., and Duret, P., "IAPAC Compressed Air Assisted Fuel Injection for High Efficiency Low Emissions Marine Outboard Two-Stroke Engines," SAE Paper 911849, *SAE Trans.*, vol. 100, 1991.

20. Monnier, G., Duret, P., Pardini, R., and Nuti, M., "IAPAC Two-Stroke Engine for High Efficiency Low Emissions Scooters," *A New Generation of Two-Stroke Engines for the Future?* (ed. P. Duret), Proceedings of the International Seminar, Nov. 29–30. Rueil-Malmaison, France, Editions Technip, Paris, 1993.

21. Duret, P., Venturi, S., and Carey, C., "The IAPAC Fluid Dynamically Controlled Automotive Two-Stroke Combustion Process," *A New Generation of Two-Stroke Engines for the Future?* (ed. P. Duret), Proceedings of the International Seminar, Nov. 29–30. Rueil-Malmaison, France, Editions Technip, Paris, 1993.

22. Duret, P., and Venturi, S., "Automotive Calibration of the IAPAC Fluid Dynamically Controlled Two-Stroke Combustion Process," SAE Paper 960363, *SAE Trans.*, vol. 105, 1996.

23. Ishibashi, Y., and Asai, M., "A Low Pressure Pneumatic Direct Injection Two-Stroke Engine by Activated Radical Combustion Concept," SAE Paper 980757, 1998.

24. Houston, R., Archer, M., Moore, M., and Newmann, R., "Development of a Durable Emissions Control System for an Automotive Two-Stroke Engine," SAE Paper 960361, *SAE Trans.*, vol. 105, 1996.

25. Fansler, T.D., French, D.T., and Drake, M.C., "Fuel Distributions in a Firing Direct-Injection Spark-Ignition Engine Using Laser-Induced Fluorescence Imaging," SAE Paper 950110, *SAE Trans.*, vol. 104, 1995.

26. Drake, M.C., Fansler, T.D., and French, D.T., "Crevice Flow and Combustion Visualization in a Direct-Injection Spark-Ignition Engine Using Laser Imaging Techniques," SAE Paper 952454, *SAE Trans.*, vol. 104, 1995.

27. Fansler, T.D., French, D.T., and Drake, M.C., "Individual-Cycle Measurements of Exhaust Hydrocarbon Mass from a Direct-Injection Two-Stroke Engine," SAE Paper 980758, 1998.

28. Petit, A., Lavy, J., Monnier, G., and Montagne, X., "Speciated Hydrocarbon Analysis: A Helpful Tool for Two-Stroke Engine Development," *A New Generation of Two-Stroke Engines for the Future?* (ed. P. Duret), Proceedings of the International Seminar, Nov. 29–30. Rueil-Malmaison, France, Editions Technip, Paris, 1993.

29. MAN and B&W, *Emission Control of Two-Stroke Low-Speed Diesel Engines*, Publication P. 331–96.12, MAN and B&W Diesel A/S, Copenhagen, Denmark, Dec. 1996.

30. Douville, B., Ouellette, P., Touchette, A., and Ursu, B., "Performance and Emissions of a Two-Stroke Engine Fueled Using High-Pressure Direct Injection of Natural Gas," SAE Paper 981160, 1998.

31. Nomura, K., and Nakamura, N., "Development of a New Two-Stroke Engine with Poppet-Valves: Toyota S-2 Engine," *A New Generation of Two-Stroke Engines for the Future?* (ed. P. Duret), Proceedings of the International Seminar, Nov. 29–30. Rueil-Malmaison, France, Editions Technip, Paris, 1993.

32. Abthoff, J., Duvinage, F., Hardt, T., Kramer, M., and Paule, M., "The 2-Stroke DI-Diesel Engine with Common Rail Injection for Passenger Car Application," SAE Paper 981032, 1998.

33. Knoll, R., "AVL Two-Stroke Diesel Engine," SAE Paper 981038, 1998.

34. Heck, R.M., and Farrauto, R.J., *Catalytic Air Pollution Control: Commercial Technology*, Van Nostrand Reinhold, New York, 1995.

35. Favennec, J., and Prigent, M., "Metal Supported Oxidation Catalyst for Two-Stroke Engines," *A New Generation of Two-Stroke Engines for the Future?* (ed. P. Duret), Proceedings of the International Seminar, Nov. 29–30. Rueil-Malmaison, France, Editions Technip, Paris, 1993.

36. Hori, M., Okumura, A., Goto, H., Horiuchi, M., Jenkins, M., and Tashiro, K., "Development of New Selective NO_x Reduction Catalyst for Gasoline Leanburn Engines," SAE Paper 972850, 1997.

37. Brogan, M.S., Clark, A.D., and Brisley, R.J., "Recent Progress in NO_x Trap Technology," SAE Paper 980933, 1998.

38. Takada, T., Hirayama, H., Itoh, T., and Yaegashi, T., "Study of Divided Converter Catalytic System Satisfying Quick Warm Up and High Heat Resistance," SAE Paper 960797, 1996.

39. Wu, H-C., Yang, S-M., Wang, A., and Kao, H-C., "Emission Control Technologies for 50 and 125 cc Motorcycles in Taiwan," SAE Paper 980938, 1998.

40. Smith, D.A., and Ahern, S.R., "Developments in the Orbital Ultra Low Emissions Vehicle," *A New Generation of Two-Stroke Engines for the Future?* (ed. P. Duret), Proceedings of the International Seminar, Nov. 29–30. Rueil-Malmaison, France, Editions Technip, Paris, 1993.

41. Henningsen, S., "Air Pollution from Large Two-Stroke Diesel Engines and Technologies to Control It," *Handbook of Air Pollution from Internal Combustion Engines*, (ed. E. Sher), pp. 477–534, Academic Press, New York, 1998.

42. Kawanami, M., Okumura, A., Horiuchi, M., Schäfer-Sindlinger, A., and Zerafa, K., "Advanced Catalyst Studies of Diesel NO_x Reduction for Heavy-Duty Diesel Trucks," SAE Paper 961129, *SAE Trans.*, vol. 105, 1996.

43. Peters, A., Langer, H-J., Joki, B., Müller, W., Klein, H., and Ostgathe, K., "Catalytic NO_x Reduction on a Passenger Car Diesel Common Rail Engine," SAE Paper 980191, 1998.

44. Annand, W.J.D., and Roe, G.E., *Gas Flow in the Internal Combustion Engine*, G.T. Foulis, Yeovil, UK, 1994.

45. Bosch, *Automotive Handbook*, 3rd ed., Robert Bosch GmbH, 1993. (Distributed by SAE, Warrendale, PA.)

46. MAN B&W, *Engines and the Environment—Noise,* MAN B&W Diesel A/S, Copenhagen, Denmark, 1989.

47. Roe, G.E., "An Empirical Approach to Motorcycle Silencing" SAE Paper 770188, 1977. (In *Two-Stroke Cycle Spark-Ignition Engines*, PT-26, SAE, Warrendale, PA, 1982.)

48. Gotoh, T., and Maeda, O., "Reduction of Disagreeable Idle Sound in Two-Stroke Engines," SAE Paper 930981, 1993.

FRICTION, LUBRICATION, AND WEAR

8-1 BASIC LUBRICATION ISSUES

In most simple crankcase-scavenged two-stroke engine designs, lubrication is carried out on a once-through-the-engine, *total-loss* basis. After lubricating the crankshaft bearings, connecting rod bearings, and piston–ring–liner interfaces, much of the lubricant burns in the combustion process and, burned or unburned, has to be removed through the exhaust system. The engine-out emissions of burned and unburned oil are among the shortcomings of this type of two-stroke engine and are evident as a blue opaque exhaust smoke. For low-cost applications, the oil is mixed with the gasoline with a mixing ratio of between 1:25 and 1:100. Higher oil levels result in a rapid increase in blue smoke emissions. A low mixing ratio is obviously desirable. The most sensitive lubrication area in a well-developed two-stroke engine of this type is not the piston–ring–liner area, but the big-end bearing of the connecting rod. With total-loss lubrication systems, and carburetted or port-fuel-injected engines, the short-circuiting gasoline (up to about 25%) carries an equal fraction of the oil (unburned) with it into the exhaust.[1]

As a consequence, motorcycle, scooter, and outboard engines often use separate oil-pumping systems. The oil pump is driven by the crankshaft so engine speed is an oil metering parameter. Usually throttle position is another input parameter, and by varying the stroke of the oil-pump piston an oil feed proportional to load can be obtained. High load and high speed require the greatest quantity of oil. With these systems the oil gasoline ratio can be reduced to 1:200, or less.[1] Direct-injection SI two-stroke engines obviously require this type of oil supply system. Even so, oil consumption rates in two-stroke SI engines are higher by a factor of 2–3 or more than oil consumption rates in modern four-stroke cycle engines. Valved two-stroke cycle engines, with an external scavenging air blower and a conventional "wet" sump, can be lubricated like their four-stroke equivalents.

When catalysts are used in the exhaust system, total-loss lubrication systems can result in oil being deposited on the catalyst. Oils and additive packages are being developed with known catalyst poisons eliminated and with good inherent lubrication properties.

For engine designs in which one or more ports are controlled by the piston motion, the piston skirt has to be much longer than that in the four-stroke engine and the piston is therefore heavier. Piston weight influences friction, and the losses due to friction between the cylinder walls and the piston skirt are more pronounced. Also, the piston rings and ports must be designed so the ring ends will not get caught in the ports. In engines with piston-controlled exhaust ports, the piston crown not only is heated by the outrushing exhaust gases, but also is in contact with a relatively hot part of the cylinder liner. For these reasons, and the higher firing frequency (once per revolution) of the two-stroke cycle engine, piston and ring temperatures are higher than in four-stroke cycle engines. To avoid the ends of the rings moving over the port, the piston rings are often pinned–held in the same position in the grooves, and there is less opportunity for the lubricating oil to penetrate the grooves and disperse any carbon that forms there. Therefore, there is a greater tendency for the rings to stick. Usually, the exhaust ports extend only partially around the cylinder periphery, so the heating will be localized, causing uneven heat flow and the potential for ring sticking in the hottest areas. In contrast, four-stroke engine pistons benefit from the removal of the pressure loading on the rings and the piston pin bearing during the exhaust and intake strokes. Because of these differences in operating conditions, the successful design of two-stroke cycle pistons is a complex process.

In larger automotive and marine engines, more sophisticated lubrication approaches are used. In prototype automobile crankcase-scavenged SI engines, greased main crankshaft roller bearings are used so a dry sump can be employed. In two-stroke diesels, pressurized-oil-fed crankshaft lubrication systems are utilized. In the larger marine engines, high-pressure oil is also used to lubricate the crosshead bearings and the cylinder liner through several strategically located oil-fed quills or slots with oil accumulators to ensure sufficient oil flow at all operating speeds. An air-flow-carried oil mist is used to lubricate the exhaust valve stem guide. Oil from the high-pressure lubrication system is used in a hydraulic "pushrod" (pipe) to actuate the exhaust valve. Oil is also often used to cool the piston in these large engines.[2]

Compared with the four-stroke cycle engine, the two-stroke engine is more susceptible to deposit formation in ports and the exhaust, formation of ash-containing deposits in the combustion chamber, piston wear, and corrosion. Oil detergency in the critical piston–ring region, and lubricant and fuel miscibility, are especially important issues also.

8-2 FRICTION LOSSES IN TWO-STROKE CYCLE ENGINES

8-2-1 Definitions and Magnitudes

The friction work per cycle or power is defined as the difference between the indicated work per cycle or power delivered to the piston from the working fluid contained within the cylinder, and the usable (brake) work per cycle or power delivered to the drive shaft: see Section 1-3-2. The friction work or power consists of the work or power needed (a) to push out the burnt gas and pull in the fresh mixture (pumping work or power), (b) to overcome the resistance to relative motion between adjacent components within the engine (mechanical or rubbing friction work or power), and (c) to drive the engine accessories. In four-stroke cycle spark-ignition engines, the pumping work or power varies most strongly with load and the rubbing power with engine speed. At higher loads, the friction from the piston and connecting rod assembly is the largest component of friction.[3] As the engine is increasingly throttled to lighter loads, the pumping power becomes comparable to the rubbing friction. In crankcase-scavenged two-stroke engines, however, the pumping work or power is due to crankcase compression and is a function of both engine load and engine speed. This results from the unsteady nature of the scavenging process, which may involve, depending on conditions, backflows through the scavenge or intake ports. When an external blower is used to compress the scavenging air, the work per cycle or power required to drive the blower is the pumping component (or it can be thought of as an accessory load).

The friction power is one of the most significant irreversible losses in small high-speed spark-ignition engines. An extensive experimental investigation of the friction in four-stroke motorcycle engines[4] showed that the friction under operating conditions close to maximum power reduces the brake output about 40% below the engine's indicated output. Friction measurements in a small aircraft two-stroke engine,[5] at high speed and load, showed that the power loss due to friction was 43%. The importance of reducing the friction losses in two-stroke cycle engines is evident. A fundamental difference in the importance of friction between two-stroke and four-stroke cycle engines is the number of crankshaft revolutions per complete engine cycle: the two-stroke only has to overcome one revolution of friction for every combustion event and power stroke, the four-stroke must overcome friction for two revolutions.

Figure 8-1 shows how the total friction mean effective pressure (tfmep) of a crankcase-scavenged two-stroke spark-ignition engine depends on engine speed, for the two extreme throttle positions, wide-open throttle (WOT) and idle.[5] From Eq. (1-9),

$$W_{c,f} = W_{c,i} - W_{c,b} \quad \text{or}$$

$$P_f = P_i - P_b \tag{8-1}$$

and

$$W_{c,f} = W_{c,p} + W_{c,rf} + W_{c,a} \quad \text{or}$$

$$P_f = P_p + P_{rf} + P_a \tag{8-2}$$

$$\text{tfmep} = \frac{W_{c,f}}{V_d} = \frac{P_f}{V_d N} \tag{8-3}$$

The tfmep of a two-stroke cycle engine differs from that of its four-stroke counterpart especially at higher load and lower engine speed conditions. As shown in Fig. 8-1, tfmep

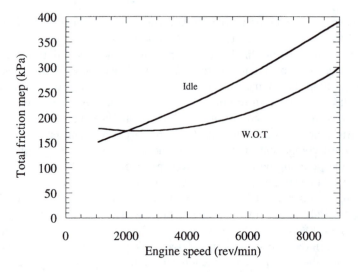

Figure 8-1 Total friction mean effective pressure (tfmep) of a crankcase-scavenged two-stroke cycle engine versus engine speed, for wide-open-throttle and idle operation.[5]

at WOT remains relatively high as engine speed decreases. In four-stroke cycle engines, WOT tfmep decreases to about one-third of its maximum power level at low speeds. In four-stroke engines, the pumping work results from the pressure difference between the in-cylinder gases and the atmosphere, over first the exhaust stroke and then the intake stroke. As engine speed decreases at WOT, the work needed to expel the exhaust gases goes down rapidly because the slower movement of the piston allows the blowdown process to be completed between exhaust valve opening and BC; thus the pressure inside the cylinder falls to close to ambient before the commencement of the exhaust stroke. During both the exhaust and the intake strokes, the pressure drops across the exhaust and the intake systems scale with the square of engine speed. In two-stroke cycle engines, the pumping work is primarily the work needed to compress the fresh charge in the crankcase. At WOT, this work increases when the engine speed declines because the slower motion of the piston allows fresh charge to fill the crankcase volume more fully before compression and, thus, the mass of fresh charge that enters the crankcase and is then compressed increases.

Because losses due to flow restrictions in the intake manifold in two-stroke cycle engines are small compared with the crankcase compression work, the pumping losses usually decrease at partial loads, whereas they increase in four-stroke engines. Figure 8-2a compares the crankcase pumping diagram (crankcase pressure versus crankcase volume) for a 1.2-liter three-cylinder modern two-stroke SI engine with the pumping loop (cylinder pressure versus cylinder volume over the exhaust and intake strokes) of a comparable-output four-stroke engine at 1900 rev/min and one-quarter load. The substantial difference in pumping work is evident. Figure 8-2b shows how pumping mean effective pressure varies with load for these two different types of SI engine.

It follows from this that the mechanical efficiency at wide-open-throttle conditions of a two-stroke cycle engine will exhibit an increasing–decreasing behavior with engine

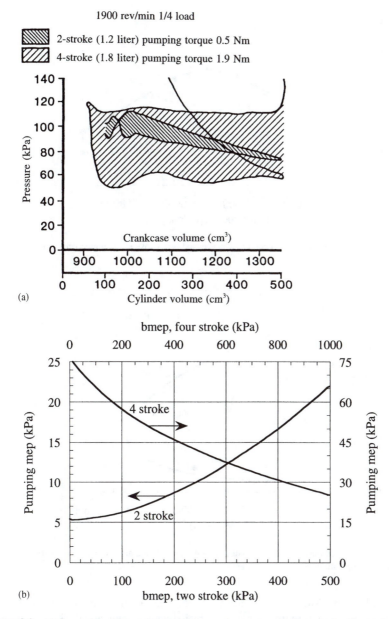

Figure 8-2 (a) Comparison of pumping work for a crankcase-scavenged two-stroke direct-injection SI engine (crankcase pressure versus volume) and for a standard four-stroke SI engine (cylinder pressure versus volume). Courtesy Orbital Engine Corp. Ltd. (b) Pumping mean effective pressure as a function of load (bmep) for crankcase-scavenged two-stroke SI engine and four-stroke cycle engine.[6]

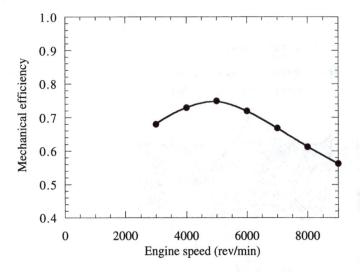

Figure 8-3 Mechanical efficiency of a crankcase-scavenged two-stroke cycle engine versus engine speed at wide-open throttle.[5]

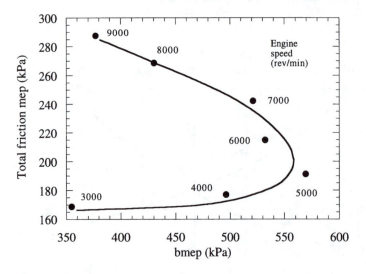

Figure 8-4 Total friction mean effective pressure (tfmep) versus brake mean effective pressure (bmep), for wide-open-throttle operation of two-stroke SI engine.[5]

speed (see Fig. 8-3). Mechanical efficiency is defined as the ratio between the brake mean effective pressure (bmep), as measured by a dynamometer, and the indicated mean effective pressure (imep):

$$\eta_m = \frac{\text{bmep}}{\text{imep}} = 1 - \frac{\text{tfmep}}{\text{imep}} \tag{8-4}$$

In contrast the mechanical efficiency of a four-stroke SI engine at WOT is highest at low

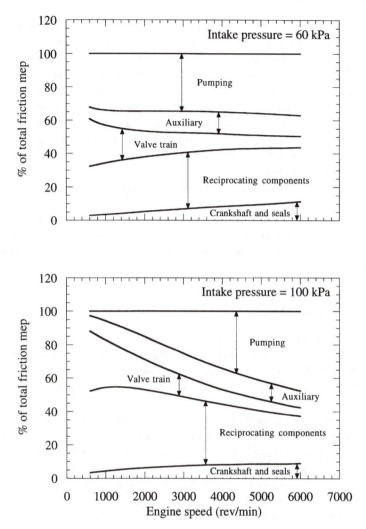

Figure 8-5 Effect of load (expressed as intake manifold pressure) and speed on the distribution of friction between major engine components for a four-stroke cycle spark-ignition engine.[8]

speed (∼0.9) and falls increasingly as speed rises to about 0.7 at maximum power. This increasing–decreasing mechanical efficiency behavior may also explain the convex-type behavior of the bmep curve typical of a simple crankcase-scavenged two-stroke engine.[7] Figure 8-4 shows how the curvature of the bmep curve relates to the variation in total friction mep. It clearly demonstrates the importance of operating a two-stroke engine as close as possible to its optimal working point.

There is a lack of detailed information in the open literature as to how the various components of the total friction losses of a two-stroke cycle engine depend on engine operating conditions. Also lacking are data on the relative contribution of each major friction component. It is, however, worthwhile reviewing the data available on how the relative contributions of individual friction components to tfmep vary with engine speed and load in four-stroke cycle engines. Figure 8-5 shows that the largest portion of tfmep at low and medium speeds is contributed by the reciprocating components—piston, rings, and connecting rod, some 40–50% of the total.[8] At higher speeds, although the reciprocating component remains large, the biggest component is pumping losses, again some 40–50% of the total. Auxiliary component losses are nearly constant at about 10% of total friction. Intake pressure has a major effect on the relative importance of the component friction contributions, primarily because changes in intake pressure and speed result in large changes in pumping losses.[8]

In two-stroke cycle engines, the crankcase (or blower) pumping mep is much lower than the four-stroke pumping work and decreases as the engine is throttled (load is reduced) rather than increasing as with the four-stroke engine. Also, with ported two-stroke engines, the valve-train component of friction is absent and the oil control ring (the bottom ring in a three-ring four-stroke engine ring pack) is not needed so its friction is absent, also.

8-2-2 Measurement Methods

A true measurement of friction in a firing engine can only be obtained by subtracting the brake power from the indicated power, determined from accurate measurements of cylinder pressure throughout the cycle. However, because of cylinder-to-cylinder differences in indicated power, and due to the difficulties in obtaining sufficiently accurate pressure data, this method is not easy to use.

The most commonly used method for evaluating friction power is the direct motoring test. The engine is motored by an external means under conditions as close as possible to those that pertain when firing. To maintain engine component temperatures close to normal values, the engine is usually switched rapidly from firing to motoring conditions. The power required to motor the engine is assumed to be close to the friction power required under firing conditions. The friction power measured by this method differs from the true (firing) friction power for the following reasons:

1. The piston is subjected to lower in-cylinder pressures during the expansion stroke. The connecting rod and crankshaft bearing loads are therefore lower, as are the transverse force exerted on the piston and the gas pressures behind the rings. The rubbing friction is therefore lower.
2. The lower temperatures of the piston, rings, and liner under motored operation result in a higher lubricant viscosity and, therefore, higher rubbing friction.
3. There is no blowdown process following exhaust valve opening and the exhaust gases have higher density. Thus the gas exchange process will be different from that under firing conditions. In four-stroke cycle spark-ignition engines, the pumping work under motored conditions is higher than under fired conditions, at the same throttle position. In two-stroke cycle engines, the net overall pumping work effect is not obvious.

How good an approximation motored engine friction data are to firing engine data is unclear, because accurate firing friction measurements are difficult and rarely done. Recent modeling studies by Patton et al.[8] of spark-ignition engine friction suggest that the net difference between motored and fired engine friction is relatively small. In four-stroke cycle engines, the three effects just summarized approximately cancel out. The extent to which this is also true for two-stroke cycle engines is unclear. Motored engine friction tests are widely used as a straightforward way to estimate actual engine friction. This technique is also used as the engine is disassembled in stages to estimate individual component engine friction contributions (e.g., the valve-train contribution can be estimated by motoring the engine after removing the cylinder head).

Recently, a modified version of the direct motoring method, the inertia method, has been proposed.[5] This method removes the need for an external motor. The engine is coupled to a flywheel having a known high moment of inertia, fired, and accelerated to the desired engine speed. The engine is kept running until steady conditions are attained. These include the cooling-water (if applicable), oil, and cylinder-head temperatures. Then, the ignition system is switched off and the rotational deceleration of the integrated system is measured. The total friction mean effective pressure at the desired engine speed (tfmep) may then be evaluated from the following relationship:

$$\text{tfmep} = \frac{2\pi T_f}{V_d} \qquad (8\text{-}5)$$

where V_d is the engine's displaced volume and T_f, the total friction torque, is determined from

$$T_f = I\alpha = I(d\omega/dt) \qquad (8\text{-}6)$$

Here I is the moment of inertia of the engine-plus-flywheel system, α is the angular crankshaft deceleration at the desired engine speed, and ω is the angular speed. Harari and Sher[5] have shown that, when compared with the friction losses determined by subtracting the brake power from the indicated power, the inertia method provides values within $\pm 3\%$ over the entire engine operating range.

8-3 MODELS FOR FRICTION LOSSES IN ENGINES

8-3-1 Friction Fundamentals

Engine friction losses can be classified into two categories depending on the type of dissipation that occurs. One category is friction between two surfaces in relative motion with a lubricant in between. The other category is viscous (turbulent) dissipation in fluid flows through parts of, or over the outside of, the engine.

In engines, friction between two lubricated surfaces in relative motion occurs in bearings and seals (crankshaft, connecting rods), at the piston-ring–liner and piston-skirt–liner interfaces, between the rings and the ring grooves, and in the valve train. The magnitude of this mechanical or rubbing friction is defined by a friction coefficient, f:

$$f = \frac{\text{tangential (friction) force}}{\text{normal (loading) force}} \qquad (8\text{-}7)$$

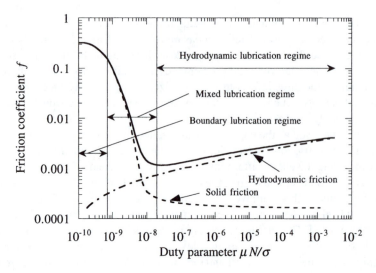

Figure 8-6 Coefficient of friction in the different lubrication regimes on a Stribeck diagram for a journal bearing. The duty parameter (the Sommerfield number) is $\mu N/\sigma$, an inverse loading parameter.

The friction coefficient depends on the properties of the surfaces (e.g., material properties, surface roughness), the lubricant (primarily through its viscosity), the normal force per unit area or the loading, and the relative motion of the two surfaces. When the two surfaces are in direct physical contact, the friction coefficient is high; this is called boundary lubrication. When the lubricant effectively separates the two surfaces in relative motion, the friction coefficient is much lower because direct physical contact is prevented. This is called the hydrodynamic lubrication regime. These different regimes of friction depend on a duty parameter, sometimes called the Sommerfield number, $\mu N/\sigma$ or $\mu U/(\sigma b)$, where μ is the lubricant kinematic viscosity, the relative sliding velocity is expressed as a rotational or linear speed N or U, and σ is the normal loading force per unit area. The length b required to make the second expression dimensionless is the length of the sliding surface in the direction of motion.

Figure 8-6 shows how the friction coefficient varies with the duty parameter $\mu N/\sigma$ as boundary lubrication (on the left) transitions through a mixed lubrication regime to hydrodynamic lubrication (on the right). This lubrication map is often called the Stribeck diagram. Under boundary lubrication conditions, f_{bl} is high and constant, depending on material and surface properties. In the hydrodynamic lubrication regime, f_{hl} is proportional to speed. In the mixed lubrication regime, the friction coefficient is given by

$$f = \alpha f_{bl} + (1 - \alpha) f_{hl} \tag{8-8}$$

where α is the surface-to-surface contact parameter, varying between unity and zero.

8-3-2 Total Engine Friction Models

Friction losses in internal combustion engines comprise a number of components, each influenced by the basic design and operating parameters of the engine.[3,8] A number of friction models has been developed for four-stroke cycle spark-ignition engines.[8-12] For two-stroke cycle engines, however, to the authors' knowledge no detailed friction model has been published in the open literature. Two historical expressions for estimating total engine friction losses have been proposed,[13,14] but these may not be accurate for modern engine designs. The two-stroke cycle engine is sufficiently different from its four-stroke cycle counterpart, both in design and operating characteristics (see Section 8-2), that the same friction model cannot be used for these two types of engine.

In any type of mechanical system, such as internal combustion engines, the friction components may be divided into three main groups where the mechanism of friction is fundamentally different in character[3]:

1. Friction components that are independent of speed (boundary lubrication), that is, where the friction between the two surfaces in relative motion is determined primarily by surface characteristics and the lubricant additive properties: This type of friction occurs when the duty parameter is very small (of order of 10^{-9}), when the lubricating film is reduced to a few molecular layers and cannot prevent metal-to-metal contact between surface asperities. Boundary lubrication occurs between engine parts during starting, stopping, and normal running at the piston-ring–cylinder-liner interfaces around top and bottom center.
2. Friction components that are proportional to speed (hydrodynamic lubrication), that is, where the friction is due to the viscosity of the lubricant: This type of friction occurs when the duty parameter is higher than 10^{-8}, when there is sufficient pressure in the lubricant film to keep the two surfaces separated, and the resistance to motion results from shear forces within this liquid film. Hydrodynamic lubrication occurs in engine bearings during normal operation (but not during starting and stopping) and at the piston-ring–cylinder-liner interfaces away from top and bottom center.
3. Friction components that are proportional to the square of the speed: This type of friction loss results from the turbulent dissipation that occurs when fluids are pumped through flow restrictions; that is, the pumping of cooling water through the cylinder block and head and the external coolant system, pumping oil through the engine lubrication system, and the fan pumping air over the engine exterior and radiator.

Total friction mean effective pressure (tfmep) data for an engine, can generally be correlated with an equation that includes a term for each of the three types of friction losses discussed above: i.e., a polynomial expression of the form

$$\text{tfmep} = a + b\left(\frac{N}{1000}\right) + c\left(\frac{N}{1000}\right)^2 \tag{8-9}$$

where N is engine speed and a, b, and c are calibrated constants. Some appropriate values for two-stroke and four-stroke cycle spark-ignition engines are given in Table 8-1, for tfmep in kPa and N in rev/min.

Table 8-1 Values for constants in total friction model for spark-ignition engines (Eq. 8-9)

Engine Volume (cm^3)	Engine Type s-stroke	No. of Cylinders	Engine Load	a	b	c	Reference
845–2000	4-s	4	WOT	97	15	5	1
4250	4-s	8	WOT	166	–2.5	3.5	3
350	2-s	2	WOT	190	–20	3.6	3
350	2-s	2	Idle	130	17	1.2	3

Based upon an analysis of an extensive number of small four-stroke cycle spark-ignition engines, and an engine speed range of up to 16,000 rev/min, Yagi et al.[4] have suggested the following:

1. That tfmep is proportional to a dimensionless coefficient which includes the engine bore B, a mean equivalent crank diameter d_c, and the piston stroke L, as follows:

$$\text{tfmep} \propto \sqrt{L d_c B} \qquad (8\text{-}10)$$

2. The constant a depends on the square of the ratio between the engine displaced volume and the effective valve open area, the constant b is zero, and the constant c depends linearly on the lubricant viscosity.

The engine friction data for two-stroke cycle engines are insufficient to validate these engine friction models for two-stroke engines. The available experience with two-stroke-cycle engines, however, does show that measured tfmep values for a particular engine may be correlated successfully with a polynomial expression of the form of Eq. (8-9).

8-4 LUBRICANTS

8-4-1 Basic Lubricant Properties

A lubricant is the agent that separates two solid surfaces that are moving relative to one another, which would otherwise come into contact. The main role of a lubricant is to prevent this contact, thereby reducing friction and wear. In addition, the lubricant may act as a sealant and as a cooling agent.

The property *viscosity* is a measure of the resistance to flow of a fluid when subjected to a shear force. The viscosity of a liquid depends on the size, shape, and chemical nature of its molecules and on the magnitude of intermolecular forces. For a series of similar compounds, the viscosity increases with the molecular mass. In liquids, the resistance arises mainly from cohesive effects and therefore the viscosity decreases with increase in temperature, and increases with increase of applied pressure. It is also known that if a lubricating oil is enriched by small amounts of high-molecular-weight organic polymers, the change in its viscosity with temperature, especially at high temperatures, decreases. Organic polymers with molecular weights of 10^3–10^4 that are used for this purpose are called viscosity index improvers. The viscosity of a two-stroke cycle engine lubricant must not exceed a prescribed value under cold ambient conditions, nor fall below the minimum allowed under full load conditions.

The most extensively used classification of engine lubricants is the SAE classification[15] (see Table 8-2). It is based on the viscosity of the oil under a defined set of operating conditions. SAE numbers followed by W (abbreviation for winter) refer to oils for use in cold climates (between $-40°C$ and $0°C$). SAE numbers without a W apply to engine oils commonly used under warmer conditions. Viscosity grades with the letter W are based on maximum low-temperature cranking and pumping viscosities, as well as a minimum viscosity at 100°C. Grades without the letter W are based on the viscosity at 100°C only. The pumping viscosity is a measure of an oil's ability to flow to the engine oil pump and provide adequate oil pressure during the initial stages of operation. Because engine pumping, cranking, and starting are all important at low temperatures, the selection of an oil for winter operation should consider both the viscosity required for successful oil flow and that for cranking and starting, at the lowest ambient temperature expected. A multiviscosity graded oil is one whose low-temperature viscosity satisfies the requirements for one of the W grades, and whose 100°C viscosity is within the prescribed range of one of the non-W grade classifications. Increasing numbers correspond to increasing viscosity. In two-stroke cycle engines where the oil is premixed and

Table 8-2 SAE viscosity grades for engine oils[15]

SAE viscosity grade	Low-temperature (°C) max viscosity (cP) at cranking	Low-temperature (°C) max viscosity (cP) at pumping	Minimum viscosity (cSt) at 100°C	Maximum viscosity (cSt) at 100°C
0W	3250 at –30	30,000 at –35	3.8	—
5W	3500 at –25	30,000 at –30	3.8	—
10W	3500 at –20	30,000 at –25	4.1	—
15W	3500 at –15	30,000 at –20	5.6	—
20W	4500 at –10	30,000 at –15	5.6	—
25W	6000 at –5	30,000 at –10	9.3	—
20	—	—	5.6	9.3
30	—	—	9.3	12.5
40	—	—	12.5	16.3
50	—	—	16.3	21.9
60	—	—	21.9	26.1

Note: 1 cP = 1 mPa-s; 1 cSt = 1 mm^2/s; for an oil density of 0.7 g/cm^3, 1 mPa-s = 0.7 cSt.

Viscosity (cP)

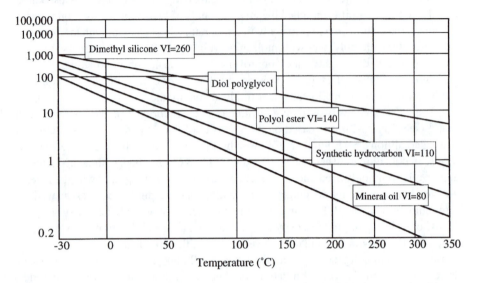

Figure 8-7 Variation of viscosity with temperature for various oils. This figure has been reproduced from a book published by Professional Engineering Publishing titled *High Temperature Lubrication* by A.R. Lansdown 1994, Fig. 11, p. 43, by permission of the Council of the Institution of Mechanical Engineers, London.[16]

(diluted) with gasoline, the viscosity grades of such oils are determined by their viscosity prior to dilution.

Viscosity is a measure of the resistance to flow. The viscosity index indicates to what extent the flow resistance changes as a result of change in temperature. A large change means a low viscosity index. The viscosity index compares the change of viscosity with temperature of the oil under discussion with that of an oil from a U.S. Gulf coast petroleum (VI = 0) and a U.S. Pennsylvania petroleum (VI = 100). The viscosity index of conventional highly-refined paraffinic oils lies generally in the range of 80–110. The actual viscosity depends on the average molecular weight of the various hydrocarbons present and on the range of molecular weights.[16] Figure 8-7 shows the variation of viscosity with temperature for various oils.[16]

Both mineral oil and synthetic-based fluids are in current use. These usually contain one or more additives that advantageously change certain important properties. Mineral-oil-based formulations normally contain brightstoke (approximately 10%) to provide proper wear protection. To avoid coke and deposit formation, the amount and quality of brightstoke must be carefully controlled. A point of growing importance is biodegradability.[17] It is well known that aleochemical esters, derived from natural oils and fats, are resistant to biodegradation processes. Synthetic liquid lubricants, such as diesters,[17] polyol-esters, polyglycols, and vegetable oils, such as jojoba,[18] are being developed as a basis for clean and environmentally benign lubricant formulations.

Additives are substances added to lubricants to improve their properties. Commonly used additives and their functions are summarized in Table 8-3. The additives

Table 8-3 Major lubricant additives, their functions and importance in internal combustion engines (extracted from O'Connor et al.[20])

Additive	Function	Importance	Example
Chemical characteristic improvers:			
Antioxidants and anti-corrosives	Retard oxidation, decrease varnish and extend oil life; reduce corrosion by adding compounds that have polar molecules that attach themselves to the surface	Lubricant oil, bearings	Zinc dithiophosphates P_2S_5 olefin or terpene products, sulfurized olefins
Detergents and dispersants	Maintain clean engine surface by keeping insoluble particles in suspension and dispersed	Rings, crankcase oil	Salts of phenolic derivatives, synthetic sulfonic acids
Antirusts	Eliminate rusting in presence of water or moisture	Fuel tank, crankcase oil	Metallic soaps, esters, ethers, and derivatives of dibasic acids
Extreme pressure (EP) agents	Increase film strength and load capacity for boundary lubrication; react with surface to form surface compound having lower shear strength	Where oil film is exceedingly thin	Chlorinated paraffin wax, chlorinated paraffin wax sulfides, chlorinated paraffinic oils
Antiwears	Protect rubbing surfaces operating with thin films; boundary lubrication	Rings	Zinc dithiophosphates
Alkaline agents	Neutralize acid from oxidation of oil so it cannot react with oil or cylinder surfaces	Upper cylinder bore	Metal salts of sulfonic acids, posphonic acids, phenols
Oiliness agents	Film-strength additive for smoothness and slipperiness; reduce friction, seizure, and wear	Where oil film is exceedingly thin	Synthetic esters of fatty acids, methyl esters, esters of phosphoric acids
Metal deactivators	Pacify, prevent, or counteract catalytic effect of metals; prevent metal ions from going into solution	Cylinder and piston surfaces, fuel tank	Organic dihydroxyphosphines, trialkyl and triaryl phosphites

Table 8-3 (Continued.)

Additive	Function	Importance	Example
Physical characteristic improvers:			
Pour-point depressants	Lower low-temperature fluidity by retarding the formation of wax crystals	Lubricant oil at low temperatures	Wax alkylate naphthalene, wax alkylated phenol, polymers of methacrylic acid esters
Viscosity index (VI) improvers	Lower rate of change of viscosity with temperature change by controlling molecular weight	Lubricant oil at high temperatures	Isobutylene polymers, acrylate copolymers
Foam inhibitors	Prevent formation of stable foam	Crankcase oil	Dimethyl silicone polymers
Tackiness agents	For greater cohesion, nondrip property	Bearings	Acrylates, polybutenes
Solid lubricants	Withstand high temperatures and/or pressures	Effective lubricant oil at very high temperatures	Graphite, molybdenum disulfide
Antiseptics	Prevent emulsion breakdown or odor from growth of bacteria	Fuel tank	Certain alcohols, phenols, organic mercuric compounds

in a specific additive package also have to be selected so that no undesirable reactions between the various lubricant components will occur. Antioxidant additives are one of the most important engine oil additives. An antioxidant strengthens the bonds between the carbon and hydrogen atoms of the oil components, thus preventing combination with oxygen.[19] At high temperatures, oil absorbs oxygen, causing new and different oxidation products to form. One of these is an acid that remains in solution in the oil and attacks metals, especially bearing metals such as cadmium, lead, nickel, and silver alloys. Antioxidants also deactivate the catalytic action of metal particles and thus weaken the deleterious effects attributed to these particles. In general, low-viscosity oils have a lower resistance to oxidation than do the heavier grades. Corrosion in engines is not only limited to bearing metals. The wire screens that retain the filtering element used in some types of oil filters may become corroded. Damage to bearing, crankshaft, and screen metals results if pieces of broken metal are carried along by the oil. The anticorrosion inhibitors act in two main ways: neutralizing acids as and when they form, and forming a protective coating over the surface of metals.[19]

All lubricating oils thicken as temperature decreases. One of the major causes of this thickening is the presence in the oil of unremoved wax and/or asphaltic compounds.[19] As paraffin-based oils cool, the wax crystals separate from the oil and form a matrix that mechanically traps and holds the fluid oil. Pour-point depressants prevent the formation of wax crystals into a matrix and prevent the formation of an adsorption gel.

All lubricants naturally possess the property of oiliness. This can, however, be improved with an additive. An important factor connected with oiliness is the polar nature of certain molecules that become attached to the metal surface as polar chain molecules. The result is a thick coating of "velvety" oiliness particles over the metal surfaces. Being chained together, the oil molecules better withstand the pumping action of pistons and are less likely to flow or be squeezed from between bearing surfaces. Oiliness also provides an explanation of surface wetting. Bearings and rings subjected to frequent stops and starts or to frequent rotation changes, require an oil rich in oiliness. Oiliness agents permit the use of lower viscosity oils. This is beneficial in high-speed bearings because these agents lower fluid friction.[19]

A detergent additive in an oil helps wash away residue from various parts as it is formed. These include ring grooves, piston skirt, under piston, crankcase, filter, and oil-insoluble contaminants. By a continuous washing action, detergents prevent the accumulation of the residue around piston rings. Detergent additives are more effective in oils operating at elevated temperatures. To avoid the residue from gathering into large clots that combine with other substances to form sludge deposits, a dispersion agent is useful. Dispersion agents act by surrounding the residue with a film that causes the residue to float. Detergent and dispersion agents are also employed to remove tarlike gummy deposits from carburetors and piston ring grooves.[19]

The churning and splashing of oil tends to trap and carry air into the main body of the oil. Oil has a greater tendency to foam when moisture is present and when the ambient pressure is low. The action of the antifoam agents is not to prevent bubbles from forming, but to destroy them by weakening the skinlike surface of the bubbles so they collapse more readily when they reach a free surface. Antifoam agents also perform a perforation function that seems to prick the bubbles.

8-4-2 Lubricant Requirements in Two-Stroke Cycle Engines

There are three common approaches to two-stroke cycle engine lubrication[21]:

1. premixing, where the lubricant is added to the fuel either at the gasoline fueling station or in the engine fuel tank;
2. line mixing, where the lubricant is metered into the gasoline as it flows from the fuel tank to the engine;
3. injection, where the lubricant is introduced into the intake manifold or other suitable points using a pump; the amount of lubricant delivered may be controlled and optimized over a wide range of operation conditions; fuel/oil ratios between 20:1 at wide-open throttle and 100:1 at idle, with 50:1 for general use, are typical.

The oil consumption of a two-stroke cycle engine is generally higher than that of an equivalent four-stroke cycle engine. Table 8-4 compares the oil consumption of small industrial two-stroke and four-stroke engines of similar power output. This two-stroke cycle engine at a fuel oil ratio of 100:1 consumes about 3.5 times as much oil as the four-stroke cycle engine. Even if the fuel consumption of the two-stroke cycle engine is reduced to that of its counterpart, the oil consumed is still higher by a factor of 2.2.

Table 8-4 Oil consumption of small industrial four-stroke and two-stroke cycle engines[22]

	Four-stroke cycle engine, 7.5 kW	Two-stroke cycle engine, 7.5 kW	
Fuel consumption (g/h)	2200	3500	
Oil consumption (g/h)	10	35	
Fuel/oil ratio	220:1	100:1	20:1
Specific fuel consumption (g/kW-h)	290	460	460
Specific oil consumption (g/kW-h)	1.3	4.6	23

Table 8-5 Important physical and chemical properties of a lubricant for two-stroke cycle engines (based on SAE[21])

Property	Importance
Miscibility and fluidity	Lubricants must have the ability to mix into gasoline and/or to flow at the prevailing ambient temperature
Rust	To prevent internal engine corrosion during shutdown
Stability and compatibility	To ensure oil homogeneity over a broad range of ambient temperatures for extended periods of time
Pour point	To ensure adequate dispensability at lower ambient temperatures
Solvent content	To ensure oil homogeneity in terms of liquid and highly volatile components
Ash content	Ash-forming lubricants are important for some air-cooled engines, where high temperatures are essentially involved.
Flash point	A safety measure that determines the flash point of solvent-diluted lubricants
Biodegradability	To enhance environmentally friendly oils
Color	To distinguish the specific oil from other different-purpose oils; blue and green are common

In total-loss or once-through lubrication systems, the lubricant is expected to protect the engine in the specified operating temperature range from wear, scuffing, ring sticking, piston deposits, and rust. At the same time, preignition, excessive plug fouling, knock, and exhaust system blockage must be avoided. Because the lubrication system is on a once-through basis, a two-stroke engine oil need not be oxidation inhibited, hold insolubles in suspension, nor resist viscosity breakdown in service. Because of the high oil consumption, however, special attention must be paid to base oil quality and to the types of additive used to reduce carbon formation in the combustion chamber and exhaust ports. The more important physical and chemical properties of a lubricant for two-stroke cycle engines are summarized in Table 8-5.

Most oils for two-stroke engines are formulated from solvent-refined high-viscosity-index base oil. They generally contain a proportion of high-viscosity base oil of high molecular weight that provides good protection against piston seizure and scuffing. Too high a proportion, however, may result in excessive deposits in the combustion chamber and exhaust.[22] Lubricants with viscosities between SAE 10W and SAE 50, selected according to the engine manufacturers' requirements and prevailing ambient temperatures, are in common use. With engines that have a separate lubrication system, the pour point and the viscosity of the lubricant are important factors in formulating the oil blend to ensure satisfactory oil flow from the tank to the pump over the entire range of operating temperatures.

The difference in lubricant temperatures between gasoline and diesel engines is modest. The operating conditions of both engines are determined by the temperature limits that engine materials and lubricants can endure. The following are typical temperatures:[23]

combustion gases, 1600–2000°C;
top piston ring, 200–260°C;
piston skirt, 140–170°C;
small-end bearings, 120–150°C;
big-end bearings, 100–150°C.

Because in two-stroke cycle engines oil is not completely burned in the combustion chamber, one expects to find carbon deposits in the scavenging ports and along the engine exhaust pipe. Most oils contain an additive package to improve engine cleanliness. This package must be essentially ashless to avoid preignition, and short-circuiting of the spark-plug gap (polarized ignition plugs attract ash). However, most lubricants today contain an ash-forming component to control ring-zone deposits at high operating temperatures. Chemically, the ashless additives are often reaction products of polyamines and long-chain fatty acids.[17] The ash-containing additives are mainly used in air-cooled engines where higher component and hence lubricant temperatures are involved. Favored compounds are the calcium sulphonates. Oils that comprise a mixture of ashless and ash-forming components (so-called low-ash additives) are in use in high-performance engines. These offer the best balance between detergency and prevention of ash formation (which can lead to preignition).

Lubricants are rated in several areas with a number of engine tests, to ensure proper engine performance and durability.[15] In each area, numerical limits are given that define whether the lubricant passes or fails. These areas include ring sticking, varnish formation (on the piston skirts, lands, and under the crown), preignition, scuffing, and exhaust system blockage.

8-5 FRICTION AND WEAR

8-5-1 Friction and Wear Regimes

The friction and wear behavior of surfaces in relative motion depends on the materials properties, surface topography, kinematics of the surfaces in contact (i.e., the direction and the magnitude of the relative motion between the surfaces in contact), and environmental conditions.[24] Under low to moderate sliding conditions, the coefficient of friction is independent of the normal load and sliding speed as discussed in Section 8-3-1. Its value for the particular system shown in Fig. 8-6 is about 0.2. Considerable asperity–asperity interaction occurs in this boundary lubrication regime, resulting in higher friction than occurs in the mixed and hydrodynamic lubrication regimes. As the sliding speed increases, the friction is less and less dominated by asperity interaction, and becomes controlled by the viscosity of the lubricant and the relative motion (hydrodynamic lubrication). The friction coefficient decreases substantially and then increases moderately as the duty parameter increases under full hydrodynamic lubrication.

The frictional behavior of materials and how much wear results can be largely explained by adhesion theory. According to this theory, the surface is assumed to consist of asperities and, under boundary and mixed lubrication, the interface is made up of asperity contact. Each pair of contacting asperities is partially welded together (microwelds) and, when a tangential force is applied to move one surface relative to another, shear forces are developed. Rabinowicz[25] has modified the adhesion theory to improve the correlation between experimental observations and theoretical predictions by assuming that the indenters of one material penetrate into the other. Based on this modified theory, he showed that the coefficient of friction is a strong inverse function of the hardness of the material (H) and may be expressed as

$$f \approx \frac{k}{H}\left(1 + K\frac{W_{ab}}{H}\right) \qquad (8\text{-}11)$$

where K is a geometric factor that depends on the geometry of the indentation asperities, W_{ab} is the surface adhesion energy, and k is the critical shear stress of the bulk material.

Under full hydrodynamic lubrication conditions, there is no opportunity for the opposing surface asperities to come into contact, and wear, if it occurs, is usually due to the intrusion of foreign bodies.[26] This is where efficient filtration of the lubricant delivered to the system is of prime importance.

Some parts of an engine, such as the main and big-end bearings, operate in the hydrodynamic lubrication regime. Some other parts, such as pistons, rings, and cylinder liners, alternate back and forth between boundary and hydrodynamic lubrication regimes.

8-5-2 Wear of Sliding and Rolling Surfaces

Lubrication is seldom complete. If it were, no wear would ever occur. There are, however, always regions where the pressure set up within the thin lubricant film is not sufficient to keep the surfaces in relative motion separated, and the surface asperities come in contact. It is common practice to distinguish between "mild" and "severe" wear. Severe wear is normally characterized by relatively low contact resistance, presence of large (about 10^{-2}-mm diameter) metallic particles, and leaves the surfaces deeply torn and rough. Mild wear results when the surfaces have high contact resistance, the debris is small (about 10^{-4}-mm diameter) and has been produced primarily by reaction with the ambient atmosphere or fluid, and generates extremely smooth surfaces.[26]

Figure 8-8 shows a wear mechanism map for an unlubricated steel surface pair of pin-on-disk configuration.[27] Here the speed of sliding is normalized by the thermal

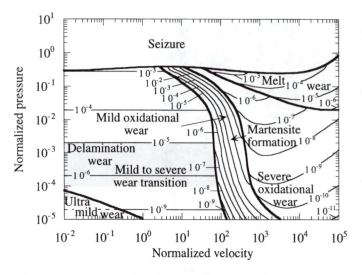

Figure 8-8 Wear mechanism map for unlubricated steel surface pair of pin-on-disk configuration. Reprinted from *Acta Metall.*, vol. 35, Lim, S., and Ashby, F.M., "Wear Mechanism Maps", p. 1, 1987,[27] with permission from Elsevier Science.

Figure 8-9 Scuff marks on the compression rings of a piston.[23]

diffusivity of the disk and the pin radius. The load is normalized by the contact area and the material hardness at room temperature. The wear rate, which is defined as the volume of material removed from the surface per unit sliding distance of one surface with respect to the other, is normalized by the area of contact. The important effects of the speed of sliding and especially the extent of the load on the wear mechanism and wear rate are apparent. The relative load seems to be the major factor leading to seizure. In engines, however, high wear may result in a dramatic increase in local temperatures, degrading the mechanical properties of the materials involved, their surface treatment,

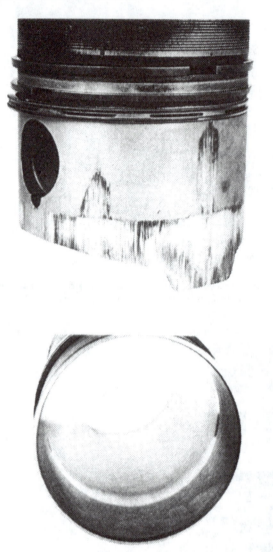

Figure 8-10 Seized piston with accompanying cylinder liner: (a) shows clearly defined seizure area at the lower end of the piston skirt; (b) shows hard contact points on the liner surface.[23]

and the lubricating oil. Under these circumstances, seizure may occur under lower load conditions.

Local fatigue stresses at the sites where solid-to-solid contact occurs due to lack of the lubricant lead to the development of localized systems of cracks, which spread and eventually cause flakes of material to become loosened, leaving pits in one or both of the surfaces in contact.[26] This process is known as *pitting*. Scuffing[16] is a form of severe adhesive wear that results in serious surface damage and often seizure or component failure. Scuffed surfaces have a welded and torn appearance with a directional trend parallel to the cylinder axis (see Fig. 8-9). The surface of the damaged parts looks softened and liquified.

Metallographically, the damaged materials often contain a hard high-carbon eutectic ferrous alloy (ledeburite), which requires temperatures over 1100°C for its formation.[16] Scuffing appears to begin near the top center position of the top ring, where the relative motion between the ring and the cylinder walls momentarily goes to zero. At this location, the oil film thickness may become too thin to prevent asperity contact, and flash temperatures at the contacts may lead to asperity welding followed by breaking of the welds.[16] This repeated welding and tearing process leads to high local friction and wear (scuffing) and, under extreme conditions, to seizure (see Fig. 8-10).

Prevention of scuffing may be achieved by using higher viscosity oil (to a limited extent), using rings with a suitably curved profile to maintain an adequate oil film between the cylinder liner and the ring, adequate cooling of the cylinder head and liners, and a more lightly loaded running-in period. Production of a suitable cylinder-liner surface structure (a very smooth surface finish tends to increase the incidence of scuffing) and using a pair of materials for piston rings and cylinder liners that do not readily adhere or weld together are important strategies for decreasing the occurrence of scuffing in the engine.

Table 8-6 Compatibility chart for various metal combinations (based on Rabinowicz[28])†

<---------- Metals, in order of increasing Young's modulus -------------->

	W	Mo	Cr	Co	Ni	Fe	Pt	Cu	Au	Ag	Al	Zn	Mg	Cd	Sn
Pb	PI	PI	I	I	I	I	C	I	PI	PC	I	I	PC	PC	C
Sn	PC		I	PC	PC	PI	C	PC	C	C	PI	PC	PI	C	
Cd			PC	PC	PI	PI	C	PC	C	C	I	C	C		
Mg		PC		PC	PI	PI		C	C	C	C	PC			
Zn		PC	C	C	C	C	C	C	C	C	C				
Al	C	PC	C	PC	C	C	PC	C	C	C					
Ag	I	PI	I	I	I	I	C	PC	C						
Au	C	PC	C	PC	C	C	C	C							
Cu	PI	I	I	C	C	PC	C								
Pt	C	C	C	C	C	C									
Fe	C	C	C	C	C										
Ni	C	C	C	C											
Co	C	C	C												
Cr	C	C													
Mo	C														

† I, incompatible; PI, partially incompatible; PC, partially compatible; C, compatible; blank boxes indicate insufficient information.

8-5-3 Friction and Wear Compatibility of Materials

The strength of adhesion between two metals is largely a function of their mutual solubility. Certain metal atoms dissolve easily in the crystal structure of other metals. A compatibility chart of possible metal pairs, in terms of preferred low-friction surfaces, is shown in Table 8-6. The chart is based on binary diagrams of the respective elements (an interpretation based on Rabinowicz[28]).

A careful selection of component materials and coatings can minimize friction and wear between the various parts of an engine (piston rings, piston skirt, and cylinder liner). Nickel atoms, for example, are completely soluble in copper, and vice versa. On the other hand, lead and tin for all practical purposes are completely insoluble in chrome and vice versa. The more soluble two metals are in each other, the more metallurgically *compatible*. Metallurgically *incompatible* materials are known as frictional compatible materials, and should be used in applications where wear is an issue.[29,30]

If, due to other design constraints, metallurgically compatible materials must be used, a metallurgically incompatible material can be placed on one of the surfaces by plating, flame spraying, or other suitable method. Cadmium, lead, tin, and silver are commonly used to coat one of the friction pair parts.

8-6 MATERIALS FOR ENGINE COMPONENTS

8-6-1 Piston Materials

The operating regime of the piston requires the following piston material properties[23]:

1. low density, to minimize inertia forces;
2. high thermal conductivity, to facilitate effective cooling;
3. good strength properties at elevated temperatures; high resistance to deformation and fatigue fracture;
4. good wear characteristics at high temperatures; wear of the piston skirt and the pin bosses is usually unimportant, but wear in the ring grooves can limit the service life of the piston;

Table 8-7 Important physical properties of materials used in piston manufacture[23]

	Density $(kg/m^3 \times 10^3)$	Thermal conductivity at $250°C$ (W/m-K)	Thermal expansion $20–200°C$ $(1/K \times 10^{-6})$
Pure aluminum	2.70	210	23.8
Pure silicon	2.33	83	4.2
Pure iron	7.86	67	11.5
Aluminum alloy	2.65–2.70	141–159	19.3–23.1
Cast iron	7.20–7.35	29–50	9–18.5
Steel	7.60–7.80	20–50	11.1–13.1

5. the thermal expansion coefficient should be close to that of the cylinder liner, to limit the running clearance.

Pistons used in spark-ignition engines are mostly made of aluminium alloys. Pistons for diesel engines are made of cast iron, which tends to be converted to wrought iron, or squeeze-cast aluminium alloys with additional measures to improve wear resistance. Several types of aluminum alloys with 8–24% silicon, 1–4% copper, and traces of nickel and magnesium are commonly used for pistons. Table 8-7 presents the important physical properties of materials used in piston manufacture.[23]

8-6-2 Piston Ring Materials

The function of the piston rings is to seal off the piston, which is the movable part of the combustion chamber, and control the amount of lubricant on the cylinder liner. In addition, a substantial part of the heat transferred from the hot in-cylinder gases to the piston passes through the rings to the cooled cylinder liner. The passage of combustion gases from the combustion chamber into the crankcase must be kept below a tolerable small amount (of order 1%), because increased blowby overheats the piston and rings and the lubrication conditions on the cylinder liner are disturbed. The sealing of the piston to prevent gas blowby is done primarily by the top or compression ring. No less important is the creation of a barrier that prevents unacceptable amounts of oil passing into the combustion chamber from the crankcase. The load-bearing surfaces of the piston, rings, and cylinder must be supplied with a quantity of lubricating oil that is as small as possible but also provides adequate lubrication under all operating conditions. This is essentially the task of the oil scraper rings.[23]

The compression ring (or rings) is normally a self-tensioning circular spring which in the initial untensioned or free state has a diameter that is larger than the cylinder bore. In this free state the end gap opens a specific distance; in the installed state the gap is reduced to the end clearance. This elastic deformation creates the spring force and thus the desired outward pressure against the cylinder liner wall. The pressure applied by the ring itself is not uniformly distributed over the entire ring circumference (see Fig. 8-11). This is usually intentional.[23] In operation, the pressure on the liner is considerably increased because of the gas pressure behind the ring. Because the tendency of a ring to flutter usually stems from the ring ends, it is a common practice in four-stroke cycle engines to increase the pressure in the ring end regions through a "pear-shaped" ring pressure pattern. In two-stroke engines, the ends of these rings would get caught in the scavenging or exhaust ports. For two-stroke engines, therefore, the increase in radial pressure is usually shifted away from the ring ends to the left and right, which leads to an "apple-shaped" pressure pattern (see Fig. 8-11b).

The following are requirements for piston ring materials[23]:

1. good running properties: the rings should run-in rapidly in a controlled manner; unavoidable wear should proceed only gradually on the rings and the cylinder after running-in; the ring should have good dry running properties to avoid seizing during a possible temporary lack of lubrication; the friction coefficient should be as low as possible, to reduce power losses;

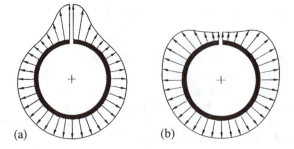

Figure 8-11 Typical radial ring-tension-generated pressure pattern of piston rings for four-stroke cycle engines (left) and two-stroke cycle engines (right).[23]

2. good elasticity and strength over a wide range of operating conditions and long periods of operation;
3. economical to manufacture.

Different kinds of cast iron such as standard gray cast iron and hardened and tempered gray cast iron have become the primary ring material. A typical mean modulus of elasticity is between 115,000 and 160,000 N/m^2, with a bending strength between 350 and 1300 N/m^2.

8-6-3 Engine Block Materials

The most widely used materials for the cylinder block are cast iron and aluminum alloys. The attractive properties of aluminum and its alloys (low density, high strength, high conductivity for heat and electricity, good corrosion properties, and versatile working properties) motivate the constantly expanding application opportunities for these alloys. During the last few decades, light-weight block materials such as high-silicon aluminum alloy 390 and several other modified alloys have been developed and used in commercial vehicles. There are several four-stroke cycle engines using aluminum blocks on the market; however, many of these use cast iron liners. This design enables a significant reduction in weight in combination with matching piston, ring, and liner materials for acceptable wear and scuffing resistance.[23]

8-7 MODIFYING ENGINE COMPONENT SURFACES

8-7-1 Coatings

The tribological behavior of a material is determined largely by the structure and composition of the material surface. When two materials are rubbed against one another, macroscopic properties such as friction and contact resistance are important. These can be modified and greatly improved by coating the material surface with a thin film. For years lubrication engineers have used this approach by blending additives into lubricants that chemically react with sliding surfaces to form thin antiwear films (see Sec-

(a)

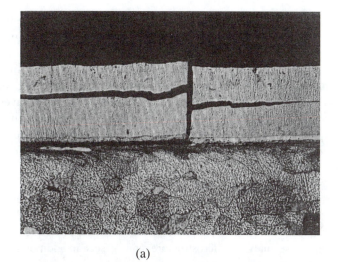

(a)

(b)

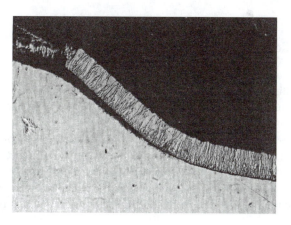

(b)

Figure 8-12 Crack nucleation site in the coating (a) or at the interface (b). Reprinted from *Wear*, vol. 100, Hintermann, H.E., "Adhesion, Friction and Wear of Thin Hard Coatings," pp. 381–397, 1984.[32]

tion 8-4-2). An alternative method for reducing friction and wear is the application of a surface coating to the material before use.

Two approaches are in common use: In the first, the surface is coated with a thin film with superior friction and wear properties, which protects the substrate material from excessive wear. In the second approach, the surface is coated with a solid lubricant that forms a sacrificial film designed to wear continuously during use. It protects the substrate by producing wear debris that provides lubrication in addition to the liquid film. This method has been successfully applied to bearing technology (babbit, an alloy used for lining bearings, is a sacrificial tribological film) and to pistons in gasoline engines (lead or tin is often used to prevent scuffing wear during engine break-in).

A coating provides a near-surface region with properties different from the bulk material. It permits a combination of properties not easily obtainable from a single or monolithic material. Coatings may be divided into two generic classes: diffusion coatings and overlay coatings. In the first, another element is diffused into the surface of the substrate (e.g., carburizing, nitriding, cyaniding, and carbonitriding; chromizing, siliconizing, sulfinuz, and boronizing[31]). In the second, a complete coating layer (or layers) of metals, alloys, refractory compounds, or polymers is deposited on top of the component surface by electroplating, porcelain enamels, flame-sprayed coatings, hard anoding coatings, and hard facings.[31] Good adhesion is secured by mechanical or interdiffusion bonding at the interface.

Coatings are deposited to improve the base material, but can also produce undesirable changes in properties. For instance, increased wear resistance may result in creating potential crack nucleation sites in the coating or at the interface (see Fig. 8-12). Therefore, the coating material and, if necessary, the interface region should be characterized in terms of composition, phase analysis, microstructure, and defects, in addition to the various types of wear tests.[33]

8-7-2 Surface Modifications for Pistons

Different kinds of surface coatings are used for pistons. These can be divided into two categories in accordance with their function. The first category includes coatings that provide protection against thermal overloading, and the second involves coatings that are intended to improve the mechanical running properties.

Anodizing or hard anodizing is used to produce oxide layers that are tightly bonded to the base material. Due to their low thermal conductivity, these layers, which are also resistant to corrosion and abrasion, provide protection to the piston crown against thermal attack by hot combustion gases. Because the emergency running properties of the hard oxide layers are inferior to those of the substrate, the piston skirt and ring belts should not be coated.

Various coatings (which range substantially in effectiveness) have been successfully applied to improve the running properties of the piston skirt. An example is graphite coating. A thin metal phosphate layer is formed in an alkaline bath (e.g., by in-bonderizing). This layer, which is approximately 1 μm thick, provides a well-prepared metallic surface for a synthetic resin graphite coating. The latter consists of fine graphite particles bonded with a phenol-resol resin. The coating, approximately 10–20 μm thick, is applied at high temperatures, is burned in (polymerized), and provides a so-called oil-compatible surface with excellent emergency running properties.

To reduce excessive wear of the piston skirt and cylinder liner during the running-in period, it is a common practice to coat the piston skirt with tin, lead, or other low-melting-point metal. The formation of tin or lead coating is based on the principle of ion exchange. The aluminum pistons are dipped in solutions of lead or tin salts. Because the two coating metals are nobler than aluminum in the electrochemical series, they are deposited on the piston surface. In the process aluminum is simultaneously dissolved until a closed surface of lead or tin is formed. The resulting 1- to 2-μm-thick metal coatings are widely used on pistons for commercial and passenger vehicle engines due to their good emergency running properties.[23]

8-7-3 Surface Modifications for Piston Rings

Engine performance and lifetime are greatly influenced by ring design, materials, and their surface modification methods. Cast iron coated with hard chromium electroplating (HCP) has been popular for piston rings. Other surface modification techniques such as plasma spray coating (PSC), and salt-bath nitriding (SN) and gas nitriding (GN) methods have also been applied with various types of steels. Ceramic coatings have been recently investigated for their high heat and wear resistance.

Figure 8-13 shows the friction coefficients of various piston ring surface coatings. Compared with chromium and molybdenum coatings, the ceramic coatings (K-ramic and IP-200) show remarkably low friction coefficients.

The running-in behavior can be improved by means of suitable protective coatings. These layers either improve the oil retention capability and thus act during wear as a polishing agent with protective action, or are quickly abraded at points of higher local pressure and thus accelerate the fitting process.

The surface of a cast iron ring is sometimes transformed into a hard, tightly adhering layer of magnetic iron oxide about 5 μm thick (ferro-oxidizing). To improve running-in properties, various processes for phosphating piston rings are widely used (see Section 8-6-2). They produce relatively soft, self-lubricating running-in surfaces to which oil adheres well. Running-in surface coatings of lead, tin, cadmium, and other low-melting-point metals used for pistons are used also for piston rings. Synthetic resin coatings, sometimes with the addition of graphite or abrasives, are also used successfully.[23]

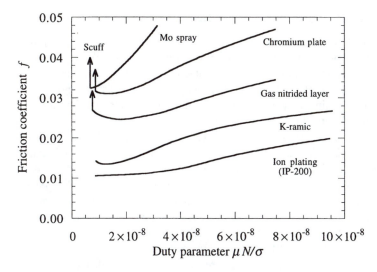

Figure 8-13 Friction characteristics of various ring coatings.[34]

Table 8-8 Wear of coatings on piston ring samples[35]

Coating	Running-in time (min)	Wear (mg/hr $\times 10^2$)		Load of seizure (kg/cm^2)	
		Cylinder	Ring	Beginning of seizure	Damage
Cast iron (bare)	125	68.1	45.0	152	195
Phosphate cold	89	72.4	49.4	160	228
Phosphate hot	40	84.0	66.9	210	310
Sn	64	58.5	36.6	195	230
Cd	78	64.2	43.5	123	226
Pb	67	66.0	41.8	135	218
Cu	86	62.1	43.7	107	201
Cu + Sn	53	57.7	38.4	220	255
Cu + MoS$_2$	72	58.0	34.1	204	271
Cr	460	49.0	26.4	160	234
Cr + Sn	263	44.5	22.5	198	260
Cr + Cu	330	42.3	20.0	176	265
Cr + Cu + Sn	280	41.0	19.4	203	280
Cr + Cu + MoS$_2$	286	40.0	18.6	230	302
Mo	160	65.3	38.1	210	265

Table 8-8 presents results of wear tests of different ring coatings for a cast iron cylinder. It suggests that the tribological properties of tin, copper + tin, and copper + MoS$_2$ coatings are superior to lead. It should, however, be noted that in selecting a suitable type of coating one should also consider the compatibility between the coating material and the cylinder liner. Other considerations such as the melting temperature of the coating material might also be important.

8-7-4 Surface Modifications for Cylinder Block Materials

All-aluminum engine blocks are widely used for small two-stroke motorcycle engines. Cylinder liner coatings of Nikasil (Ni–SiC composite plating) and nickel-based composite coating (NCC) are in common use. Hypereutectic Al–Si alloys such as 390 and a modified group (Si $= 16$–26%)[36] have excellent wear and heat resistance properties. Their castability and machinability, however, are rather poor compared with cast iron and hypereutectic aluminum alloys.

Beside the hypereutectic Al–Si alloys such as 390, Mercosil and other modified alloys may become applicable with the development of superior surface modification methods such as Nikasil and new ceramic coatings (NCCs).[37] Nikasil plating has been applied to Corvette, Porsche, and Mercedes-Benz Formula One engines. New NCCs (SiC, BN) have been applied to Suzuki, Yamaha, and Kawasaki motorcycle, snowmobile, and marine engines. Both Nikasil and NCC electroplating films contain high-hardness, fine ceramic particles (such as silicon carbide) to improve wear properties, or boron nitride to give excellent self-lubricating properties. The nickel-based electroplating films have excellent corrosion and wear resistance and can withstand corrosive conditions induced by high-percentage-methanol fuels, for example.

The hardness of these films under high-temperature conditions depends on the film composition and the phosphor content.[37] NCC film which contains phosphor is hardened under elevated-temperature operation, whereas a film that does not contain phosphor is softened. The most important property of NCCs, besides hardness, is low friction coefficient. The friction coefficient of NCCs under unlubricated conditions is far lower than that of hard chromium plating. NCC–BN film can further reduce the friction coefficient because of the self-lubricating property of the boron nitride particles. The scuffing resistance of NCCs, especially one containing boron nitride, is far superior to that of the conventional aluminum alloys.

Recent trends toward better wear and seizure resistance, along with less noise and low friction, are accelerating the investigation of various surface modification methods. Applications of composite polymer coatings (CPC) containing solid lubricant particles have been developed to achieve further reductions in wear and friction, operating noise, and fuel consumption.[38]

REFERENCES

1. Laimböck, F.J., "The Potential of Small Loop-Scavenged Spark-Ignition Single Cylinder Two-Stroke Engines," SAE Paper 910675, *SAE Trans.*, vol. 100, 1991.
2. Brown, D., "Modern Sulzer Marine Two-Stroke Crosshead Diesel Engines," Report R e/22.18.07.40, Sulzer Brothers Ltd., 1986.
3. Heywood, J.B., *Internal Combustion Engine Fundamentals*, McGraw-Hill, 1988.
4. Yagi, S., Ishibasi, Y., and Hirashi, S., "Experimental Analysis of Total Engine Friction in Four-Stroke S.I. Engines," SAE Paper 900223, *SAE Trans.*, vol. 99, 1990.
5. Harari, R., and Sher, E., "Measurements of Engine Friction Power by Using Inertia Tests," SAE Paper 950028, *SAE Trans.*, vol. 104, 1995.
6. Onishi, S., Hong Jo, S., Do Jo, P., and Kato, S., "Multi-Layer Stratified Scavenging (MULS)—A New Scavenging Method for Two-Stroke Engine," SAE Paper 840420, *SAE Trans.*, vol. 93, 1984.
7. Sher, E., "Flattening the Torque Curve of a Two-Stroke Engine with a Fluid Diode Installed at the Scavenge Port," SAE Paper 830092, 1983.
8. Patton, K.J., Nitschke, R.G., and Heywood, J.B., "Development and Evaluation of a Friction Model for Spark-Ignition Engines," SAE Paper 890836, *SAE Trans.*, vol. 98, 1989.
9. Bishop, I.N., "Effect of Design Variables on Friction and Economy," SAE Paper 812A, *SAE Trans.*, vol. 73, pp. 334–358, 1965.
10. Goenka, P.K., Paranjpe, R.S., and Jeng, Y.R., "FLARE: An Integrated Software Package for Friction and Lubrication Analysis of Automotive Engines—Part 1: Overview and Applications," SAE Paper 920487, 1992.
11. Thring, R.H., "Engine Friction Modeling," SAE Paper 920482, 1992.
12. Hamai, K., Masuda, T., Goto, T., and Kai, S., "Development of a Friction Prediction Model for High Performance Engines," presented as an STLE Paper at the ASME/STLE Triboloty Conference in Toronto, Ontario, Canada, Oct. 8–10, 1990: (Also in *J. Lubrication Eng.*, vol. 47, no. 7, pp. 567–573, 1991.
13. Taylor, C.F., *The Internal Combustion Engine in Theory and Practice*, vol.1, MIT Press, Cambridge, MA, 1985.
14. Blair, G.P., "Prediction of Two-Cycle Engine Performance Characteristics," SAE Paper 760645, *SAE Trans.*, vol. 85, 1976.
15. SAE, "Engine Oil Viscosity Classification," SAE J1300, Feb. 1991.
16. Landsdown, A.R., *High Temperature Lubrication*, The Institution of Mechanical Engineers, London, 1994.
17. Kenbeek, D., and van der Waal, G., "High Performance Ester-Based Two-Stroke Engine Oils," *Engine Oils and Automotive Lubrication* (ed. W.J. Bartz), Marcel Dekker, New York, 1993.

18. Gupta, M., Pandey, N.K., Mishara, G.C., and Singhal, S., "Development and Performance Aspects of Jojoba Based Lubricant Formulations For Two-Stroke Engines," SAE Paper 932795, 1993.

19. *Fundamentals and Application of Fuels and Lubricants*, compiled and edited by the Editorial Staff of the Petroleum Educational Institute, 1953.

20. O'Connor, J.J., Boyd, J., and Avallone, E.A., *Standard Handbook of Lubrication Engineering*, McGraw-Hill, New York, 1968.

21. SAE, "Lubricants for Two-Stroke-Cycle Gasoline Engines," SAE J1510, Oct. 1988.

22. Bostock, P.B., and Simpson, D., "The Two-Stroke Engine: Recent Developments and Future Potential," BP Report MKG-MISC330, British Petroleum, 1974.

23. MAHLE, *Piston Manual*, MAHLE GmbH, Stuttgart, Germany (Bad Cannstatt), 1988.

24. Suh, N.P., *Tribophysics*, Prentice-Hall, Englewood Cliffs, NJ, 1986.

25. Rabinowicz, E., *Friction and Wear of Materials*, Wiley, New York, 1965.

26. Quinn, T.F.J., *Physical Analysis for Tribology*, Cambridge Univ. Press, 1991.

27. Lim, S., and Ashby, M.F., "Wear Mechanism Maps," *Acta Metall.*, vol. 35, p.1, 1987.

28. Rabinowicz, E., "The Determination of the Compatibility of Metals through Static Friction Tests," presented at the ASME/ASLE Lubrication Conference, Cincinnati, OH, Oct. 13–15, 1970, American Society of Lubrication Engineers. (Also in *ASLE Trans.*, vol. 14, pp. 198–205, 1971.)

29. Brady, B.G., "Fretting Wear," SAE Paper 901786, *SAE Trans.*, vol. 99, 1990.

30. Garbar, I.I., "Structural Regularities of Metal Wear and Tribological Compatibility," D.Sc. Thesis, Moscow, 1988.

31. Lansdown, A.R., and Price, A.L., *Materials to Resist Wear*, Pergamon Press, Elmsford, NY, 1986.

32. Hintermann, H.E., "Adhesion, Friction and Wear of Thin Hard Coatings," *Wear*, vol. 100, pp. 381–397, 1984.

33. Bunshah, R.F., "Selection and Use of Wear Tests for Coatings," ASTM STP 769 (ed. R.G. Bayer), pp. 3–16, American Society for Testing and Materials, 1982.

34. Yoshida, H., Kusama, K., and Sagawa, J., "Effects of Surface Treatments on Piston Ring Friction Force and Wear," SAE Paper 900589, *SAE Trans.*, vol. 99, 1990.

35. Friction, *Wear and Lubrication*, Reference Book (eds. I.V. Kragelsky and V.V. Alisin), vol. 2, Mashinostroenie, Moscow, 1979. (In Russian.)

36. Myers, R.J., *Mercosil Engine Block Technology*, T91–035, North American Die Casting Assoc., Detroit, 1991.

37. Tareda, H., Hirose, M., Takama, M., and Konagai, N., "Industrialization of Ni-P-BN Dispersion Coating for Cylinders," *J. Automotive Eng. Japan*, vol. 43, no. 5, 1985.

38. Wada, A., "Trends of Surface Finishing Technologies in Toyota Motor," NEPTEC'92 Nagoya, Oct. 29, 1992.

CHAPTER NINE

OPERATING CHARACTERISTICS OF TWO-STROKE ENGINES

This chapter reviews the operating characteristics of the common types of two-stroke spark-ignition and compression-ignition engines. The effects of changes in the major design and operating variables on their operating characteristics are then related to the more fundamental engine processes such as scavenging, fluid flow, thermodynamics, combustion, heat transfer, and friction, which we have examined in earlier chapters. The intent is to provide data on, and an explanation of, actual engine operation—performance, efficiency, and emissions.

9-1 ENGINE PERFORMANCE PARAMETERS

The practical engine performance parameters of interest are power, torque, brake specific fuel consumption, and emissions. Following the definitions in Section 1-3, the brake power and the torque of a reciprocating engine can be expressed in terms of the brake mean effective pressure (bmep) as follows[1]:

$$P_b = \begin{cases} \text{bmep } A_p \overline{S}_p/2 & \text{(for two-stroke cycle)} \\ \text{bmep } A_p \overline{S}_p/4 & \text{(for four-stroke cycle)} \end{cases} \tag{9-1}$$

and

$$T = \begin{cases} \text{bmep } V_d/(2\pi) & \text{(for two-stroke cycle)} \\ \text{bmep } V_d/(4\pi) & \text{(for four-stroke cycle)} \end{cases} \tag{9-2}$$

Thus, for a well-designed engine, where the maximum values of bmep and mean piston speed are airflow- and/or stress-limited, power is proportional to piston area and torque to displaced volume. Introducing the definitions of trapping efficiency and delivery ratio from Section 2-1, and of mechanical and fuel conversion efficiencies from Section 1-3, bmep can be expressed in terms of other basic parameters as follows:

$$\text{bmep} = \eta_m (\eta_{tr} \Lambda) \eta_{f,i} \rho_{a,i} \frac{Q_{HV}}{\lambda (A/F)_S} \tag{9-3}$$

Note that Eqs. (9.1)–(9.2) can be written for the indicated power and torque in terms of indicated mean effective pressure. The equivalent to Eq. (9-3) for imep is identical, except that the mechanical efficiency η_m is omitted.

Equations (9-1)–(9-3) indicate that engine power and torque depend linearly on each of the following parameters: mechanical efficiency, trapping efficiency and delivery ratio, indicated fuel conversion efficiency, inlet air density, and amount of energy stored in the trapped fuel. Similar expressions can be obtained for four-stroke engines, for which the product of the delivery ratio and trapping efficiency is replaced by the volumetric efficiency.[1] Brake specific fuel consumption is related to indicated fuel conversion efficiency by Eq. (1-19a):

$$\text{bsfc} = \frac{1}{\eta_m \eta_{f,i} Q_{HV}} \tag{9-4}$$

The concentrations of gaseous emissions in the engine exhaust gases are usually measured as mole fractions. Normalized indicators of emissions levels are more useful (see Section 7-1-3). Specific emissions are the mass flow rate of pollutant per unit power output (usually in grams per kilowatt-hour); thus

$$\begin{aligned} \text{sNO}_x &= \frac{\dot{m}_{NO_x}}{P} & \text{sCO} &= \frac{\dot{m}_{CO}}{P} \\ \text{sHC} &= \frac{\dot{m}_{HC}}{P} & \text{sPart} &= \frac{\dot{m}_{Part}}{P} \end{aligned} \tag{9-5}$$

Emission rates can also be normalized by the fuel flow rate, and an emission index (EI) is commonly used: for example,

$$\text{EI}_{CO} = \frac{\dot{m}_{CO} \; (g/s)}{\dot{m}_f \; (kg/s)} \tag{9-6}$$

with similar expressions for NO_x, HC, and particulates.

In the following sections, the practical engine performance parameters of interest for different types of two-stroke engines are discussed. The value and relative importance of each of the preceding parameters vary with the operating conditions of the engine.

It is instructive to use typical numbers for full-load and part-load operation to compare the performance of equivalent two-stroke and four-stroke cycle engines. For typical modern spark-ignition engines of the same displaced volume, at the same fuel/air ratio, the ratio of their maximum torques is given by

$$\begin{aligned} \frac{T_{2\text{-stroke}}}{T_{4\text{-stroke}}} &= \frac{[\eta_m (\eta_{tr} \Lambda) \eta_{f,i}/2]_{2\text{-stroke}}}{[\eta_m (\eta_v) \eta_{f,i}/4]_{4\text{-stroke}}} \\ &\approx \frac{[0.85 \times (0.8 \times 0.9) \times 0.3/2]}{0.8 \times (0.95) \times 0.34/4]} = 1.4 \end{aligned} \tag{9-7}$$

This is consistent with typical maximum bmep values of about 850 kPa for this type of two-stroke engine compared with the 1200-kPa value for a four-valve-per-cylinder modern four-stroke SI engine. This gives a torque ratio [Eq. (9-2)] of

$$\frac{T_{2\text{-stroke}}}{T_{4\text{-stroke}}} = \frac{2 \times \text{bmep}_{2\text{-stroke}}}{\text{bmep}_{4\text{-stroke}}} \approx \frac{2 \times 850}{1200} = 1.4 \qquad (9\text{-}8)$$

Note that if the bmep of the two engines could be the same (not likely to occur in practice), then the torque ratio would be 2.

A comparison of maximum powers from these two different cycle engines is less straightforward because the maximum power in the two-stroke depends strongly on the scavenging performance of the engine. This depends on the scavenging system, the design of the ports, and especially on the sophistication of the exhaust system and its tuning (see Chapter 5). For two-stroke SI engines the specific power can vary between about 30 and 80 kW/liter, with the upper value typical of high-performance motorcycle engines. High-performance four-stroke passenger car SI engines average about 50 kW/liter, with the highest values being about 70 kW/liter.

In a similar way, for equivalent engines at the same operating conditions, we can estimate the ratio of the brake specific fuel consumptions to be

$$\frac{\text{bsfc}_{2\text{-stroke}}}{\text{bsfc}_{4\text{-stroke}}} = \frac{(\eta_m \eta_{f,i})_{4\text{-stroke}}}{(\eta_m \eta_{f,i})_{2\text{-stroke}}} \approx \frac{(0.6 \times 0.38)}{(0.75 \times 0.36)} = 0.85 \qquad (9\text{-}9)$$

Because the pumping losses under part-load conditions in two-stroke engines are smaller than in four-stroke engines and the friction of the four-stroke's gas exchange crank revolution is absent, the mechanical efficiency is higher and, although the indicated fuel conversion efficiency is somewhat lower, for the same fuel/air ratio and engine displacement volume, the brake specific consumption at typical part-load operating conditions of a direct-injection two-stroke SI engine is lower than that of a standard modern four-stroke SI engine by about 15%. Note that if the two-stroke SI engine is carburetted or port-fuel-injected, due to fuel short-circuiting this benefit essentially disappears.

9-2 ENGINE PERFORMANCE CHARACTERISTICS

Because of its lower pumping and friction losses, the two-stroke crankcase-scavenged engine has inherent theoretical advantages over its four-stroke counterpart in terms of efficiency. Realizing this advantage requires minimizing fuel short-circuiting and improving combustion quality. Direct fuel-injection technologies can satisfy these requirements. Because of the inherent high two-stroke-engine residual fraction, low NO_x emissions are another intrinsic advantage. Figure 9-1 shows a comparison among several engine technologies for a 1000-kg-weight vehicle in terms of the trade-off between fuel consumption and NO_x emissions.[2] This assessment indicates that the diesel and direct-injection SI four-stroke engines have similar or slightly better fuel consumption than the direct-injection two-stroke engine but higher NO_x levels. The direct-ignition two-stroke SI engine comes close to the fuel-economy potential of the best diesel engines while

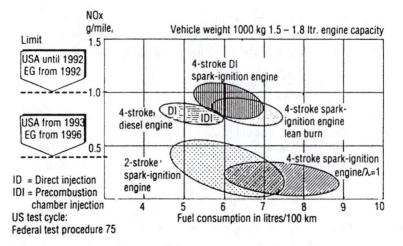

Figure 9-1 Ranges of fuel consumption and NO_x emissions levels, for different types of IC engine.[2] Reproduced by permission of the Council of the Institution of Mechanical Engineers.

having the advantage of engine NO_x emissions levels compatible with California's most stringent light-duty vehicle (ULEV†) levels without the need of NO_x catalysts.

9-2-1 Carburetted Spark-Ignition Engines

While the application of new direct-injection fuel injection technology to spark-ignition engines provides substantial performance and emissions benefits, it increases engine cost. A likely consequence of this is that direct-injection will be increasingly used in larger-sized higher-performance applications where the additional cost can be justified. Carburetted or manifold fuel-injected two-stroke SI engines are likely to be restricted to specific applications where simplicity and cost requirements are paramount. Table 9.1 shows typical results for exhaust gas composition and fuel consumption for carburetted two-stroke and four-stroke marine engines designed for similar applications.[3] The two engines were tested according to the ISO 8178–4 modified E4 cycle procedure, which calls for engines to be run at set percentages of rated (maximum) speed and torque. The results shown in Table 9-1 correspond to the following operating conditions [engine speed (%), engine torque (%), weighting factor]: (100, 100, 0.06), (91, 71.6, 0.14), (80, 46.5, 0.15), (63, 25.3, 0.25), and (idle, 0, 0.40). The two-stroke engine is inferior to the four-stroke engine in CO and HC emissions levels, while superior in NO_x emission level. The first two are mainly attributed to the lubrication method of the two-stroke engine: the lubricant is mixed with the fuel and introduced through the fuel system. The lower NO_x emission level of the two-stroke engine is attributed to its inherent high residual fraction due to the imperfect scavenging process. Short-circuiting of the fuel

---

†California's ultra-low-emission vehicle (ULEV) levels for light-duty vehicles are HC(NMOG) 0.04, CO 1.7, NO_x 0.2 g/mile for 50,000 miles.

Table 9-1 Comparison between exhaust gas composition and fuel consumption of two- and four-stroke two-cylinder carburetted marine engines.[3]

Engine type	Two-stroke Johnson OMC-J10RCSE		Four-stroke Honda BF9.9A	
Rated power (kW)	7.4 at 5000 rev/min		7.4 at 5000 rev/min	
Displacement (cm^3)	216		280	
Fuel/oil ratio	50:1		N/A	
Fuel consumption (g/kW-h)	909		549	
CO (g/kW-h)	519	EI = 571	368	EI = 670
CO_2 (g/kW-h)	1299	EI = 1429	1052	EI = 1916
HC (g/kW-h)	235.7	EI = 259	31.8	EI = 57.9
NO_x (g/kW-h)	0.73	EI = 0.80	3.39	EI = 6.17

through the exhaust port cannot be avoided, however, and higher HC emissions and fuel consumption are inevitable.

Because of its simple construction, compactness, high power-to-weight ratio, and low cost, the small carburetted two-stroke cycle engine is widely used for handheld equipment. As illustrated, its most serious drawback is its high HC emission and low thermal efficiency due to short-circuiting of fuel to the exhaust port. In a successful attempt to conform to the California Air Resources Board 1999 or EPA Tier 2 standards, a commercial Schnürle-type carburetted engine has been modified at Komatsu Zenoah Co.,[4] without sacrificing its mechanical simplicity and light weight, into a much cleaner engine by using a stratified-scavenging system. During the scavenging process, burnt gas is first scavenged by pure air, which is introduced into the scavenging port through a reed valve. The air–fuel mixture is then introduced into the cylinder later, toward the end of the scavenging process. The lubricating oil is mixed with the gasoline at a ratio of 1:40. Table 9-2 shows the details of this engine. Figure 9-2 shows the wide-open-throttle performance of the original and modified versions, for an air/fuel ratio of 14. Power output, brake fuel conversion efficiency, and bsHC, bsCO, and bsNO_x emissions are shown as a function of engine speed. At rated speed, the modified engine provided 80% lower HC and 51% higher brake efficiency with a 7% lower power output than the original engine. These results are due to the significant reduction in the fuel short-circuiting that cannot, in practice, be avoided in the original engine. The lower power output is attributed to the lower charging efficiency of the modified engine.[4] This engine also provides 24% lower CO and 11% higher NO_x at the rated speed, which indicates a more efficient combustion process than in the original version. The declining levels of NO_x with increasing engine speed at least partly result from the increasing combustion duration (in crank angle degrees) when engine speed increases, which reduces the maximum cylinder pressure and temperature.

The emission tests were conducted in accordance with SAE J1088 test cycle C, a weighted steady-state procedure based upon raw exhaust gas sampling. The test is composed of two modes: 100% torque at rated speed, and 0% torque at idle. Mode

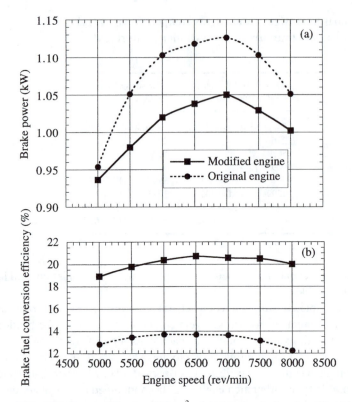

Figure 9-2 Performance of Komatsu 33.6-cm³ original (carburetted) and modified (stratified charging) engine: (a) brake power output, (b) fuel conversion efficiency, and (c)–(e) specific HC, CO, and NO$_x$ emissions, at wide-open-throttle conditions, as a function of engine speed; air/fuel ratio = 14.[4]

Table 9-2 Details of the Komatsu air-cooled carburetted two-stroke engine[4]

	Current	Modified
Engine type		
Scavenging type	Schnürle scavenging	Stratified scavenging
Inlet method	Piston valve	Crankcase reed valve
Maximum power	110 kW at 7000 rev/min	
Bore/stroke	36/33 mm	
Displacement	33.6 cm³	
Effective compression ratio	7.5	

weights for this test cycle, which has been adopted by EPA and CARB, are 90% for 100% torque and 10% for 0% torque. Table 9-3 summarizes the test results at an air/fuel ratio of 14.

Because short-circuiting is expected to be more significant when a higher amount of fresh mixture is introduced to the cylinder, HC emissions from the original engine tend to be much higher under high-load conditions. The charge stratification in the modified

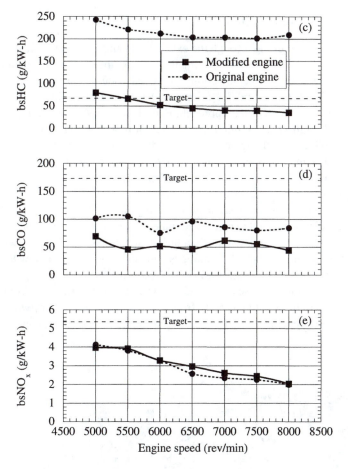

Figure 9-2 (Continued).

Table 9-3 Emission test results for the Komatsu air-cooled carburetted two-stroke engine[4]

	Original engine			Modified engine			
	Idle (g/h)	Full load (g/h)	Weighted (g/kW-h)	Idle (g/h)	Full load (g/h)	Weighted (g/kW-h)	CARB (g/kW-hr)
HC	46.5	230.3	208.4	34.9	42.9	44.6	67
CO	38.2	97.1	89.7	27.4	68.3	67.9	174
NO_x	0.11	2.71	2.41	0.12	2.80	2.68	5.36

engine significantly reduces the amount of fuel that otherwise would short-circuit and, therefore, the HC emissions level. The modified engine with its charge stratification conforms to the CARB or U.S. Federal Tier 2 emission standards.

9-2-2 Direct Fuel-Injected SI Engines

Elimination of fuel short-circuiting through the exhaust port and charge stratification at lighter loads can be achieved by directly injecting the fuel into the combustion chamber. Several different approaches to "direct injection" have been discussed in earlier chapters (see Sections 2-3-6, 3-3-2, 6-3-2, and 7-3): direct injection into the cylinder with air-assisted fuel injection (e.g., the Orbital approach, Figs. 1-24 and 9-3); direct injection into the cylinder with high-pressure liquid fuel injection (e.g., the Piaggio approach, Fig. 7-12; the Toyota S-2, valved engine, Fig. 1-26); direct mixture injection with fuel injected into a chamber in a secondary transfer duct with a separate valve to control *mixture* entry into the cylinder (e.g., the IFP IAPAC concept, see Fig. 2-25). We now review the more promising of the concepts.

An important example of a combustion system using air-assist fuel injection technology is the Orbital direct-injected two-stroke engine. Figure 9-3 illustrates the major components of the engine. The engine features a turbulent combustion chamber, which allows the use of high dilution rates and encourages rapid mixing at high and low loads. It also features an exhaust-port scavenge-flow control valve that is controlled by the engine electronic control unit to increase low speed torque and assist in emissions control. The lubricant is supplied to the engine parts from a pressurized oil system, which is separated from the fuel supply system. The fuel is introduced by a direct low-pressure

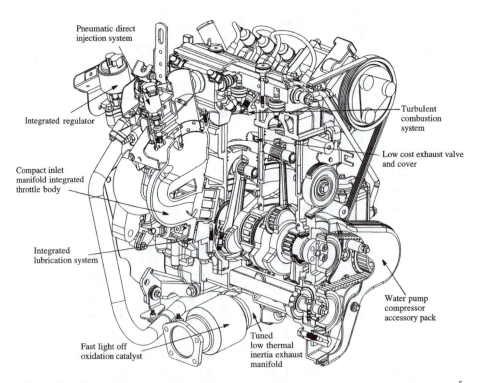

Figure 9-3   The Orbital 1.2-liter three-cylinder supercharged two-stroke direct-injected gasoline engine.[5]

air-assisted injection system, which uses air at a pressure of 550 kPa to effect the injection and atomization of fuel, allowing the use of a low-pressure (620 kPa) fuel supply. This approach to direct fuel injection provides a very fine spray having a mean diameter of about 8 μm (a typical diesel engine system provides sprays having a mean diameter of 15–20 μm). See also the discussion in Sections 6-3-2, 6-4-1, and 7-3-1.

Owing to these advanced features, the minimum brake specific fuel consumption of the engine is low, about 290 g/kW-h. The wide-open-throttle characteristics of a 1200-cm^3 Orbital engine (with forced induction) are shown in Fig. 9-4. The maximum specific power of the engine is 67.5 kW/liter and its maximum torque is 150 N-m. Figure 9-5 shows a comparison of the WOT performance of two versions of this two-stroke Orbital engine with several other commercial engines. Because a combustion process in two-stroke engine occurs each crank revolution (rather than every other revolution as in four-stroke engines) the naturally aspirated two-stroke engine reaches a bmep value of 655 kPa, which is higher by some 10% than that of its best four-stroke counterpart (bmep$_{4\text{-stroke}}$ = 1200 kPa, equivalent to 600 kPa for a two-stroke). The forced induction two-stroke engine of Fig. 9-3 reaches a bmep value of 785 kPa (which corresponds to the maximum torque of 150 N-m).

Owing to direct injection and the fast combustion process resulting from its turbulent combustion chamber, the bsHC emissions of the Orbital two-stroke engine are around 2.5 g/kW-h, which is an extremely low value relative to carburetted two-stroke engines (see Table 9-1). Figure 9-6 shows how the HC emission index of the engine correlates with its NO$_x$ emission index. For comparison, a typical HC engine-out emissions index for a modern four-stroke spark-ignition engine is about 2% (see also Fig. 7-17). In general, as the NO$_x$ emission index increases, the HC emission index decreases; as the air/fuel ratio and residual fraction are varied, a higher NO$_x$ emission indicates a higher in-cylinder burned gas temperature, which enhances combustion efficiency and

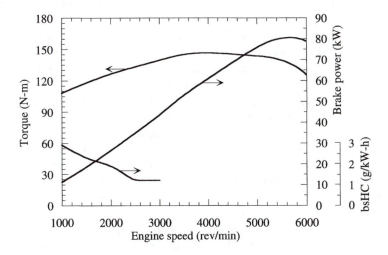

Figure 9-4 Wide-open-throttle characteristics of the Orbital 1.2-liter two-stroke direct-injected gasoline engine in Fig. 9-3.[5]

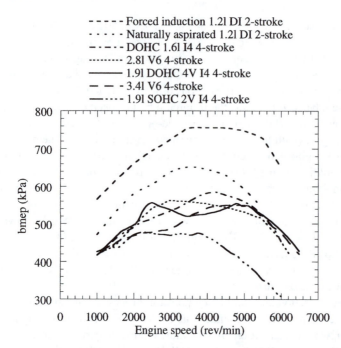

Figure 9-5 Wide-open-throttle brake mean effective pressure, as a function of speed, for boosted and naturally aspirated Orbital two-stroke gasoline engines and several commercial four-stroke port-injected SI engines.

HC consumption. Owing to the nature of the gas exchange process in the two-stroke engine, high residual fraction at partial load is an inherent property, and lower engine NO_x emissions compared with the four-stroke engine are expected. However, owing to the turbulent combustion system, which allows the use of high dilution rates and encourages rapid mixing at high and low loads, the Orbital two-stroke engine demonstrates low emission-index values over the entire range of engine load.

An alternative to direct *fuel* injection to achieve charge stratification is direct *mixture* injection. A successful example of a modern two-stroke mixture injection SI engine is the IAPAC concept developed by the Institut Français du Petrole (IFP)[6] (see Section 2-3-6). In one study they modified a 1230-cm³ cross-scavenged two-stroke engine[7] to include an intake fuel injection system in one version, and an air-assisted mixture injection system (IAPAC) in the other (see Table 9-4).[6] The air needed for the latter version was provided by the crankcase, where the air was compressed to the required pressure. A conventional port fuel injection system was installed in each cylinder head just ahead of an intake valve (see Fig. 9-7). During the valve-open period, a rich, low-pressure, fuel–air mixture is delivered directly into the combustion chamber. To minimize fuel short-circuiting, pneumatic injection takes place separately from the scavenging airflow. Figure 9-8 shows the wide-open-throttle performance of the two modifications. It is important to mention that the IAPAC engine uses the same configuration as the intake fuel injected engine except for the exhaust port open duration.

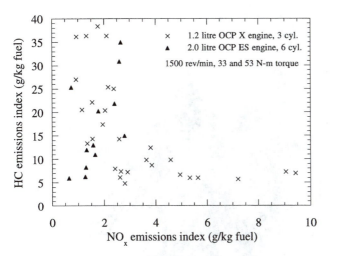

Figure 9-6 HC emission index versus NO_x emission index for the Orbital two-stroke direct-injected engine over the U.S. Federal Test Procedure driving cycle test.[5]

Table 9-4 Engine specifications for Figs. 9-7 and 9-8[6,7]

Engine type	Marine three-cylinder SI engine
Scavenging type	Cross-scavenging
Maximum power	53 kW at 5000 rev/min
Bore/stroke	85.7/71.1 mm
Displacement	1230 cm^3
Total scavenging opening duration	128°
Total exhaust opening duration	158°

This has been slightly reduced in order to obtain the best trade-off between acceptable levels of torque and power at WOT, and high efficiency and low emissions at partial load. Because, in the pneumatic system, air has to be compressed (although to a modest degree), the intake fuel injected engine produces a higher net power (44.7 kW/liter, compared with 40.7 kW/liter for the IAPAC version). The relatively high bmep (around 600 kPa) of both engines at low engine speeds is explained by the high trapping efficiency achieved with the cross-flow scavenging system, which is estimated to be some 20% higher than that achieved by a conventional loop-scavenged engine.[6] The low bsfc and HC emission is attributed to the elimination of the fuel portion that short-circuits through the exhaust port with intake fuel injection. Figure 9-8 indicates that, for the pneumatic fuel injected engine, a typical value for the HC emission index about is 5 compared with about 350 for the port injected version (see Table 9-1 for a comparison with carburetted two-stroke engine values).

Toyota has developed a supercharged two-stroke gasoline engine with poppet valves, which is based on a commercial four-stroke engine design.[8] In this engine, the intake and exhaust valves were used instead of ports in the cylinder wall, thus avoiding

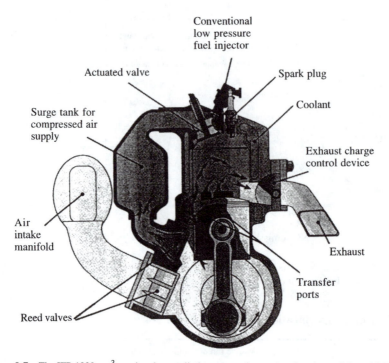

Figure 9-7 The IFP 1230-cm^3 marine three-cylinder two-stroke pneumatic mixture-injected IAPAC gasoline engine.[6]

the lubrication and piston-ring durability problems usually associated with the traditional piston-port-controlled two-stroke approach and allowing asymmetric port timing and charge boosting. A full water jacket and oil jets from the cylinder block to cool the underside of the piston were also employed to reduce piston distortion due to asymmetric heating, and thus reduce the probability of piston seizure. Figure 5-35 shows the arrangement and geometry of the ports, valves, and fuel injector in the cylinder head for this engine, and the in-cylinder scavenging flow these produce. Figure 9-9 shows the engine's wide-open-throttle performance; Table 9-5 gives the engine's specifications. Nomura and Nakamura[8] found that, compared with the four-stroke gasoline engine of equal displacement, maximum power increased by approximately 20% and maximum torque increased about 70%. The minimum fuel consumption is almost the same as that of the four-stroke engine, but has the potential to be superior at low engine loads due to lower pumping work. NO_x emissions at low engine loads were less, by a factor of 2 (0.3 g/mile in a light-duty vehicle), because a higher amount of residual gas remains in the cylinder, especially at partial load. Figure 9-9 indicates that the maximum specific power of the engine is 58 kW/liter, and its maximum bmep is 922 kPa.

Performance measurements in manifold-injected and direct-injected Schnürle-scavenged Honda engines[9] (see Fig. 9-10), under wide-open-throttle conditions, are presented in Figure 9-11. In this engine (see Table 9-6 for engine details) an "acti-

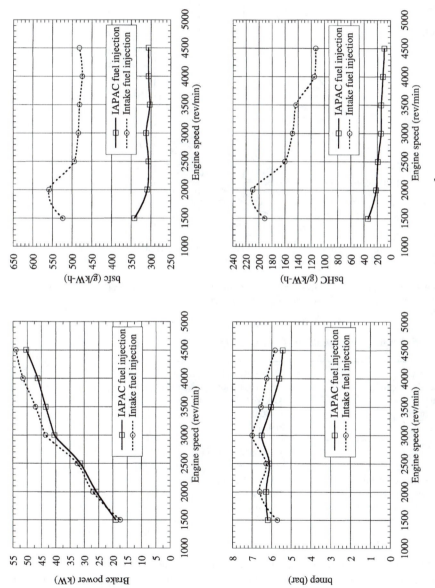

Figure 9-8 Comparison of wide-open-throttle characteristics of two versions of the IFP 1230-cm^3 marine three-cylinder two-stroke gasoline engine: (solid lines) air-assisted mixture injection system (IAPAC); (dashed lines) intake fuel injection system.[6] See Table 9-4 for engine details and Fig. 9-7 for a drawing of the engine.

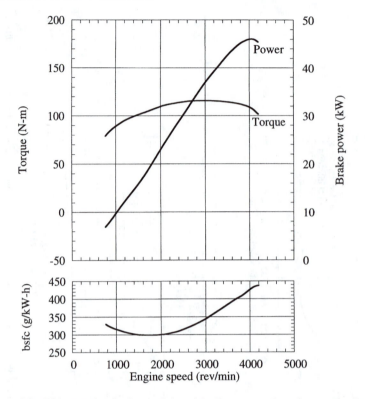

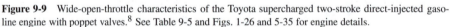

Figure 9-9 Wide-open-throttle characteristics of the Toyota supercharged two-stroke direct-injected gasoline engine with poppet valves.[8] See Table 9-5 and Figs. 1-26 and 5-35 for engine details.

Table 9-5 Engine specifications for Fig. 9-9[8]

Engine type	Toyota two-cylinder in-line SI engine
Scavenging approach	Belt-driven roots-type supercharger
Maximum power	47 kW at 4000 rev/min
Bore/stroke	80/80 mm
Displacement	804 cm^3
Geometric compression ratio	8.5
Inlet valve-open/close	50° BBC/70° ABC
Exhaust valve-open/close	70° BBC/50° ABC
Inlet/exhaust valve lift	7.2 mm/7.2 mm
Fuel injection pressure	10 MPa
Sauter mean diameter of fuel spray	25 μm

vated radical combustion" concept and a low-pressure pneumatic direct-injection system were employed. The activated radical (AR) combustion concept has been suggested by Ishibashi and Tsushima[10] to control the combustion-produced in-cylinder pressure rise (see Section 6-3-1). This design is based on their finding that control of the timing

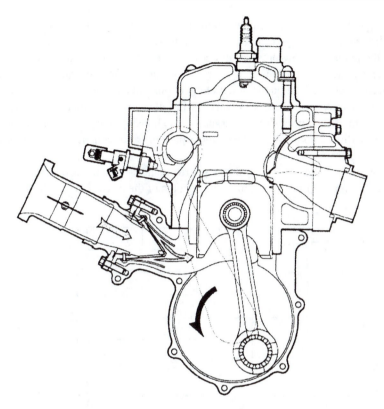

Figure 9-10   The Honda experimental two-stroke engine installed with pneumatic direct fuel injection and activated radical combustion systems.[9]

Table 9-6 Engine details for Figs. 9-10 and 9-11[9]

Engine type	Honda single-cylinder SI engine
Scavenging type	Five-port Schnürle
Maximum power	33 kW at 6900 rev/min
Bore/stroke	80/80 mm
Displacement	402 cm^3
Trapped compression ratio	6.1
Scavenge port open	123° ATC
Exhaust port open (with throttling)	86° ATC
Fuel injection pressure	250 kPa

of spontaneous autoignition (analogous to controlling spark timing) could be obtained by controlling the in-cylinder pressure at the time the exhaust port closes, according to the engine speed and the load. Throttling the exhaust is accomplished with an exhaust flow control valve (see Section 5-5-4). Ishibashi and Asai[9] claimed that the active radical combustion could achieve remarkable reductions in fuel consumption and HC emissions at part-throttle operation, due to the reduction in incomplete combustion cycles and the reduction in cycle-by-cycle burn rate variables. In the engine shown in Fig. 9-10, the pneumatic low-pressure direct fuel injection system consists of a rotary valve driven by the crankshaft with a cogged belt, an air chamber where in-cylinder pressure accumulates for the pneumatic fuel injection, and a conventional low-pressure fuel injector for metering. The rotary valve is installed in the cylinder, supported by bearings on both ends, with a slight clearance to the housing. An oil seal on the pulley end is the only sliding point. When the cylinder pressure and temperature are high, this valve is protected by the piston.

A comparison of the wide-open-throttle performance of the manifold-injected and direct-injected engines (Fig. 9-11) shows the benefits of the direct-injected fuel system.

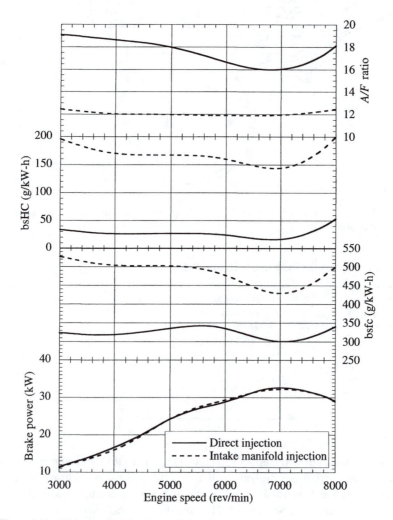

Figure 9-11 Wide-open-throttle characteristics of intake-manifold-injected and direct-injected engines.[9]

The engine can operate lean and match the power output of the manifold-injected engine, which operated rich. Losses due to short-circuiting are much smaller. Thus HC emissions and fuel consumption are significantly lower. Figure 9-11 indicates that the maximum specific power of the engine is 82 kW/liter, and its maximum bmep is around 750 kPa.

9-2-3 Small-Bore Compression-Ignition Engines

While the four-stroke gasoline engine is superior in terms of power density, particulate emission, sound, and driving comfort, the diesel engine is well known to be superior in the areas of fuel consumption and CO_2, CO, and HC emissions levels. For these reasons, the small-bore four-stroke diesel engine has found a major market in light-duty vehicles. Because the small-bore two-stroke gasoline engine has traditionally been employed in applications where high power density and simplicity are paramount requirements, the diesel engine has never been a serious competitor. The strictness of air pollution regulation has lead to a more bulky spark-ignition engine, however, and with likely future pressures for CO_2 emissions reduction, the small-bore two-stroke diesel engine is being reconsidered.

Recently, Daimler-Benz[11] has examined the question of what a promising two-stroke direct-injection small-bore diesel concept should be. A common rail injection system was applied to ported loop- and uniflow-scavenged engines. The uniflow-scavenged configuration was operated with a swirl level comparable to that of four-stroke direct-injected diesel engines. No swirl motion was realized with the ported loop-scavenged arrangement. Table 9-7 provides details of these two engines.

The full-load operating characteristics of both these engines are presented in Fig. 9-12. For these experiments, a smoke number of 2.5 was defined as the full-load limit. Indicated mean effective pressures up to 570 kPa and an engine speed of 1600 rev/min were achieved for the ported loop-scavenged engine, which is half the maximum value of 1100 kPa at 3000 rev/min for the uniflow-scavenged engine. The air utilization of the ported loop-scavenged engine becomes significantly worse than that of the uniflow-scavenged engine as engine speed increases. Also, although the relative air/fuel ratio is acceptable, it is lower by about 0.3 units. The volumetric efficiency with the loop-scavenged arrangement was noted to be from 30% to over 50% lower than for the uniflow-scavenged engine. The indicated specific fuel consumption measured for the ported loop-scavenged engine averages about 10% higher than that of the uniflow-scavenged chamber. Due to the unfavorable combustion conditions as well as the higher

Table 9-7 Daimler-Benz direct-injected two-stroke diesel engines[11]

Engine type	Ported loop scavenging	Uniflow scavenging
Bore/stroke	80/70 mm	80/98 mm
Displacement	350 cm^3	493 cm^3
Combustion chamber	Bowl in cylinder head	Bowl in piston
Scavenging air	Crankcase	External blower

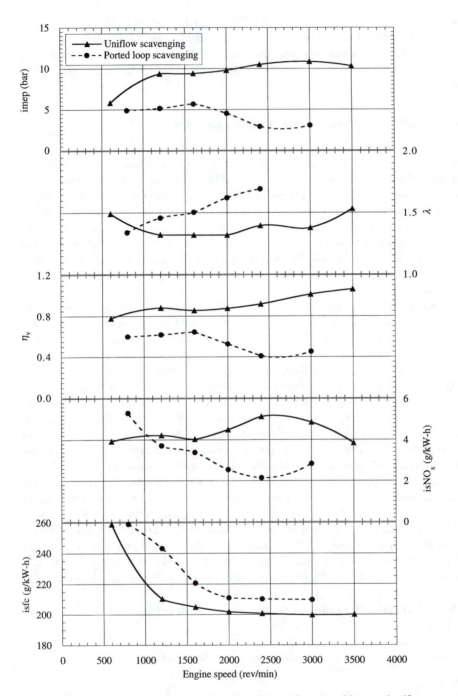

Figure 9-12 Full-load operating characteristics of the Daimler-Benz ported loop- and uniflow-scavenged two-stroke direct-injected diesel engines.[11]

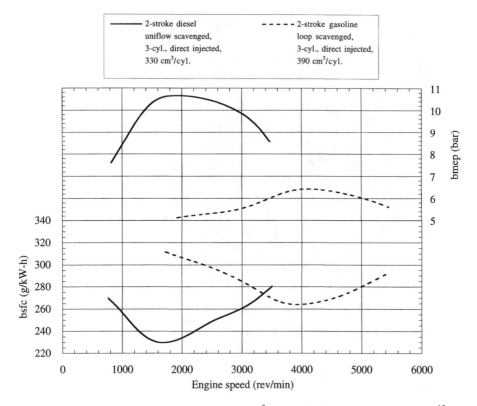

Figure 9-13 Full-load characteristics of the AVL 980-cm^3 two-stroke direct-injection diesel engine.[12] Also shown are data from a two-stroke loop-scavenged three-cylinder 1170-cm^3 direct-injected gasoline engine, for comparison.

exhaust residual, the NO_x emissions become lower as speed increases, with the ported loop-scavenged design. After allowing for friction losses and the power required for the scavenging blower, the specific power output for the uniflow-scavenged engine is more than 45 kW/liter. This is achieved by the high volumetric efficiency and the higher fuel–air mixing rates achieved with the swirl-producing scavenging flow. Due to its low scavenging quality, only 20 kW/liter could be achieved with the ported loop-scavenged engine.

AVL has developed a prototype small-bore direct-injection two-stroke diesel for passenger-car applications.[12] The engine has three water-cooled cylinders in-line with a total displaced volume of 980 cm^3. The cylinders are uniflow scavenged through four exhaust valves, and the engine is supercharged by a combined mechanical charging and turbocharging system to obtain high torque at low and at high engine speeds. Figure 9-13 shows the operating characteristics of this engine at full load. Also shown are the characteristics of a two-stroke loop-scavenged, three-cylinder direct-injected 1170-cm^3 gasoline engine. In this AVL diesel design, a high charge swirl level was created to control the fuel burning rate, almost approaching a constant-pressure combustion process.

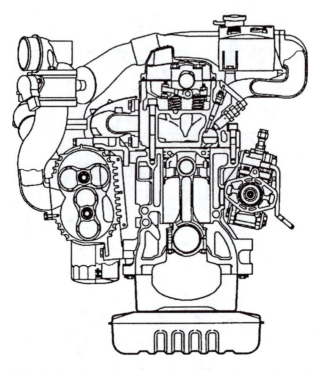

Figure 9-14   The Toyota supercharged two-stroke indirect injection diesel engine with poppet valves.[8] (See also Fig. 6-25.)

Table 9-8 Details of Toyota S-2 engine, Figs. 9-14 and 9-15[8]

Engine type	Four-cylinder in-line CI engine
Scavenging type	Supercharging
Maximum power	100 kW at 3400 rev/min
Maximum power	320 N-m at 1800 rev/min
Bore/stroke	96/86 mm
Displacement	2,489 cm^3
Geometric compression ratio	18
Inlet valve-open/close	45° BBC/50° ABC
Exhaust valve-open/close	75° BBC/65° ABC
Inlet/exhaust valve lift	7.2 mm/8.6 mm
Minimum fuel consumption	274 g/kW-h

It was reported[12] that the high power density of 50 kW/liter and, especially, the torque values of almost 170 N-m/liter (a bmep of 1068 kPa) could be reached at a maximum cylinder pressure of only 12 MPa. This enables a very light power-train construction of 70–80 kg, which is comparable to the weight of an equivalent two-stroke gasoline engine.

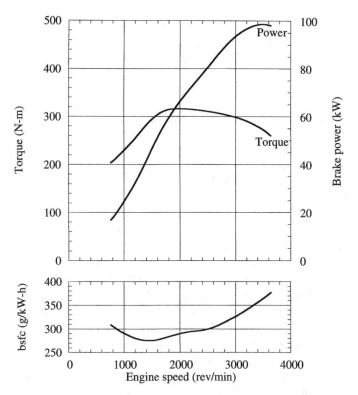

Figure 9-15 Full-load characteristics of the Toyota supercharged two-stroke diesel engine with poppet valves.[8]

Toyota[8] has developed a supercharged two-stroke indirect injection diesel engine with poppet valves, which is based on a commercial four-stroke engine design (see Figs. 9-14 and 6-25 and Table 9-8). Owing to the larger quantity of hot residual gas typically remaining in the cylinder in a two-stroke engine, the compression ratio of the engine could be lowered to 18. Consequently, peak pressure and temperature have been lowered and NO_x emissions, particulate emissions, and combustion noise have been noticeably reduced. Because of the lowered compression ratio, running-up time in cold weather conditions is shortened. In this engine, intake and exhaust valves were used instead of ports on the cylinder wall, thus avoiding the lubrication and piston-ring durability problems usually associated with traditional piston-port-controlled two-stroke engines. Using poppet valves also allows asymmetric port timing and boosting. A full water jacket and oil jets from cylinder block to cool the underside of the piston were also employed to reduce piston distortion due to asymmetric heating, reducing the probability of piston seizure. Figure 9-15 shows the engine's full-load performance curves. It was reported[8] that, compared with a four-stroke diesel engine of equal displacement, maximum power shows an increase of about 25%, and maximum torque an increase of about 40%. The minimum brake specific fuel consumption is almost the same as that of the four-stroke engine, but has the potential to be superior at low engine loads due to lower pumping work. NO_x emissions at low engine loads are less by a factor of 2 (0.3 g/mile), because a higher amount of residual gas remains in the cylinder. The data in Fig. 9-15 give a maximum specific power of 40 kW/liter, and maximum bmep of 808 kPa.

9-2-4 Large-Bore Compression-Ignition Engines

Large-bore diesel engines are mainly used as power units for generating electricity and for marine propulsion. The designs of these engines are optimized for a specific engine speed to match the requirements of the electrical power system, or for a specific power–engine-speed curve corresponding to the propeller performance. The engine's performance is therefore evaluated in terms of normalized parameters such as brake mean effective pressure, brake fuel conversion efficiency (or bsfc), and specific emissions rather than in terms of the engine torque or power curve. The reduction of the fuel consumption without jeopardizing the engine reliability and lifetime has been the key issue in the development of such engines in recent years. Operation at low specific power, 2–6 kW/liter, allows long cylinder lifetimes of 50,000–70,000 hours. Modern engines of this type are now available with bsfc of 162–170 g/kW-h, which corresponds to a brake fuel conversion efficiency of up to 54%, and lubricant oil consumption is around 0.9–1.2 g/kW-h. These engines are all turbocharged, and charge air cooled with an intercooler, and therefore achieve high bmep values of some 1700 kPa. They can also be turbocompounded at higher loads, which increases output and improves bsfc and efficiency by about an additional 5 g/kW-h or 3%. Table 9-9 gives the geometric details and performance characteristics of two examples of these large-bore two-stroke diesel engines. Figure 9-16 shows how the operating characteristics of these engines vary over their load range. As these engines approach their maximum power, injection is retarded to hold the maximum cylinder pressure below the design limits of the engine. We will discuss these variations in more detail, later, in Section 9-5-2.

Table 9-9 Geometric and performance details for two large-bore two-stroke engines†

Engine	Sulzer RTA 84	MAN-B&W L90MC
Number of cylinders	12	12
Bore/stroke	840/2400 mm	900/2916 mm
Displacement	16.0 m^3	22.3 m^3
Engine speed	90 rev/min	82 rev/min
Maximum power	39.7 MW	51.7 MW
Maximum mean piston speed	7.20 m/s	7.97 m/s
Specific power	2.49 kW/liter	2.32 kW/liter
Torque	4295 kN-m at 90 rev/min	6,023 kN-m at 82 rev/min
Brake mean effective pressure	1691 kPa	1700 kPa
Brake specific fuel consumption	171 g/kW-h	166 g/kW-h
Specific oil consumption	N/A	1.0–1.4 g/kW-h

†Data supplied by MAN and B&W, and by Wärtsilä NSD and Sulzer.

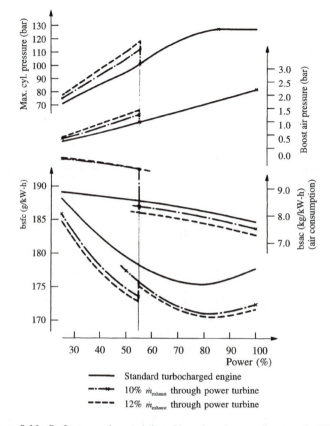

Figure 9-16 Performance characteristics of large-bore low-speed two-stroke RTA58 Sulzer diesel engine, operating in standard turbocharged mode, and turbocompounded with 10 or 12% of the exhaust flowing through a power turbine coupled to the drive shaft above 55% load.[13].

The levels of pollutants in the exhaust gas are a function of engine design and operating parameters and fuel characteristics, and the means employed to control such emissions so that they meet regulatory requirements. Figure 9-17 shows the exhaust levels of the main pollutants as a function of engine load for a large-bore low-speed diesel using gas oil and a typical commercial heavy fuel. The amounts of carbon monoxide and remaining unburned hydrocarbons are a function of the excess air ratio and the burned gas temperature (see Section 7-4-1). Because these types of engines use a high excess air ratio, and because the speed of these engines is very low, combustion is highly efficient and the CO and HC emissions are much lower than those of small-bore engines, both CI and SI.

This highly efficient combustion process of the direct injection low-speed diesel engine produces high local combustion temperatures (typically above 2400°C). At this temperature, oxygen and nitrogen will inevitably react sufficiently rapidly to form substantial amounts of nitric oxide (NO) (see Section 7-1-2). Later in the process, during expansion and in the exhaust system, part of the NO (about 6%) will convert to NO_2

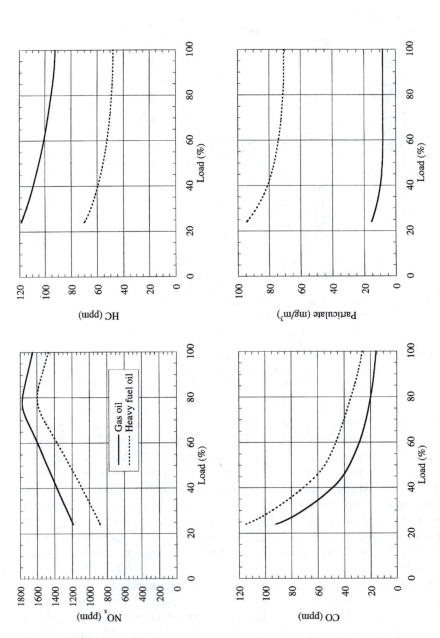

Figure 9-17 Exhaust concentrations of regulated pollutants, as a function of engine load, for a large-bore low-speed diesel using gas oil and a typical commercial heavy fuel. From data provided by MAN and B&W, and Wärtsilä NSD and Silzer.

(5%) and N_2O (1%). Because low-speed two-stroke diesel engines are optimized for high efficiency, the NO_x emissions from uncontrolled engines are high. The particulate emissions from these engines are mainly unburned carbon from the fuel oil, with various trace metals from the fuel, also. Calcium particles from the lubricating oil are also emitted. The nuclei around which the particles form in the engine cylinder are largely tiny carbon particles formed in fuel-rich regions of each fuel spray, during the diesel combustion process (see Section 6-4-2). These nuclei grow as additional increasingly dehydrogenated carbon compounds are deposited, and metal compounds from the fuel oil ash become attached, also. Substantial particulate oxidation and burn up occurs within the cylinder, prior to exhaust. Then, in the exhaust system, additional higher molecular weight hydrocarbons, sulfate from the sulfur in the fuel, and inorganic compounds from the fuel and the oil are absorbed. Because of the high sulfur content of heavy large-engine diesel fuels, the absorbed sulfate is especially significant. See Section 7-4-1 for additional discussion, and see Henningsen.[14]

9-3 OPERATING VARIABLES THAT AFFECT TWO-STROKE SI ENGINE PERFORMANCE, EFFICIENCY, AND EMISSIONS

The major operating variables that affect two-stroke spark-ignition engine performance, efficiency, and emissions at any given load and speed are spark timing, fuel injection parameters in direct-injection engines, charge composition (relative air/fuel ratio, residual gas fraction, and amount of the exhaust gases that are recycled), exhaust throttling, and ambient conditions. The effect of these variables will be reviewed in this section. Design and operating considerations that relate to the gas exchange process, such as the geometry of the ports and crankcase volume, are discussed in Chapter 5. The effect of major design parameters, which are not specific to two-stroke cycle engines, such as compression ratio, combustion chamber shape, and variables that affect heat transfer, can be found elsewhere (e.g., Heywood[1]).

9-3-1 Load and Speed

A convenient way to present the operating characteristics of an internal combustion engine over its full load and speed range is to plot contours of parameters such as bsfc or emission level on a graph of brake mean effective pressure (bmep) versus engine speed. The relationships between the bmep, torque, and engine power can be found in Section 9-1 and in Chapter 1.

Figures 9-18 and 9-19 show examples of such performance maps for carburetted and manifold-fuel-injected (IAPAC) engines. The upper envelop of each map is the wide-open-throttle (WOT) performance curve. Points below this curve define the part-load operating characteristics. Maximum bmep occurs in the mid-speed range, whereas the minimum bsfc region is located around the same engine speed and at 80–90% full load. The maximum bmep curve reflects the variation with speed of the delivery ratio and trapping efficiency (the product of these two is equivalent to the four-stroke engine's volumetric efficiency) due to the changing effectiveness of the gas exchange process, the decrease in the mechanical efficiency as engine speed increases, and the decrease in

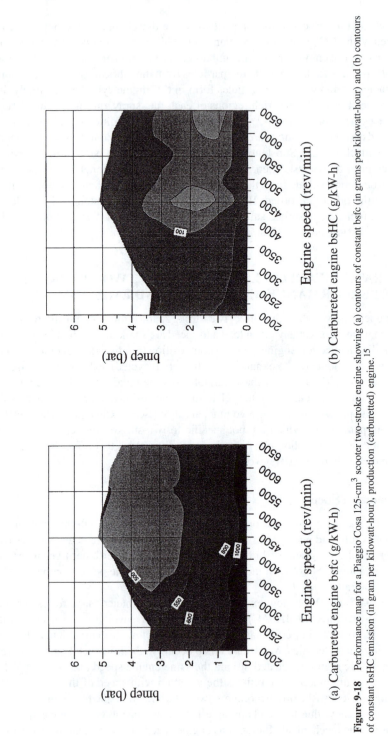

Figure 9-18 Performance map for a Piaggio Cosa 125-cm^3 scooter two-stroke engine showing (a) contours of constant bsfc (in grams per kilowatt-hour) and (b) contours of constant bsHC emission (in gram per kilowatt-hour), production (carburetted) engine.[15]

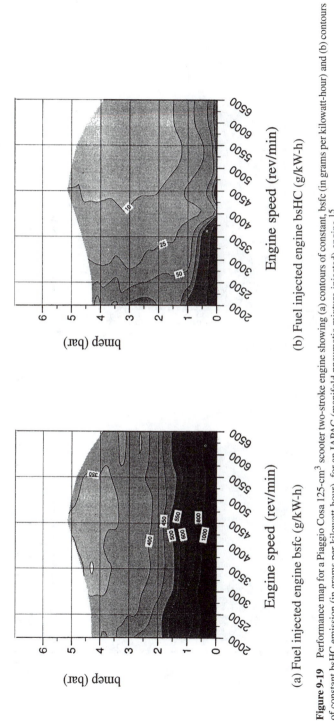

(a) Fuel injected engine bsfc (g/kW-h)

(b) Fuel injected engine bsHC (g/kW-h)

Figure 9-19 Performance map for a Piaggio Cosa 125-cm^3 scooter two-stroke engine showing (a) contours of constant, bsfc (in grams per kilowatt-hour) and (b) contours of constant bsHC emission (in grams per kilowatt-hour), for an IAPAC (manifold pneumatic mixture-injected) engine.[15]

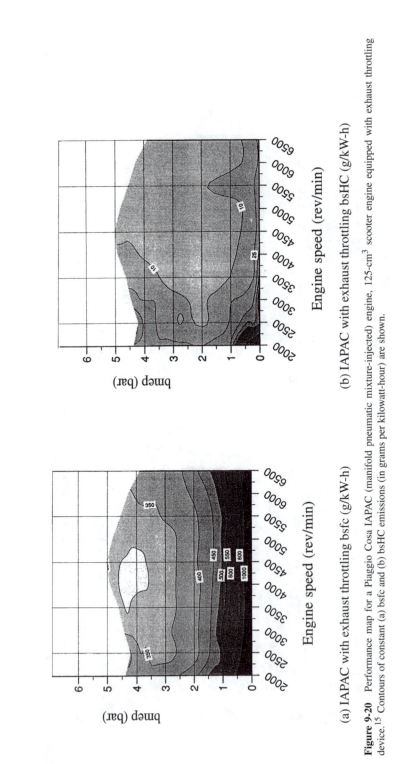

(a) IAPAC with exhaust throttling bsfc (g/kW-h)

(b) IAPAC with exhaust throttling bsHC (g/kW-h)

Figure 9-20 Performance map for a Piaggio Cosa IAPAC (manifold pneumatic mixture-injected) engine, 125-cm^3 scooter engine equipped with exhaust throttling device.[15] Contours of constant (a) bsfc and (b) bsHC emissions (in grams per kilowatt-hour) are shown.

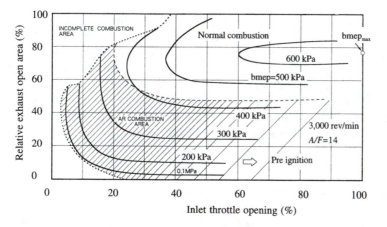

Figure 9-21 The effect of degree of exhaust throttling (throttled exhaust port open area divided by un-throttled open area) and amount of inlet throttling on the bmep of a 246-cm^3 carburetted Honda engine at 3000 rev/min.[10] The cross-hatched area shows where controlled autoignition (active radical combustion) occurs.

the heat losses to the cylinder wall on a per-cycle basis as engine speed increases. The bsfc contours may be explained by the same arguments, while the possibility of short-circuiting of some fuel through the exhaust port should also be considered for both the carburetted and manifold-fuel-injected engines.

Figure 9-19 shows that a significant improvement in torque in the lower engine speed range occurs when the carburetor is replaced by the IAPAC mixture injection system. This is attributed to the use of inlet reed valves instead of the rotary valve in the carburetted engine.[15] The torque in the high engine speed range of the IAPAC engine is less mainly because of the increase in friction losses of the added valve train. In spite of these higher friction losses, the full-load bsfc of the IAPAC engine is always between 20 and 50% better. Also as a result of the minimization of the short-circuiting losses in the IAPAC engine, bsHC emissions (Fig. 9-19b) are reduced by more than 70% at full load. At partial loads and low engine speeds, a significant level of HC emission still persists with the IAPAC engine. This is attributed in part to misfiring or poor combustion.[15] Under these operating conditions, exhaust throttling can significantly improve the combustion process, and thus reduce HC emission.[15,16,10] Figure 9-20 shows how exhaust throttling affects the bsfc and bsHC emission with the IAPAC engine. The low-bsfc regions are greatly extended, whereas bsHC emissions are substantially reduced over the entire operating range.

Exhaust port throttling, by controlling the in-cylinder pressure according to the engine speed and the load, can cause a different type of combustion process (termed active radical combustion) to occur; see Section 6-3-1. This is a spontaneous, rapid enough but not too rapid, autoignition process occurring throughout the fuel, air, and residual gas mixture region that is both more complete and more repeatable, cycle-by-cycle, than the normal light-load two-stroke combustion process. Under certain operating conditions, controlled exhaust port throttling can continuously provide the proper exhaust pressure to obtain active radical combustion, and thereby achieve substantial reductions in fuel consumption and HC emissions at part load. The effect of the degree of exhaust throttling, at various engine loads, on the bmep of a 246-cm^3 carburetted Honda engine[10] at 3000 rev/min is shown in Fig. 9-21. Controlled autoignition (active radical combustion) takes place in the cross-hatched area. The combination of low exhaust throttling (little reduction in exhaust port open area) and low engine load (low inlet throttle opening) causes incomplete combustion, whereas a high degree of exhaust throttling and high engine loads cause preignition. Because the bmep maps depend also on the engine speed,[10] load should be controlled by selecting the optimum combination of inlet and exhaust throttling for the specific engine speed.

9-3-2 Mixture Composition

Mixture composition during combustion is most critical, because this determines the development of the combustion process. A robust combustion process—one that is sufficiently fast and repeatable—is essential for achieving satisfactory engine operation. The mixture inside the cylinder prior to ignition is composed of air, fuel (mainly vapor with a small fraction of tiny droplets), and burned gas. The burned gas fraction in small two-stroke engines is usually the residual gas; recycled exhaust gas, which is commonly used in four-stroke engines for NO_x control, is normally used only in larger (car) two-stroke engines.

Figure 9-22 shows how the overall air/fuel ratio affects the power, efficiency, and emission characteristics of a Schnürle-type carburetted crankcase-scavenged engine at wide-open throttle.[4] In these tests, the fuel–air mixture was prepared in two different ways: (1) with the normal carburetor; (2) with a carburetted stratified-scavenging system (see Section 9-2-1 for engine details). The engine power peaks slightly rich of stoichiometric ($A/F \approx 14.6$). Due to dissociation at the high temperatures following combustion (in particular, dissociation of H_2O and CO_2), molecular oxygen is present in the burned gases under stoichiometric conditions, so some additional fuel can be added and partially burned. This increases the temperature and the number of moles of the burned gases in the cylinder. These effects increase the pressure to give increased power and mep. At the same time, the incomplete oxidation of fuel molecules results in increased HC and CO emissions. For the standard engine design, the charge mixture is more homogeneous, oxygen utilization is more complete, and a higher engine power can be obtained. However, because of the homogeneity of the incoming charge, a part of the fuel is inevitably short-circuited to the exhaust during the gas exchange process, significantly increasing fuel consumption and HC emissions.

For mixtures lean of stoichiometric, the fuel conversion efficiency increases as the fuel/air equivalence ratio decreases below 1.0 (relative air/fuel ratio increases above 1.0). Combustion of lean mixtures produces products at lower temperature, and with less dissociation. Thus the fraction of the chemical energy of the fuel that is released as sensible energy is greater, and a larger fraction of the burned gas sensible energy is transferred as work to the piston during expansion. As the mixture becomes leaner still, cycle-to-cycle pressure fluctuations and the total duration of the burning process increase and become the dominant factor in the degradation of engine efficiency.

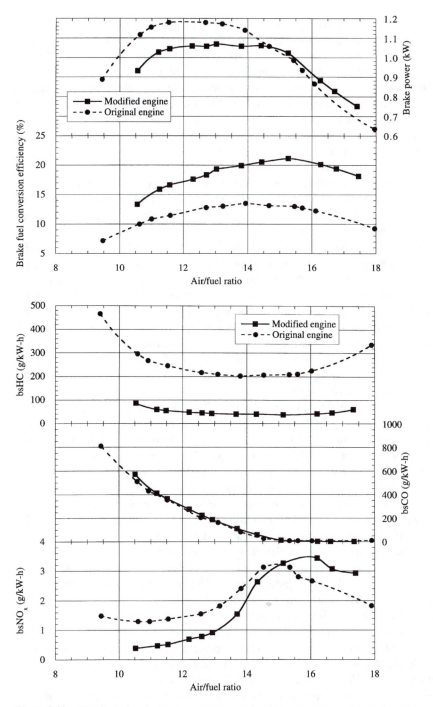

Figure 9-22   The effect of air/fuel ratio on the power, efficiency, and emissions characteristics of a Schnürle-type carburetted crankcase-scavenged engine at wide-open throttle.[4] Engine details are given in Table 9-2.

The critical factors affecting emissions that are governed by the relative fuel/air or air/fuel ratio are the oxygen concentration and the temperature of the burned gases. The maximum burned gas temperatures occur slightly rich of stoichiometric early in the expansion stroke. Excess oxygen is available in the burned gases lean of stoichiometric. The formation of NO depends on the gas temperature and oxygen concentration. Although maximum burned gas temperatures occur at a relative air/fuel ratio of around 1.1, at this equivalence ratio oxygen concentrations are low. As the mixture is leaned out, increasing oxygen concentration initially offsets the falling gas temperatures and NO emissions peak at a fuel/air equivalence ratio of 0.9 (see Sections 7-1-2 and 7-2-2).

The modified (stratified) engine provides 80% lower HC and some 50% higher thermal efficiency in spite of 7% lower power output than the current engine. These results are due to the significant reduction in the short-circuiting that is unavoidable in the current standard engine design. The lower power output is attributed[4] to the lower charging efficiency of the modified engine.

Exhaust gas recycle (EGR) is a commonly used technique for control of four-stroke SI engine NO_x emissions. A fraction of the exhaust gases (typically 5-15% is recycled through a control valve from the exhaust to the engine intake system. EGR acts as an additional diluent in the unburned gas mixture, thereby reducing the peak burned gas temperature and NO formation rate. The retained burned gas in the cylinder at the commencement of the compression stroke thus comprises both residual gas from the previous cycle and exhaust gas recycled to the intake. In four-stroke engines, typical values of the residual mass fraction vary between about 5% at WOT and 30% at idle, and depend on the engine load and speed. A higher speed and lower load increase the residual mass fraction. In simple loop and cross crankcase-scavenged two-stroke engines, the scavenging efficiency is typically low (70–80%), the residual mass fraction is high (about 10–30%), and EGR is not employed. In uniflow-scavenged engines, in direct injection engines, and when an external-scavenging blower is used, the scavenging process is more efficient, and EGR is more frequently employed.

The effect of exhaust gas recycle on engine performance and efficiency, for lean and stoichiometric mixtures, is similar to the addition of excess air. Both EGR and excess air dilute the unburned mixture, and hence reduce the peak burned gas temperature. Figure 7-17 shows the effect of increasing EGR and air–fuel ratio on bsfc, $bsNO_x$, and bsHC emissions, at constant engine torque in a direct injection two-stroke gasoline engine.[5] In this engine the cylinder charge is stratified, and the recycled exhaust gas was introduced in two ways: directly to the crankcase through piston-skirt-controlled ports; and into the intake manifold. The latter is more effective. In these experiments, ignition timing was set to MBT (optimum) timing. In the figure, line AB shows the effect of leaning out the overall air/fuel ratio to the point B where the efficiency is maximized— the largest torque for a given fuel flow is achieved. Then B–C and B–D show the effect of increasing EGR, at this (fixed) air/fuel ratio. Initially, EGR significantly reduces NO_x with negligible change in fuel consumption. At the higher EGR levels, however, combustion starts to deteriorate and as a result fuel consumption increases. EGR does have an adverse effect on bsHC emissions (see Fig. 7-17b), presumably because burned gas temperatures are reduced, which would increase bulk quenching in the lean outer edges of the fuel–air mixture region and reduce the afterburning of hydrocarbons that escape the normal combustion process (see Section 7-2-4). Note that EGR is a part-load emission control strategy. At full load, EGR would replace air in the cylinder so the amount of fuel that can be burned, and thus the torque produced, would be reduced. At partial load in this direct-injection engine there is excess air, so whether the EGR is additional to or replaces air is not critical (and it undoubtedly does both). In four-stroke SI engines, use of EGR at partial load with good (fast) combustion systems can improve bsfc due to lower burned gas temperatures (increasing fuel energy released as sensible energy and the relative amount of sensible energy extracted by the moving piston as work, and reducing heat losses to the walls), and due to reduced pumping work (due to the higher intake pressure, at fixed torque, with EGR). In the two-stroke engine only the former effect is present (because the effect of EGR on pumping work is small). The effect of EGR on the scavenging flow, and thus on mixture preparation, has yet to be explored.

9-3-3 Spark Timing

The effect of changes in spark timing on the performance characteristics of a two-stroke SI engine is similar to that in four-stroke engines. If combustion starts too early in the cycle, the work needed to compress the cylinder contents during the compression stroke is large, as also are heat losses from the burned gas; if combustion starts too late, the peak cylinder pressure is reduced and occurs later, and the work produced during the expansion stroke is lower. There exists a particular spark timing for which the indicated work, that is, the enclosed area on a cylinder-pressure–cylinder-volume diagram, reaches a maximum value. This timing gives the maximum engine torque at a given engine speed, and mixture composition. Maximum brake torque (MBT) and minimum brake specific fuel consumption are both achieved at this point. This optimum ignition timing is referred to as MBT timing.

As engine speed increases, the combustion duration in crank angle degrees increases, and the spark has to be advanced to maintain optimum timing. Optimum spark timing also depends on load. As load is decreased, the intake pressure decreases, the combustion duration increases due to the increasing amount of residual, and the spark timing has to be further advanced to maintain optimum engine performance.

Spark timing affects peak cylinder pressure and therefore peak unburned and burned gas temperatures. Retarding spark timing from the optimum reduces peak pressure and temperature, and increases exhaust temperatures because the engine is less efficient and extracts less expansion stroke work. The increase in gas temperature during the latter part of expansion and in the exhaust increases the amount of hydrocarbons oxidized during expansion and exhaust and thus reduces HC emissions. Retarding spark timing also brings the ignition point closer to TC, where the conditions for avoiding misfire are more favorable. Retarding spark timing is, therefore, customarily used to (i) reduce NO_x emissions, (ii) avoid knock, (iii) reduce hydrocarbon emissions, and (iv) avoid misfire at engine idle.

9-3-4 Fuel Injection Parameters

Fuel injection timing essentially controls the process of preparing the fuel–air mixture for combustion. In SI engines, the fuel is customarily sprayed into air at atmospheric or lower pressure before the commencement of the compression process, thus allowing the use of low-pressure low-cost fuel injection systems. Intake manifold injection, crankcase injection, and port injection arrangements are in common use. With these arrangements, a relatively long time is available for the fuel to evaporate and mix before combustion takes place, and larger fuel droplets can be tolerated. The injection timing may be synchronized with the incoming air flow to produce charge stratification. In direct-injection engines, the fuel is sprayed into the cylinder a relatively short time before combustion starts. During this short period, the droplets must evaporate and mix with the appropriate amount of air. A fine spray with appropriate penetration and spread, and small (10–20-μm diameter) drops is needed. This requires either a high-pressure liquid fuel injection system (injection pressures of order 50 bar) or a lower pressure (of order 5 bar) air-assisted fuel injection system.

An experimental comparison between an air-assisted injection system and a high-pressure liquid injection system in the same engine has been carrried out at Hyundai.[17] The spark plug and fuel injector were installed in the cylinder head with the injector located on the exhaust-port side and directed toward the opposite side of the cylinder to reduce short-circuiting and enrich the fuel–air mixture in the vicinity of the spark plug. The combustion chamber was a hemisphere in the cylinder head, with a squish region on the exhaust-port side. Additional engine details are given in Table 9-10.

Figure 7-14 shows how varying the start of injection timing affects the bsfc, bsHC, and bsCO of the engine with both the high-pressure injection system (HPFI), and with the air-assisted fuel injection system (AAFI).[17] The tests were performed at a part-load bmep of 200 kPa and an engine speed of 1200 rev/min. Spark timing was set at MBT. In general, advancing direct injection timing is limited by the onset of fuel short-circuiting.

Table 9-10 Details of Hyundai direct-injection SI engine[17]

Engine type	Single-cylinder Schnürle loop-scavenging with an external blower	
Displacement	399 cm^3	
Bore/stroke	82.3/75 mm	
Geometric compression ratio	10.8	
Trapped compression ratio	6.9	
Scavenge port open	67°BBC	
Exhaust port open	87°BBC	
Lubrication system	Wet sump	
Fuel injection	High pressure	Air-assisted
Cone angle	55°	70°
SMD, partial load/full load	15/6 μm	Less than 6 μm
Pulse width, partial load/full load	0.3/1.7 ms	3.0/3.5 ms
Injection pressure	7 MPa	0.75 MPa

Retarding injection timing is limited by insufficient time for fuel vaporization and fuel–air mixing prior to combustion. An increase of the engine speed therefore requires a more advanced injection timing. The steep increase in bsfc and bsHC as injection timing is retarded with the high-pressure liquid injection system suggests that vaporization, mixing, and mixture region location relative to the spark plug create major mixture preparation problems. The experimental results show that the air-assist injection system allows a later injection timing, indicating that smaller fuel droplets and consequently faster vaporization and more appropriate mixture region geometry are achieved, compared with the performance of the HPFI system. The bsCO emission, however, tends to increase as the injection timing is retarded, as a result of more fuel-rich regions due to less time for vaporization and mixing. Note that because the cone angle of the liquid fuel injector is smaller than that of the air-assisted fuel injector, the HPFI allows earlier start of injection timing without noticeable short-circuiting losses.

9-3-5 Ambient Conditions

The crankcase-scavenged two-stroke engine is most attractive where low weight, small bulk, and high power output are the paramount requirements. These qualities are especially important in small pilotless aircraft and in meteorological observation balloons where a small electrical generator unit is needed.[18] In both cases the performance at high altitudes is determined primarily by the ability of the engine to produce the desired power there. Several correlations have been suggested in the literature to estimate the effect of changes in atmospheric conditions on the indicated power of a reciprocating spark-ignition engine. The SAE[19] and Bosch[20] handbooks recommend a correction factor (not specifically for a two-stroke engine) of the form

$$CF = \frac{(p - p_w)}{(p - p_w)_s} \left(\frac{T_s}{T} \right)^{1/2} \tag{9-10}$$

where p and p_w are the atmospheric pressure and the partial pressure of the water vapor, respectively, T is the atmospheric temperature, the subscript s denotes standard atmosphere conditions at sea level, and the correction factor CF is defined by

$$P_i = CF P_{i,s} \tag{9-11}$$

where P_i is the indicated power of the engine. Obert[21] suggests that if all variables other than atmospheric conditions are held constant, the correction factor for the *indicated* performance of an engine should be directly proportional to the mass of dry air inducted. The volumetric efficiency is not constant but increases as $T^{1/2}$ at medium and high engine speeds because the airflow through the engine's intake system behaves similarly to gas flow through an orifice or flow restriction. This would explain the $1/2$ power exponent in Eq. (9-10). The DIN 6270 standard, as well as the JIS, adopted a similar correction factor for general-purpose internal combustion engines, but preferred a 0.75 exponent for (T_s/T), rather than 0.5. Watanabe and Kuroda[22] studied experimentally the effect of the atmospheric temperature on the power output of a two-stroke crankcase-compression gasoline engine and found that the power output varied inversely with the absolute temperature raised to an exponent from 0.5 to 0.9. They explained their

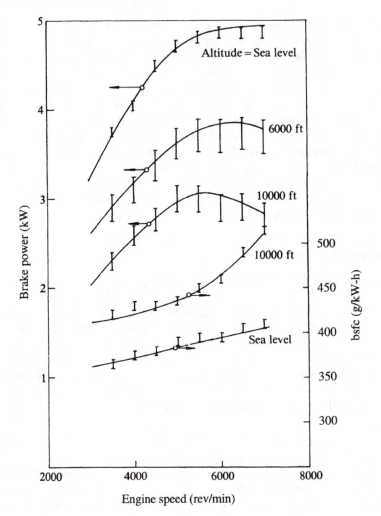

Figure 9-23 The effect of altitude (see Table 9-11) on the performance of a crankcase-scavenged Schnürle-type loop-scavenged spark-ignition two-stroke engine.[18] Reproduced by permission of the Council of the Institution of Mechanical Enginneers.

result by the fact that the pressure ratio of the crankcase compressor decreases as the inlet air temperature increases. These authors also indicated that the temperature of the cooling air was kept constant in their experiments, but in practice the cooling function of the crankcase deteriorates when air temperature increases. It is conceivable, therefore, that the power output will decrease more sharply with increasing air temperature than Eq. (9-10) indicates.

The effect of changes in atmospheric pressure and temperature on the performance of three different types of 111-cm^3 crankcase-blown Schnürle-type loop-scavenged spark-ignition two-stroke engines is shown in Fig. 9-23.[18] The change in the brake

Table 9-11 Standard atmosphere

Altitude above sea level		Pressure	Temperature
(ft)	(m)	(kPa)	(K)
0	0	101.0	288
1,000	305	97.4	286
2,000	610	93.9	284
3,000	914	90.5	282
4,000	1219	87.2	280
5,000	1524	84.0	278
6,000	1829	80.9	276
7,000	2134	77.9	274
8,000	2438	75.0	272
9,000	2743	72.2	270
10,000	3048	69.4	268
11,000	3353	66.8	266
12,000	3658	64.2	264
13,000	3962	61.7	262
14,000	4267	59.3	240
15,000	4572	57.0	258
16,000	4877	54.7	256
17,000	5183	52.5	254
18,000	5487	50.4	252
19,000	5791	48.4	250
20,000	6096	46.4	248
21,000	6402	44.4	246

power at high altitude was mainly attributed to the deterioration of the charging efficiency because the delivery ratio only increases to a moderate extent with increasing ambient pressure and temperature.

The following empirical correlation factors have been suggested[18] for predicting the indicated power and specific fuel consumption of an engine at any pressure and temperature using information obtained at standard atmospheric conditions (Table 9-11):

$$\frac{P_i}{P_{i,s}} = \left[\frac{(1 - \tilde{x}_w)}{(1 - \tilde{x}_{w,s})} \cdot \frac{p}{p_s} \right]^{9/8} \left(\frac{T_s}{T} \right)^{0.8} \tag{9-12}$$

and

$$\frac{isfc}{isfc_s} = \left[\frac{(1 - \tilde{x}_w)}{(1 - \tilde{x}_{w,s})} \cdot \frac{p}{p_s} \right]^{1/8} \tag{9-13}$$

It was also concluded[18] that the optimum inlet port timing is independent of altitude level and only depends modestly on engine speed. A higher altitude level and a higher engine speed require a smaller crankcase-to-swept-volume ratio to obtain the maximum available indicated power.

Another study of the effect of the ambient pressure on the brake power of an airborne 350-cm^3 two-cylinder engine[23] has suggested that for altitudes higher than 3000 m (10,000 ft), the correction factor is proportional to the ambient pressure raised

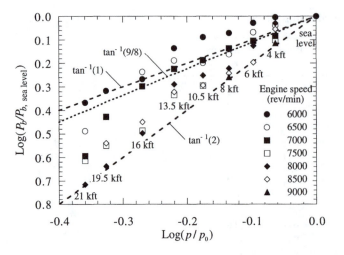

Figure 9-24 Maximum engine power as a function of ambient pressure, on logarithmic coordinates, for various engine speeds.[23] 350-cm^3 two-cylinder crankcase-scavenged two-stroke SI engine, 1 kft = 1000 ft.

to a power that depends on engine speed: 1 for low, and 2 for high engine speed. Figure 9-24 shows the results. It was concluded that the decrease of brake power with decreasing air pressure results from the decreasing charging efficiency, due to the deterioration of both the delivery ratio and the scavenging efficiency. The delivery ratio decreases mainly due to backflows from the crankcase to the ambient that occur at low engine speeds. The scavenging efficiency deterioration was attributed to charge losses by short-circuiting. The exhaust gas purity, therefore, increases with increasing altitude.

9-4 OPERATING VARIABLES THAT AFFECT TWO-STROKE CI ENGINE PERFORMANCE, EFFICIENCY, AND EMISSIONS

The major operating variables that affect naturally aspirated two-stroke compression-ignition engine performance, efficiency, and emissions are engine speed, engine load, fuel injection parameters, and ambient conditions. The effect of changes in altitude is mainly due to changes in the gas exchange process (deterioration of both the delivery ratio and the scavenging efficiency). Therefore, two-stroke diesel engine performance changes with ambient conditions will be similar to those of an SI engine.[22] This topic has been covered in Section 9-3-5. Design considerations that relate to the gas exchange process (such as the port geometries and crankcase volume) can be found in Chapter 5. The effects of other major design parameters that are not specific to two-stroke engines, such as compression ratio, combustion chamber shape, and the variables that affect heat transfer, are reviewed elsewhere.[1] The effects of the engine speed, engine load, and fuel injection parameters will be discussed in this section.

9-4-1 Load and Speed

Two-stroke diesel engines have, for many years, been mainly used as power plants in large-capacity seacraft and electric generators. For these applications, it is convenient to present the operating characteristics such as bsfc and bsCO as a function of engine load (Figs. 9-16 and 9-17). The thermodynamic efficiencies of these large-bore two-stroke engines are much higher than those of SI engines; typical values for brake fuel conversion efficiency are above 50% [e.g., from Fig. 9-16: at 25% load, a bsfc of 175 g/kW-h corresponds to $\eta_{f,b} = 0.51$; at 50% load, bsfc $= 167$ g/kW-h and $\eta_{f,b} = 0.53$, at 80% load bsfc is a minimum (165 g/kW-h) and $\eta_{f,b} = 0.54$]. As maximum power is approached, injection is retarded to control the maximum cylinder pressure, so bsfc and fuel conversion efficiency worsen slightly.

The CO and HC emissions from these large engines are low: the low engine speeds and high excess air ratios at which these engines operate lead to high combustion efficiencies and low levels of these pollutants. These same operating conditions lead, however, to high NO_x emissions. Smoke and particulate emissions are of importance, too.

The effects of changes in engine load and speed on NO_x emissions are shown in Fig. 9-25. This large marine engine produces 2.8 MW, has a bore and stroke of 450 and 750 mm, respectively, and has a maximum bmep of 1.02 MPa. The engine was used to drive a matched propeller. The figure shows relative NO_x emissions (ppm), NO_x emission rate, brake specific NO_x emissions, and NO_x emissions index. The relative NO_x emissions (Fig. 9-25a) rise with a reduction in the engine load and show high values in the low engine load range following the propeller characteristics. When engine load decreases while keeping engine speed constant, relative NO_x emissions go down. Relative emissions do not provide a measure of the total quantity of NO_x emitted because exhaust gas mass flow rate increases as the engine load increases. When the NO_x emission per unit time is plotted against the engine load, it exhibits a monotonic increase with the load (Fig. 9-25b). Decreasing the engine load by a factor of 3, from 75% to 25%, results in only about a 50% reduction in NO_x emissions per unit time. The brake specific NO_x emissions (Fig. 9-25c) and the NO_x emission index (Fig. 9-25d), however, show a behavior similar to the relative emission because these emissions parameters have been normalized by power and fuel flow rate, respectively.

These data indicate that the magnitude of this engine's NO_x emissions increase with an increase in the engine load, and decrease with an increase in the engine speed at the same engine load. This can be understood by considering the variation of the cylinder pressure, mean in-cylinder gas temperature, and rate of heat release as a function of crank angle shown in Fig. 9-26. Assuming that injection timing depends mainly on engine speed and is therefore approximately constant in this figure, the combustion duration becomes longer with an increase in the engine load, the amount of fuel burned per cycle increases the maximum cylinder pressure and the mean gas temperature in the cylinder rises. Consequently, the amount of NO_x emitted per unit time increases. As engine speed increases at constant load, the amount of NO_x emitted decreases since injection and combustion now occupy longer crank angle intervals, so the peak cylinder pressure occurs later after TC (and will be lower); hence peak cylinder gas temperatures are later and lower. Maeda et al.[24] showed that the amount of NO_x emitted per unit time

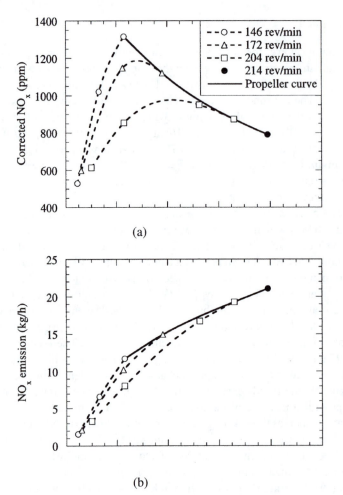

Figure 9-25 The effects of engine load and speed on (a) relative NO_x emissions (ppm in exhaust corrected to 13% oxygen), (b) rate of NO_x emission, (c) brake specific NO_x emission, and (d) NO_x emissions index, of a large-bore two-stroke marine diesel engine.[24]

correlates well with the maximum temperature in the cylinder irrespective of engine speed.

We now discuss small-bore ($\sim$80 mm) two-stroke cycle diesel engine performance. Figure 9-27 shows how the operating characteristics of two small-bore direct injection diesels depend on engine load. Results are shown for ported loop-scavenged and uniflow-scavenged engines. Engine details are given in Table 9-7. From both engines, as engine load increases, the air/fuel ratio decreases as expected, isHC emissions decrease, and the smoke number (SN) and isNO$_x$ emissions both increase. The indicated specific fuel consumption reaches a minimum value at about 80% of the maximum load. The increase in isCO emissions when load decreases below 50% of its maximum value

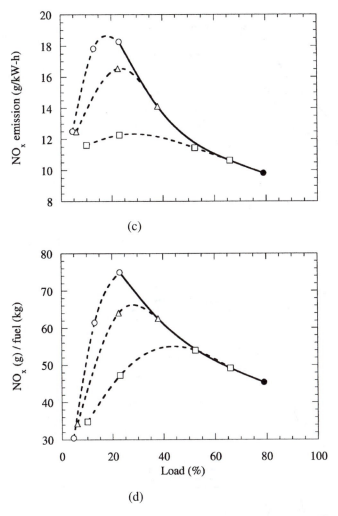

(c)

(d)

Figure 9-25   (Continued).

indicates that poor combustion is occurring in some regions inside the cylinder due to excessively lean mixtures. With the ported loop-scavenged engine, the smoke number increases significantly even at low loads: at an imep of 300 kPa, the SN exceeds a value of 1.0 and at imep = 570 kPa, the full load limit of a SN of 2.5 is reached. With the uniflow-scavenged engine, the values of the smoke number are significantly lower. Up to an imep of 700 kPa, the smoke number remains below 0.5 and the full load limit (SN = 2.5) is reached at an imep of 880 kPa. As a consequence, the fuel injection pressure requirement at the same mean effective pressure is some 50% lower than for the ported loop-scavenged engine. The significant difference[11] in the smoke number as load increases show that satisfactory mixture preparation cannot be achieved with a nonswirling scavenging flow, in spite of high injection pressures and extremely small

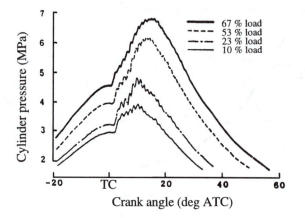

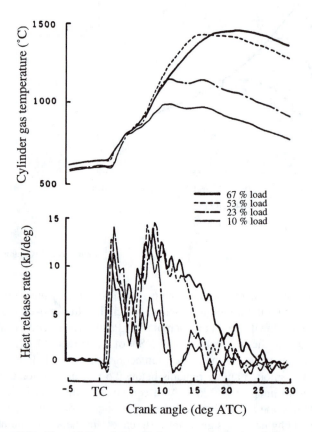

Figure 9-26 The variation of the cylinder pressure, mean in-cylinder gas temperature, and rate of heat release with crank angle, at various load levels, for a large-bore two-stroke diesel engine.[24]

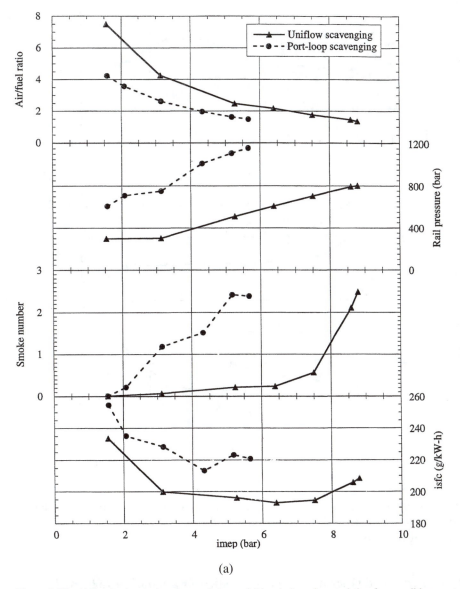

(a)

Figure 9-27 (a) Part-load operating characteristics, and (b) emissions characteristics, for a small-bore ported loop-scavenged and a uniflow-scavenged two-stroke diesel engine.[11]

nozzle hole diameters (see Fig. 9-28). The poor combustion quality with ported loop-scavenged engine is also reflected by the increases in specific CO and HC emissions (Fig. 9-27b).

The fuel consumption at partial load is 10% higher for the ported loop-scavenged engine. This is attributed in part to the unfavorable surface-to-volume ratio resulting form the approximately 30% shorter stroke. The specific NO_x emission values are sim-

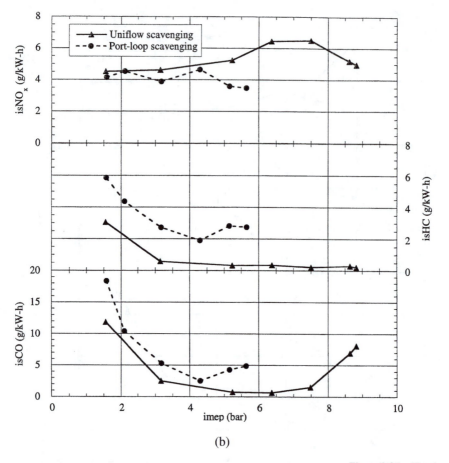

Figure 9-27 (Continued).

ilar to or slightly below the values with the uniflow-scavenged engine. In part, this is the result of the higher residual gas fraction in the cylinder, resulting from the poorer scavenging behavior.

9-4-2 Part-Load Operating Characteristics for a Small-Bore Ported Loop-Scavenged and a Uniflow-Scavenged Two-Stroke Diesel Engine[11]

In diesel engines, the fuel injection parameters control the characteristics of the combustion process, such as the crank angle at which combustion starts, combustion duration, rate of cylinder pressure rise, maximum gas pressure and temperature, and levels of emissions. The fuel injection parameters, which include injection pressure, injection nozzle design, and injection timing, all affect the characteristics of the diesel fuel spray (its penetration, rates of vaporization and mixing with air in the combustion chamber) and hence the combustion process.

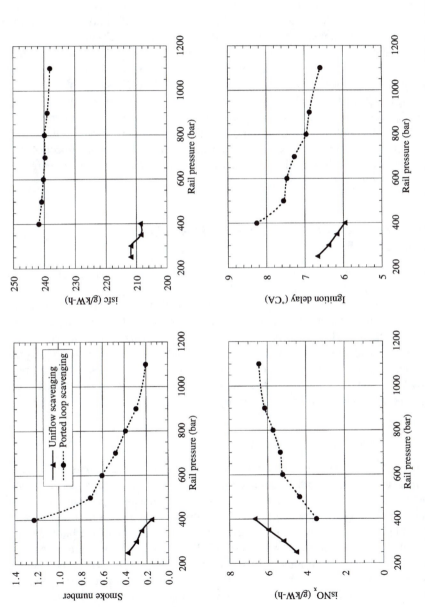

Figure 9-28 The effect of injection pressure in a common-rail injection system, on the smoke number, isNO$_x$ isfc, and ignition delay, for small-bore direct-injected ported loop-scavenged and uniflow-scavenged two-stroke diesels.[11]

Figure 9-28 shows the effect of the injection pressure in a common-rail injection system on the smoke number, isNO$_x$, isfc, and ignition delay, in the small-bore ported loop-scavenged and uniflow-scavenged engines discussed in Section 9-2-3. Engine details are given in Table 9-7. As expected, the performance of the non-swirling ported loop-scavenging system is inferior to the uniflow-scavenging system. It was found[11] that, to achieve comparable smoke numbers with the non-swirling scavenging system, it was necessary to double the rail pressure (Fig. 9-28) and significantly reduce the size of the injection hole diameter, from 0.122 to 0.085 mm, to compensate for the poorer evaporation and mixing of the fuel. The isfc in the ported loop-scavenged engine is higher by 10–15% than that of the uniflow-scavenged engine [due to the reasons noted previously: worse surface-to-volume ratio, less lean overall A/F, poorer (slower) combustion, higher CO and HC emissions]. Note that as rail pressure is increased by a factor of 3, isfc decreases only slightly (about 1%). A more profound effect of rail pressure is observed with the uniflow-scavenged engine, indicating an improvement in the rate at which the combustion process releases the fuel energy due to higher fuel injection rate and spray quality. This results also in higher combustion temperatures and, therefore, higher isNO$_x$ emissions.

9-5 SUPERCHARGED AND TURBOCHARGED ENGINE PERFORMANCE

The power and torque that an internal combustion engine produces are proportional to the mass of the air induced per cycle. This depends on the density of the trapped charge at the beginning of the compression stroke. It follows that the performance of an engine of a given displacement can be increased by compressing the inlet charge prior to entry to the cylinder. The more common methods of achieving higher charge density include mechanical supercharging, turbocharging, and pressure-wave supercharging. Fundamental consideration of these devices and a more extensive discussion are provided by Heywood[1] and Watson and Janota.[25] In this section, we examine the effects on engine performance of boosting air density.

9-5-1 Two-Stroke Cycle SI Engines

In the simple crankcase-scavenged two-stroke SI engine, the fresh charge is compressed in the crankcase prior to its introduction to the cylinder. The fresh charge is either combustible mixture (in manifold-injected or carburetted engines) or pure air (in direct-injection engines). The fresh charge must be compressed to a pressure that is higher than the exhaust system pressure to achieve effective scavenging of the burned gases (typical values are 130–150 kPa). An excessively high pressure, however, results in additional energy losses due to unnecessary compression. In engines where the opening and closing of the ports are controlled by the piston, the exhaust port closes after the scavenging port, and the pressure of the trapped charge is determined by the pressure of the exhaust system. The degree of supercharging that can be achieved is therefore limited.

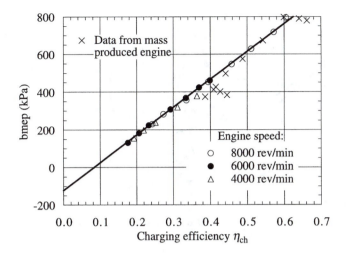

Figure 9-29 Relationship between brake mean effective pressure and charging efficiency at various engine speeds and loads, for a Honda 125-cm^3 displacement single-cylinder two-stroke SI engine installed with a reed valve.[26]

Table 9-12 Engine details for Fig. 9-31[10]

Engine type	Honda single-cylinder SI engine
Scavenging type	Schnürle loop-scavenged
Maximum power	29 kW at 8000 rev/min
Maximum torque	36 N-m at 6500 rev/min
Bore/stroke	66/72 mm
Displacement	246 cm^3
Trapped compression ratio	6.4
Scavenge port opens	124° ATC
Exhaust port opens (with throttling)	86°–114° ATC
Scavenging/exhaust port area	15/13 cm^2

The engine power and torque are also strongly dependent on the extent to which the displaced volume is filled with fresh mixture, that is, on the charging efficiency η_{ch} [see Eq. (2-2)]. The fuel consumption depends on both the charging efficiency and trapping efficiency η_{tr} [Eq. (2-5)]. Figure 9-29 shows how bmep depends on the charging efficiency, for a Honda single-cylinder engine with a displaced volume of 125 cm^3 installed with a reed valve.[26] The charging efficiency increases from about 0.2 for low engine load to about 0.65 at high engine load. The linear dependence of bmep on the fresh mass retained over the entire range of operation is clear. Figure 9-30 shows how the trapping efficiency of this engine varies with increasing delivery ratio Λ, at several engine speeds and loads. The delivery ratio increases from about 0.3 at idle conditions to 1.0 at wide-open throttle and high engine speed. Lines of constant charging efficiency, which equals $\Lambda\eta_{tr}$, are shown. The figure shows that the trapping efficiency decreases, for all engine speeds, with increasing mass of delivered charge (or delivery ratio). The

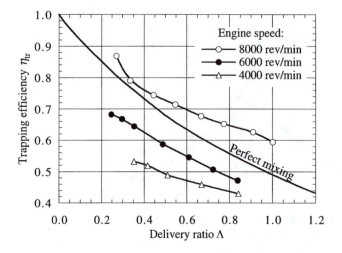

Figure 9-30 Trapping efficiency as a function of delivery ratio, at several engine speeds and loads, for Honda 125-cm³ displacement single-cylinder two-stroke SI engine installed with a reed valve[26]: data are lines of constant charging efficiency.

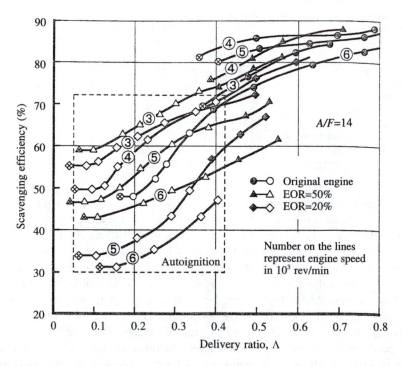

Figure 9-31 Scavenging efficiency as a function of delivery ratio, engine speed, and exhaust throttling degree, for a 246-cm³ displacement single-cylinder Honda two-stroke SI engine.[10] Engine details are given in Table 9-12.

slope of these lines of constant charging efficiency are steeper for low values of η_{tr}, because less mixing between the fresh charge and the cylinder contents occurs when the mass of the delivered fresh charge is small so a larger fraction of this charge is retained inside the cylinder. Note that for the data shown, as the speed increases, the trapping efficiency increases. For an engine speed of 8000 rev/min, values above the perfect mixing curve are obtained.

Figure 9-31 shows how the scavenging efficiency η_{sc} [Eq. 2-3)] depends on the delivery ratio (which is a function of the engine load), engine speed, and degree of exhaust throttling, for a 246-cm^3-displacement single-cylinder engine. The engine details are given in Table 9-12.

9-5-2 Two-Stroke Cycle CI Engines

The two-stroke engine is not self-aspirating hence the addition of a turbocharging system will significantly affect the scavenging process. Because in two-stroke engines both inlet and exhaust ports are open during the scavenging process and the exhaust gases are pushed out by the incoming fresh charge, the port timing and pressure difference across the cylinder must be carefully designed to ensure an efficient gas exchange process with minimum waste of pumping energy.

Large two-stroke cycle CI engines are used in marine power stations, and some automotive applications. These engines are usually turbocharged to achieve high brake mean effective pressures and specific outputs. Pulse and constant-pressure turbocharging systems† and, more recently, pulse converters, are in common use.[1,25]

The two-stroke engine can run successfully with a pulse exhaust system[25] provided that the exhaust manifold design is such that positive pressure waves do not arrive at the exhaust port during the course of the scavenging process. For most of the scavenging process the exhaust pressure has to be kept low. However, a rise in pressure at the end of the scavenging process can help by increasing the trapped mass of the fresh charge in the cylinder. With the pulse exhaust system, effective operation is only achieved over a limited speed range, so these engines are sometimes fitted with auxiliary compressors to aid scavenging at low speeds. Such two-stroke engines running without an auxiliary compressor are usually started from compressed air bottles.[25] Turbocharging with a pulse system provides substantial pressure energy at the turbine inlet during the blowdown period and effective scavenging at partial load. The latter can be achieved by using a small manifold volume and short pipes. Pulse reflections that interfere with the scavenging process may occur.

When a two-stroke engine is turbocharged with a constant-pressure system, the exhaust turbine inherently increases the pressure in the exhaust manifold, and an ap-

---

†Pulse turbocharging systems use short small-cross-section pipes to connect the exhaust ports to the turbine so that much of the kinetic energy of the gas in the exhaust blowdown process can be utilized. Constant-pressure turbocharging systems use an exhaust manifold of sufficiently large volume to damp out the mass flow and pressure pulses so the flow into the turbine is essentially steady. Although some of the exhaust gas kinetic energy is thereby dissipated, higher turbine efficiencies can be obtained.

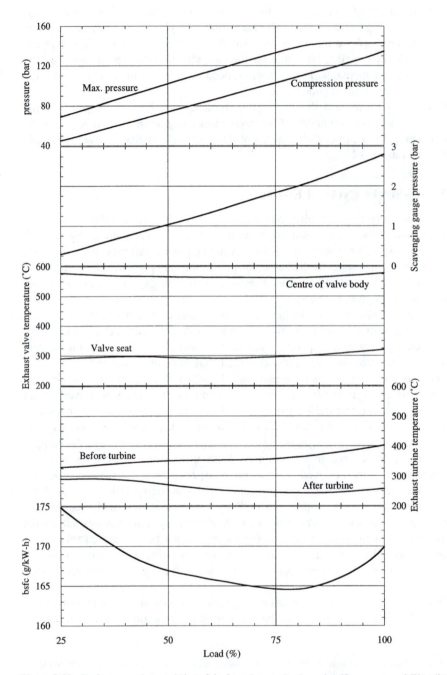

Figure 9-32 Performance characteristics of the large-bore turbocharged uniflow-scavenged Wärtsilä NSD 11RTA96C marine diesel engine.

propriate higher boost pressure must be provided. However, because the exhaust temperature in a two-stroke engine is typically low (about 400°C at full load) compared with that in four-stroke engines (typically 500°C at full load), the turbine power may not be sufficient to generate the required boost, and an auxiliary compressor is then used. Using a rotary compressor, using a simple fan, and using a reciprocating pump (connected either in series or in parallel with the turbocharger compressor) are each common approaches. With all types of reciprocating scavenging pumps, the unsteady nature of the flow may involve surge problems in the centrifugal turbocharger compressor, and an adequate margin between the average mass flow rate at the operating condition and surge limit must be maintained. A reasonably large receiver between compressor and pump is therefore required.[25] Figure 9-32 shows some measurements in a large-bore turbocharged uniflow-scavenged two-stroke diesel engine. The engine has 11 cylinders, which produce a total brake power of 54 MW at 90 rev/min. The bmep of the engine is 1.82 MPa. The engine performance curves show flat gas temperature characteristics before and after the turbine over the full engine load range. The temperature level is relatively low to extend engine life expectancy and reliability. The part-load bsfc levels are low due to very lean operation achieved by using an elevated scavenging pressure and appropriate exhaust valve timing. A slightly reduced geometrical compression ratio was used in this engine to increase the distance between the piston crown and injected fuel sprays to achieve low piston surface temperatures of about 375°C, at full load. The following are some other component temperatures of interest: cylinder cover temperature of around 315°C up to a maximum of 350°C near the joint with the liner; cylinder liner temperature 285°C maximum and 240°C at TC of top cylinder ring; exhaust valve surface temperature below 585°C in center and 540°C at outer circumference; and exhaust valve seat temperature of 315°C.

Medium- and low-speed uniflow-scavenged two-stroke engines operating over a narrow speed range, mainly for large marine and stationary power plants, are able to run turbocharged with no additional scavenging aids. With some control of the energy supplied to the turbine achieved by early opening of the exhaust valve, these engines can operate successfully with a pulse turbocharging system alone. However, as engine load increases, the loss of piston work due to early opening of the exhaust valve cannot be recovered by the additional work provided by use of the turbocharger. Figure 9-33 illustrates the relationship between exhaust valve timing, piston work losses, the pressure pulse at the turbine inlet, and the required scavenging pressure.[27] There is a crossover point[25] where use of constant-pressure turbocharging becomes more advantageous than use of pulse turbocharging. With a turbocharging efficiency of 62%, the changeover point may occur at a bmep as low as 1.2 MPa. Figure 9-34 compares the potential of providing the required turbine shaft power with pulse and constant-pressure operation.[27] Figure 9-35 shows how the turbocharger efficiency affects the engine bsfc and the after-turbine temperature. For an increase of 4% in the turbocharger efficiency, the after-turbine temperature decreases by some 24°C, which makes additional exhaust gas energy recovery difficult, whereas the bsfc decreases by some 0.3%. Some manufacturers prefer to use under-piston pumps in series with the turbocharger to ensure good scavenging over a wider range of engine

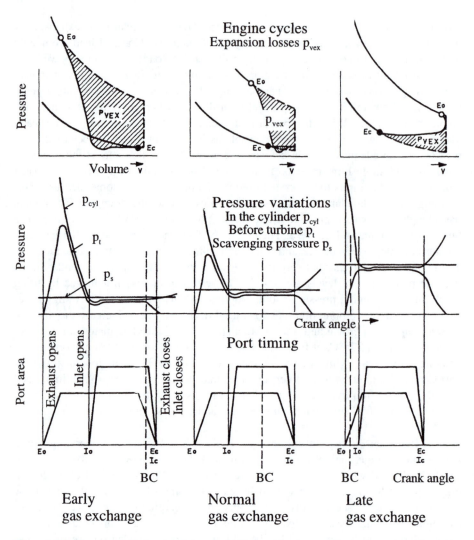

Figure 9-33 The relationship between exhaust valve timing, lost piston work, pressure pulse at the turbine inlet, and the required scavenging pressure, for early, normal, and late gas exchange.[27]

speed and load. When this is done, a constant-pressure turbocharging system is preferred.

For engines that are required to operate over a wide speed and load range, for example, in automotive applications, a separate mechanically driven scavenge blower is required. Some small two-stroke CI engines use Lysholm (roots) compressors as scavenge pumps, driven directly from the crankshaft. These typically have a low efficiency (55–65%), which falls further to around 50% as the pressure ratio increases (mainly due to leakage problems). Screw-type compressors are also sometimes used. These are

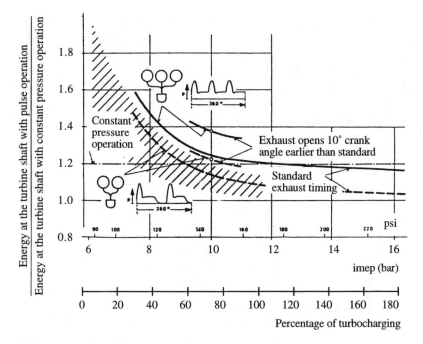

Figure 9-34 A comparison of the potential of providing the required turbine shaft power with pulse and constant-pressure turbocharger operation.[27]

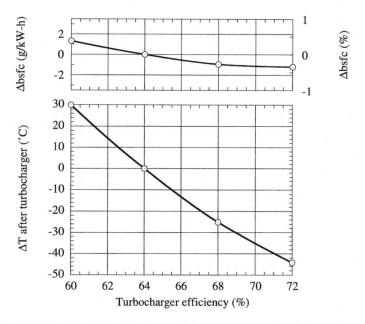

Figure 9-35 The effect of the turbocharger efficiency on engine bsfc and after-turbine temperature. From data provided by MAN and B&W, and Wärtsilä NSD and Sulzer.

quieter and more efficient than the Lysholm-type compressors over a wider range of pressure ratios (up to 3), but are more expensive to manufacture. A smaller scavenging pump, better bsfc at low speed and partial loads, and a better low-speed torque characteristic are the main advantages of using the pulse system. Constant-pressure turbocharging provides much simpler and more reliable exhaust manifolding,[25] especially on an engine with a large number of cylinders, and longer life of the turbocharger turbine. Pulse converters can be used successfully on two-stroke engines with cylinder numbers unfavorable to pulse turbocharging.

REFERENCES

1. Heywood, J.B., *Internal Combustion Engine Fundamentals*, McGraw-Hill, New York, 1988.
2. Anon., "Engine Control: Ignition and Fuel Injection Management, Petrol and Diesel," *Automotive Engineer*, vol. 18, no. 1, pp. 38–41, Feb./March, 1993.
3. Barton, P.J., and Fearn, J., "Study of Two and Four Stroke Outboard Marine Engine Exhaust Emissions Using a Total Dilution Sampling System," SAE Paper 972740, 1997.
4. Sawada, T., Wada, M., Noguchi, M., and Kobayashi, B., "Development of a Low Emission Two-Stroke Cycle Engine," SAE Paper 980761, 1998.
5. Data provided by Orbital Engine Corporation Ltd. Also Houston, R., Archer, M., Moore, M., and Newmann, R., "Development of a Durable Emissions Control System for an Automotive Two-Stroke Engine," SAE Paper 960361, 1996.
6. Duret, P., Venturi, S., and Carey, C., "The IAPAC Fluid Dynamically Controlled Automotive Two-Stroke Combustion Process," *A New Generation of Two-Stroke Engines for the Future?* (ed. P. Duret), pp. 77–98, Proceedings of the International Seminar held at IFP, Rueil-Malmaison, France, Nov. 29–30, Editions Technip, Paris, 1993.
7. Davis, R.A., "A Five-Cylinder Outboard Engine," SAE Paper 891751, 1989.
8. Nomura, K., and Nakamura, N., "Development of a New Two-Stroke Engine with Poppet-Valves: Toyota S-2 Engine," *A New Generation of Two-Stroke Engines for the Future?* (ed. P. Duret), pp. 53–62, Proceedings of the International Seminar held at IFP, Rueil-Malmaison, France, Nov. 29–30, Editions Technip, Paris, 1993.
9. Ishibashi, Y., and Asai, M., "A Low Pressure Pneumatic Direct Injection Two-Stroke Engine by Activated Radical Combustion Concept," SAE Paper 980757, 1998.
10. Ishibashi, Y., and Tsushima, Y., "A Trial for Stabilizing Combustion in Two-Stroke Engines at Part Throttle Operation," *A New Generation of Two-Stroke Engines for the Future?* (ed. P. Duret), pp. 113–124, Proceedings of the International Seminar held at IFP, Rueil-Malmaison, France, Nov. 29–30, Editions Technip, Paris, 1993.
11. Abthoff, J., Duvinage, F., Hardt, T., Kramer, M., and Paule, M., "The 2-Stroke DI Diesel Engine with Common Rail Injection for Passenger Car Application," SAE Paper 981032, 1998.
12. Knoll, R., "AVL Two-Stroke Diesel Engine," SAE Paper 981038, 1998.
13. Lustgarten, G.A., "The Latest Sulzer Marine Diesel Engine Technology," SAE Paper 851219 in SP-625 *Marine Engine Development*, SAE, Warrendale, PA, 1985.
14. Henningsen, S., "Air Pollution from Large Two-Stroke Diesel Engines and Technologies to Control It," *Handbook of Air Pollution from Internal Combustion Engines: Pollutant Formation and Control* (ed. E. Sher), pp. 477–534, chap. 14, Academic Press, New York, 1998.
15. Monnier, G., Duret, P., Pardini, R., and Nuti, M., "IAPAC Two-Stroke Engine for Efficiency Low Emissions Scooter," *A New Generation of Two-Stroke Engines for the Future?* (ed. P. Duret), pp. 101–111, Proceedings of the International Seminar held at IFP, Rueil-Malmaison, France, Nov. 29–30, Editions Technip, Paris, 1993.
16. Sher, E., Hacohen, Y., Refael, S., and Harari, R., "Minimizing Short-Circuiting Losses in 2-S Engines by Throttling the Exhaust Pipe," SAE Paper 901665, 1990.

17. Yoon, K.J., Kim, W.T., Shim, H.S., and Moon, G.W., "An Experimental Comparison between Air-Assisted Injection System and High Pressure Injection System at 2-Stroke Engine," SAE Paper 950270, 1995.
18. Sher, E., "The Effect of Atmospheric Conditions on the Performance of an Airborne Two-Stroke Spark-Ignition Engine," *Proc. Inst. Mech. Eng. Part D*, vol. 198, no. 15, pp. 239–251, 1984.
19. SAE, recommended practice engine test code, SAE J816b, *SAE Handbook*, SAE, Warrendale, PA, 1995.
20. Bosch, *Automotive Handbook*, Robert Bosch GmbH, 1993.
21. Obert, E.E., *Internal Combustion Engines and Air Pollution*, Harper & Row, 1973.
22. Watanabe, I., and Kuroda, H., "Effect of Atmospheric Temperature on the Power Output of a Two-Stroke Cycle Crankcase Compression Gasoline Engine," SAE Paper 810295, 1981.
23. Harari, R., and Sher, E., "The Effect of Ambient Pressure on the Performance Map of a Two-Stroke SI Engine," SAE Paper 930503, 1993.
24. Maeda, K., Yasunari, M., Hikasa, S., and Mirishita, S., "Characteristics and Reduction Methods of NO_x Emission from Two-Stroke Marine Diesel Engine," *Bull. M.E.S.J.*, vol. 22, no. 1, pp. 1–7, 1994.
25. Watson, N., and Janota, M.S., *Turbocharging the Internal Combustion Engine*, Macmillan, New York, 1982.
26. Hashimoto, E., Tottori, T., and Terata, S., "Scavenging Performance Measurements of High Speed Two-Stroke Engines," SAE Paper 850182, 1985.
27. Gyssler, G., "Problems Associated with the Turbocharging Large Two-Stroke Diesel Engines," CIMAC Report 18–65–1, pp. 1047–1078, 1965.

INDEX

Active radical combustion 236–239, 400–404
  emissions 297–299
  regions of 236–237, 417–418
Air/fuel ratio:
  burnt gases 88
  definition 13
  determination 99–103
  engine data 102, 280, 315, 331, 404, 418–420
  stoichiometric 13, 14, 418
Altitude:
  standard atmosphere 425
  effects on performance 61, 73, 203, 423–426
Ambient condition effects (see also Altitude effects) 423–426
  correction factor 423, 424–425
Applications, two-stroke engines:
  automotive 358, 405–410, 439–440
  chain saws 326, 329
  externally blown 40, 41
  marine 35, 358, 410
  mopeds/scooters 326, 357
  outboard engines 357
  spark-ignition 38, 267
    carburetted 18, 326–330
    loop-scavenged 16
    valved 39–41, 357

  special 17–23
  summary 14–23

Bore to stroke ratio 9, 35, 39, 51
Burn angles:
  flame development 218
  main/rapid burn 218
Burnt gas composition 100

Carbon monoxide emissions:
  catalyst performance 322–323, 329, 331, 334
  data 279, 295, 310, 316, 328–329, 334
  diesels 308, 316
  spark-ignition engines:
    air/fuel ratio effects 278–281
    formation mechanism 271, 278–281
    short-circuiting effects 278–280
    throttle effects 281
Catalysts:
  ceramic monolith 317–318
  chemical energy release 324
  configurations 320–321
  construction 317–319
  conversion efficiency 292, 321, 326–329, 331–332
  diesels 336–340

445